Computational Biology

Volume 33

Endorsed by the *International Society for Computational Biology,* the *Computational Biology* series publishes the very latest, high-quality research devoted to specific issues in computer-assisted analysis of biological data. The main emphasis is on current scientific developments and innovative techniques in computational biology (bioinformatics), bringing to light methods from mathematics, statistics and computer science that directly address biological problems currently under investigation.

The series offers publications that present the state-of-the-art regarding the problems in question; show computational biology/bioinformatics methods at work; and finally discuss anticipated demands regarding developments in future methodology. Titles can range from focused monographs, to undergraduate and graduate textbooks, and professional text/reference works.

More information about this series at https://link.springer.com/bookseries/5769

Hannes Hauswedell

Sequence Analysis and Modern C++

The Creation of the SeqAn3 Bioinformatics Library

Hannes Hauswedell ⓘD
Reykjavik
Iceland

ISSN 1568-2684   ISSN 2662-2432   (electronic)
Computational Biology
ISBN 978-3-030-90992-5   ISBN 978-3-030-90990-1   (eBook)
https://doi.org/10.1007/978-3-030-90990-1

This Springer imprint is published by the registered company Springer Nature Switzerland AG
The registered company address is: Gewerbestrasse 11, 6330 Cham, Switzerland

Preface

This is a book about software engineering, bioinformatics, the C++ programming language and the SeqAn library. In the broadest sense, it will help the reader create better, faster and more reliable software by deepening their understanding of available tools, language features, techniques and design patterns.

Every developer who previously worked with C++ will enjoy the in-depth chapter on important changes in the language from C++ 11 up to and including C++ 20. In contrast to many resources on Modern C++ that present new features only in small isolated examples, this book represents a more holistic approach: readers will understand the relevance of new features and how they interact in the context of a large software project and not just within a "toy example". Previous experience in creating software with C++ is highly recommended to fully appreciate these aspects.

SeqAn3 is a new, re-designed software library. The conception and implementation process is detailed in this book, including a critical reflection on the previous versions of the library. This is particularly helpful to readers who are about to create a large software project themselves, or who are planning a major overhaul of an existing library or framework. While the focus of the book is clearly on software development and design, it also touches on various organisational and administrative aspects like licensing, dependency management and quality control.

The field that SeqAn3 provides solutions for is sequence analysis or, in a broader sense, bioinformatics. Readers working in this domain will recognise many of the discussed problems. However, almost all content is useful to software engineers in general and research software engineers in particular; no background in biology or previous experience with the SeqAn library is required.

This book is based on a dissertation, so the general style is more reminiscent of a "story" than might be typical for a computer science book. Some readers will enjoy reading it cover to cover while others will want to jump to sections of interest directly. The original preface of the dissertation is given on the following page as the acknowledgements section. In addition to the persons mentioned there, I would like to thank Martin Vingron who was part of my defence committee and suggested this book project. I would also like to thank Susan Evans and the team at Springer Nature for helping it become reality.

Reykjavik, Iceland   Hannes Hauswedell

The original version of this book was revised: Revised book has been uploaded to Springerlink. The correction to this book is available at https://doi.org/10.1007/978-3-030-90990-1_13.

Acknowledgements

The SeqAn library is a very active project with a long history. Over the last more than 10 years, it has had different core developers and many people who contributed features and fixes. Although SeqAn3 contains almost no code from SeqAn1/2, the experience of working on and with previous versions was invaluable in the development of SeqAn3. I feel that it is therefore only proper to mention Andreas Gogol-Döring, David Weese, Enrico Siragasu and Manuel Holtgrewe at this point, all of whom contributed significantly to SeqAn1/2. Of course Knut Reinert has always guided and does until today lead the project. His experience is the main pillar of its continued success.

This thesis introduces a new and radically different version of the SeqAn library. The scope of this project is huge, and it certainly would not have been possible to create the library single-handedly in this time. I do, however, credit myself with its inception, the vision behind the project and the endurance to pursue a complete rewrite of the library when most people called it infeasible. The design process, the overarching goals and the technical decisions are overwhelmingly my work—that is the foundation of this thesis. On the practical side, I have also written and changed more code than the next most important contributors combined, but I want to state clearly that relevant parts of SeqAn3 have also been implemented by people other than myself.

René Rahn has shared the responsibility of leading the project with me on a social and administrative level. Since the early beginnings of SeqAn3, I relied strongly on his counsel. Later, we assembled the SeqAn *core team* to discuss design and strategy matters on a regular basis. This included Svenja Mehringer, Marcel Ehrhardt and Enrico Seiler. All members of the core team have left their mark in some way on the library, and I am confident that SeqAn3 is in good hands after I leave the project.

I would like to thank everyone who contributed to SeqAn3, but more generally I want to also thank everyone for the great time at Freie Universität and the unforgettable SeqAn retreats! Special thanks go to Sara Hetzel and Felix Heeger who provided very helpful comments on a draft of this dissertation. Sara will also continue work on Lambda, an application presented later in this thesis.

On a professional and personal level, my sincere gratitude goes to Knut Reinert who has been my mentor now for so many years. None of this would have been possible without him. I would also like to express my sincere gratitude to Stefan Kurtz who agreed to co-supervise this (quite comprehensive) thesis although we had not worked together previously.

Attending the meetings of and contributing to the ISO C++ committee has had the most profound influence on my understanding of C++ and has thus helped greatly with creating SeqAn3. I would like to thank Fabio Fracassi and Nico Josuttis from the DIN Arbeitskreis Programmiersprachen as well as Corentin Jabot and JeanHeyd Meneide for helping me find my way around WG21.

Before working at Freie Universität, my studies were funded through a stipend of the Max-Planck-Gesellschaft. I additionally received a fellowship by the Hans-Böckler-Stiftung which allowed me to attend various extracurricular activities, for which I am very grateful.

Finally, I would like to thank my parents for supporting me during my youth and my early university studies. I am privileged to have had access to computers as a child and to grow up in an environment that fostered my curiosity in science and technology. I am grateful for the support of my friends and especially Romy and Betti. I look forward to spending more time with everyone again!

Contents

Part I
Background

The first part of this book lays the foundation for the remaining parts. It briefly introduces the reader to sequence analysis, a central field in current bioinformatics research. It then covers the design and implementation of the SeqAn library prior to the release of version 3 and discusses in how far it was successful in achieving its set goals. Finally, this part devotes a large chapter to explaining the recent and not-so-recent developments in the C++ programming language and how they might enable us to solve the current challenges in sequence analysis in more elegant and/or more efficient ways.

Chapter 1
Sequence Analysis

Sequence analysis is a domain in bioinformatics which encompasses all computer-aided studies of biological sequence data. This data is produced from molecules such as DNA and RNA, which store a cell's genetic information, and proteins, which are the "machines" of a cell and provide a myriad of functions including signalling, metabolism and immune response. While these biological molecules (especially proteins) exhibit complex three-dimensional structures, they can also be represented as linear polymer *sequences* of their molecular building blocks.[1] These molecular building blocks in turn are nucleotides (in the case of DNA/RNA) or amino acids (in the case of proteins). They are the basic units of information in sequence analysis, and the types of these units are referred to as "alphabets" in the context of computer science.[2]

The type of analysis performed on such data varies greatly: it ranges from functional analysis (e.g. "what is the purpose of this gene in the cell?") over comparative analysis (e.g. "how is sequence X related to sequence Y in another or the same organism?") to quantitative analysis (e.g. "what does the frequency of this RNA transcript indicate regarding the activity of the cell?"). Subject of research may be a single short sequence like a gene, the entire genome or transcriptome, or even all genetic material in some sample. The latter is called *metagenomics* and is becoming increasingly common.

Scientific domains that perform sequence analysis or that make use of sequence analysis tools are even more diverse. They include most areas of modern biological research, because the need to understand genetics and evolutionary processes has become pervasive. But sequence analysis has also come to influence fields such as ecology where it is used to assess the microbial diversity and its response to certain perturbations (Mackelprang et al., 2011). This research in turn has far-reaching

---

[1] The three-dimensional structure as well as the connection between sequence representation and three-dimensional structure is the subject of *structural bioinformatics*.

[2] Chapter 6 discusses them in detail.

© The Author(s), under exclusive license to Springer Nature Switzerland AG 2022
H. Hauswedell, *Sequence Analysis and Modern C++*, Computational Biology 33,
https://doi.org/10.1007/978-3-030-90990-1_1

implications for other fields such as climatology. In the realm of medical research, sequence analysis is central to identifying genetic markers for hereditary diseases (Liu et al., 2019) as well as cancer (Banerji et al., 2012). It is becoming more and more important for analysing the human microbiome (Turnbaugh et al., 2007) and its contribution to human health. And it is also part of infectious disease research and treatment, both, for detecting contagion in a sample (Ho & Tzanetakis, 2014) and in developing vaccines (Maiden, 2019). Through its role in developing genetically modified organisms (GMOs), sequence analysis contributes to further fields such as agriculture, industrial processing and energy production.

The substance of all sequence analysis is the sequence data. This data is generated by different (bio-)technological methods and the properties of these techniques have a profound effect on the types of analysis technically possible and economically feasible. Especially, the technological leaps in DNA/RNA sequencing have dwarfed progress in other scientific domains:

> [T]he first whole human genome sequencing in 2000 [. . .] cost over $3.7 billion and took
> 13 years of computing power. Today, it costs roughly $1000 and takes fewer than three days.
> With trillions of genomes waiting to be sequenced, both human and otherwise, the genomic
> revolution is in its infancy. (Bannon, 2014)

The decline in cost over the years for sequencing one human genome is displayed in Fig. 1.1. This is given as a general indicator for the trend of sequencing costs although—as noted above—attaining a genome is not always the goal and other forms of sequencing are even cheaper, e.g. species identification through the so-called *barcoding*. While the price curve has flattened in recent years, new sequencing technologies promise to produce longer sequencing *reads* which improve the quality of some and enable new research areas (Pollard et al., 2018). It is important to note the logarithmic scale of the Y-axis in Fig. 1.1 and the expected progress suggested by Moore's Law which vaguely indicates development of computing power in the same time.[3]

This connection between progress in sequencing technologies and computing power is very important, because decreasing prices imply increasing availability of sequencing data and corresponding growth of sequence databases. Many problems, like searching for all homologues ("related sequences") of a given sequence, grow in computational complexity with the size of the database. Often this relationship is even super-linear, i.e. searching a database twice the original size is more than twice as difficult for the computer. And, as Fig. 1.1 indicates, sequence data grows at orders of magnitude faster than the capabilities of computer *hardware*, so solving well-known problems becomes more and more costly over time. To counter this trend, high-performance sequence analysis *software* needs to be developed that reduces complexity on an algorithmic level.

The increasing diversification of research areas using sequence analysis and the progress of sequencing technologies have led to many new research questions for which equally many new applications have been published. Developing these

---

[3] More on Moore's Law and why effective speed-ups may even be lower in Sect. 2.4.1.

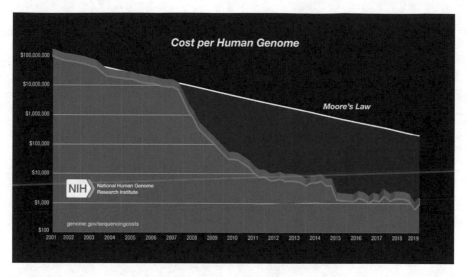

Fig. 1.1 Decline in the sequencing cost of a human genome. Note the log-scale Y-axis and the expected decline based on Moore's Law. Public domain image. Courtesy: National Human Genome Research Institute

sequence analysis applications is a scientific area of its own[4] and significant resources go into developing novel applications—either to solve new problems or to solve existing problems more efficiently. However, the main algorithmic steps in most of these applications are very similar (Gogol-Döring, 2009), e.g. the reading and writing of common file formats, the indexing of large databases and the computation of sequence alignments.

Thus, *software libraries* can help reduce the cost of creating new applications. Software libraries are pre-written program code, mostly algorithms and data structures, that can be used by applications to perform such frequent tasks for them. Since library code is shared between many applications and often reused, more time is invested into quality control and performance optimisations; this leads to better applications. And because the full implementation of complex algorithms can be hidden behind a simple interface, using libraries enables less-versed programmers to solve difficult problems. This is especially important in bioinformatics where application developers are often domain specialists but not experts in software engineering.

This book is about SeqAn, a software library written in C++, that covers the most important areas of sequence analysis and enables bioinformaticians to create high-performance solutions to existing and new challenges.

---

[4] Many consider it part of *research software engineering*, see also Gesellschaft für Forschungssoftware (2018).

Chapter 2
The SeqAn Library (Versions 1 and 2)

This chapter gives a brief overview of the SeqAn library, important design goals and programming principles, as well as an analysis of in how far these were reached. I will discuss all aspects that I deem necessary to understanding the design and development process of SeqAn3, but I strongly recommend reading the original SeqAn publication (Döring et al., 2008), the publication documenting the second major release of SeqAn (Reinert et al., 2017) and the doctoral thesis of Andreas Gogol-Döring (Gogol-Döring, 2009) that explain the original motivation and design choices in detail.

2.1 History

The SeqAn library is being developed primarily in Knut Reinert's groups at Freie Universität Berlin and Max Planck Institute for molecular Genetics, but it has contributors from many other research groups in Berlin and around the world. Since moving to a public Git repository in 2015, the number of contributions from individuals not affiliated with Knut Reinert's lab or cooperation partners has grown steadily. The most important events are shown in Table 2.1.

The author of this book knows the library since his undergraduate thesis in 2009 and has worked with it in different roles since. As of SeqAn-2.1, he is the shared project lead and release manager (together with René Rahn). He is the main architect of the SeqAn3 library.

Since SeqAn is a software library, SeqAn's history is of course also the history of the applications built with SeqAn. It is beyond the scope of this work to cover all these applications, but several noteworthy examples are Bowtie (Langmead et al., 2009), TopHat (Trapnell et al., 2009), DELLY (Rausch et al., 2012), FLEXBAR (Dodt et al., 2012), RazerS (Weese et al., 2012) Mason (Holtgrewe, 2010), Stellar (Kehr et al., 2011) and SLIMM (Dadi et al., 2017). Part III discusses

H. Hauswedell, *Sequence Analysis and Modern C++*, Computational Biology 33, https://doi.org/10.1007/978-3-030-90990-1_2

Table 2.1 A brief history of important SeqAn events

Year	Event
2008	SeqAn-1.0; first publication (Döring et al., 2008)
2009	Doctoral thesis of Gogol-Döring (2009)
2013	SeqAn-1.3; with significant changes
2015	SeqAn-2.0; move to GitHub
2016	SeqAn-2.1; follows semantic versioning
2017	Second publication (Reinert et al., 2017)
2018	SeqAn-2.4; last feature release of 2.x series
2019	SeqAn-3.0
2020	*planned:* SeqAn-3.1 (stable) and third publication

Lambda (Hauswedell et al., 2014), an application developed by the author based on SeqAn.

2.2   Design Goals

In his dissertation, Gogol-Döring (2009) defined the over-arching goals of the library as being an instrument of engineering ("Enabling the rapid development of efficient *tools* [.]") and as being academic/instructive ("Promoting the design, comparison and testing of algorithms[.]").

He formulated the concrete design goals as (all quotes by Gogol-Döring, 2009)

Performance   "[. . .] designed to produce code that runs as fast as possible".
Simplicity   "All parts [. . .] are constructed and applicable as simple as possible. [sic]".
Generality   "All parts [. . .] are applicable in as many circumstances as possible".
Refineability   "Whenever a specialization is reasonable, it is possible to integrate it easily[.]".
Extensibility   "[It] can always be extended without changing already existing code".
Integration   "[It] is able to work together with other libraries and built-in types".

2.3   Programming Techniques

To achieve the previously defined goals, Gogol-Döring describes the following programming techniques:

C++   "We decided to implement SeqAn in C++, because *performance* is among our main goals [. . .] and the extended features of C++, namely *templates*[. . .], are well suited to an excellent *library design*." (Gogol-Döring (2009))

Generic Programming   "Generic programming designs algorithms and data structures in a way that they work on all types that meet a minimal set of requirements
   [...] [, it] promotes the *generality* of the library." (Gogol-Döring (2009))
Template Subclassing   This term is used by Gogol-Döring to describe a kind of
   polymorphism based on partial template specialisation and function overloading
   using partially specialised template parameters.
Global function interfaces   Functions declared at namespace scope (instead of as
   members of a class) are called *global functions* by Gogol-Döring, otherwise
   often also known as *free functions*. The use of free functions **for polymorphism**
   is required by generic programming, but SeqAn extends this approach even to
   object interfaces.
Metafunctions   A "metafunction" is described by Gogol-Döring as an entity that
   "returns" for a given type or constant another type or constant (at compile-time).
   SeqAn1/2 uses metafunctions not only as observers of the properties of a type
   but also as modifiers of these properties.

All of these points are elaborated on in the doctoral thesis of Gogol-Döring. I will
cover the C++ programming language extensively in Chap. 3 but want to guide the
reader through the remaining techniques in the following sections as it is important
to understand the specifics of SeqAn1/2 to comprehend (and appreciate) the changes
in SeqAn3.

2.3.1 Generic Programming

Generic programming is a paradigm that became popular in the C++ community
later than object-oriented programming (OOP) and its goal is to overcome some
(performance) problems of OOP (Duret-Lutz et al., 2001). It is facilitated through
the use of function and class templates and it is strongly associated with *static polymorphism* (see below). Beside performance, the main goal of generic programming
is the reuse of code within a code base and interoperability with user-defined types:

> Generic programming recognizes that dramatic productivity improvements must come from
> reuse without modification, as with the successful libraries. Breadth of use, however, must
> come from the separation of underlying data types, datastructures, and algorithms, allowing
> users to combine components of each sort from either the library or their own code. (Dehnert
> & Stepanov, 2000)

2.3.2 Template Subclassing

Polymorphism is a key feature in most programming languages and is part of
different programming paradigms. Bjarne Stroustrup defines it as "providing a

single interface to entities of different types".[1] In Snippet 2.1,[2] I present an example (adapted from the example in Gogol-Döring (2009)):

```
 2   // base class                              2   // base template
     struct IntContainer                            template <typename TSpec>
     {                                              struct IntContainer
 4       // virtual member function            4   {
         virtual size_t find(int i)                     /* ... */
 6       {                                      6   };
             /* find index of i by linear scan */
 8       }                                      8   // most generic overload
                                                    template <typename TSpec>
10       /* ... */                            10   size_t find(IntContainer<TSPec> & c, int i)
     };                                             {
12                                            12       /* find index of i by linear scan */
                                                    }
14                                            14
                                                    // tag for derived type
16                                            16   struct MapSpec;
     // derivation via inheritance                  // derived type via specialisation
18   struct IntMap : IntContainer           18   template <>
     {                                              struct IntContainer<MapSpec>
20       // member function override          20   {
         virtual size_t find(int i)                     /* ... */
22       {                                     22   };
             /* find index of i by binary search */
24       }                                     24   // refined overload
                                                    size_t find(IntContainer<MapSpec> & c, int i)
26       /* ... */                            26   {
     };                                                 /* find index of i by binary search */
28                                            28   }

30                                            30   // polymorphic interface
     // polymorphic interface                       template <typename TSpec>
32   void print_idx_of(IntContainer & c,    32   void print_idx_of(IntContainer<TSpec> & c,
                       int elem)                                      int elem)
34   {                                        34   {
         std::cout << c.find(elem);                     std::cout << find(c, elem);
36   }                                        36   }
```

Code snippet 2.1: Polymorphism in object-oriented programming vs template subclassing. Adapted from "Listing 2" in Gogol-Döring (2009). Neither is valid SeqAn code

- Given a container of integers, there shall be a `find()` function that finds the position of the first occurrence of a given integer in that container.
- The trivial solution is to do a linear-time scan over the container.
- But for containers that are ordered, such a search can be performed in logarithmic time; for these containers, a *more refined* algorithm should be selected.
- Furthermore, a polymorphic interface should be able to handle objects of base type and the derived type.

In object-oriented programming, polymorphism is implemented via inheritance and virtual member functions; *derived* classes inherit from *base* classes. Pointers

---

[1] http://www.stroustrup.com/glossary.html.

[2] Please see Sect. A.1.2 for notes on how to read code snippets in this thesis.

and references to the base type can also bind objects of the derived type, so one can pass an object of type `IntMap` to `print_idx_of()` in Snippet 2.1. When the `find()` member function is invoked, a virtual function lookup selects the most refined implementation **at runtime**. Because the selection happens at runtime, this form of polymorphism is also called *dynamic polymorphism.*

In generic programming on the other hand, polymorphism is implemented via templates and (free) function overloading. The selection of the best/most refined implementation happens **at compile-time**, so it is called *static polymorphism.* Since it happens at compile-time, static polymorphism is notably faster than dynamic polymorphism (Driesen & Hölzle, 1996), which is the reason SeqAn prefers it.

Template subclassing is one "style" of static polymorphism (there are others). Instead of through inheritance, a base template is defined and derived classes are modelled as template specialisations of that template. The so-called *tag types* are often used to denote such specialisations.[3] Generic functions are then also implemented as free/global function templates with some template parameters "fixed". If an overloaded free function is invoked, the overload that is most refined is picked by the compiler.

Both of the mentioned styles have in common that one can refine arbitrarily often/"deep" (in the case of template subclassing by making the tags also be templates that are further specialised). They also share that the polymorphism is restricted to one's own types, i.e. one needs to explicitly inherit from the respective base class (dynamic polymorphism) or specialise the respective base template (template subclassing); one cannot plug in foreign types, e.g. from a different library (more on this in Sect. 3.4).

2.3.3   Global Function Interfaces

As previously explained, generic algorithms have to be implemented as free functions in the generic programming paradigm. This is, however, not true for all functions. For a long time, the C++ standard library has provided algorithms as free functions, but it still implemented most other functions (that are related to the properties of an object more closely) as member functions. For example, `std::find()` is a generic free function that can be called with different containers (or more precisely their iterators) as arguments, but `.size()` is a member function of the respective container.

In later revisions of the C++ standard (C++ 11, C++ 17), the standard library picked up free function wrappers for many of these member functions, e.g. `std::begin()`, `std::end()`, `std::size()`, `std::empty()`. The reasoning is that although the

---

[3] They have no other purpose and are usually optimised out of the final code entirely.

functions are seen as accessing properties of the object and not as free-standing components, working with a free function is more flexible in a generic programming context.

If for example a generic algorithm needs to know an object's size, it would previously always look for a `.size()` member function. This works if all input types of the algorithm are designed together with the algorithm, but it will fail if a user provides a type from a different library which happens to provide a `.length()` member and not `.size()`. If one's algorithm instead looks for a free function `size(obj)`, the user of the library can provide a custom wrapper around the other library's type so that it will satisfy the requirements of the algorithm without needing to be changed ("reuse without modification"; Dehnert & Stepanov, 2000).

SeqAn has used this style since its inception, however in a more radical fashion where practically all functions are free functions. They are not even wrappers around member functions but directly access the state of an object (e.g. `seqan::length()` directly accesses respective data members). This is a notable difference to the standard library that provides encapsulation on an implementation level (the actual functionality is implemented as members) and only exposes these member functions via free function wrappers.

The implications of this for the general library design are important to note. On the one hand, the users are able to overload implementation details that might otherwise be considered `private`, granting a higher level of extensibility/refineability; on the other hand, this can introduce subtle changes in other parts of the library that rely on the previously defined behaviour. In effect, the definition of how a type behaves becomes highly non-local, because essential functions can practically be overridden from anywhere in the library or even in application code.

2.3.4 Metafunctions

> What we need therefore is a mechanism that returns an output type (e.g. a value type) given an input type (e.g. the string) [...]. Such a task can be performed by *metafunctions*, also known as type traits [...]. A *metafunction* is a construct to map some types or constants to other entities like types, constants, or objects at compile-time. (Gogol-Döring, 2009)

Especially the last sentence of the quote articulates well the mechanism behind metafunctions/type traits. Note that I would not equate the terms *metafunction* and *type trait* entirely, and I prefer using the latter (see also Sect. 3.3).

Following a similar argument as in the previous section, Gogol-Döring argues that it is beneficial to have "global" type metafunctions (e.g. `seqan3::Value<T>::Type`) over relying on a type's member types (e.g. `T::value_type`). The C++-standard adopted this style much later and a transformation type trait that does exactly what Snippet 2.2 does will be included in C++ 20 under the name `std::ranges::range_value_t<T>`. Note that this is a wrapper and that, similarly to the free functions and in contrast to SeqAn1 and SeqAn2, the

actual implementation is provided by the type as a member, i.e. in most cases
`::value_type`.

A notable difference of the style used in SeqAn and the (modern) standard
library is that in SeqAn metafunctions are not only used as *accessors* but also as
modifiers.[4] This means they do not simply expose certain (type) properties but can
be specialised/overloaded to change the properties that are exposed for existing
type(s):

```
    template <typename T> class Value;
2
    template <typename TValue, typename TSpec>
4   class Value < Container<TValue, TSpec> >
    {
6       typedef TValue Type;
    };
8
    template <typename T, size_t I>
10  class Value < T[I] >
    {
12      typedef T Type;
    };
14
    template <typename T>
16  void swapvalues(T & container)
    {
18      typedef typename Value<T>::Type TValue;
        TValue help = container[0];
20      container[0] = container[1];
        container[1] = help;
22  }
```

Code snippet 2.2: "Listing 4: meta functions [sic] example" from Gogol-Döring
(2009)

SeqAn offers the metafunction `Size` [...]. This type is by default `size_t`, and it is
hardly ever changed by the user, so it is not worth to specify it in another template argument.
Nevertheless [...][,] it is possible to overwrite the default with a new type [...] by defining
a new specialization of the metafunction `Size`. (Gogol-Döring, 2009)

The quote suggests that this "feature" was initially reserved for manipulating only
obscure properties of types, and however later the design was adopted throughout
the library and is even taught in the beginner's tutorial for working with suffix arrays:

All Indices in SeqAn are capable of indexing Strings [...] up to 2^{64} characters. [...][If] the
text to be indexed is shorter, e.g. it does not exceed 4.29 billion (2^{32}) characters[...], one
can reduce the memory consumption of an Index by changing its internal data types, with
no drawback concerning running time. [...]
    In order to change the size type of the suffix array entry we simply have to overload the
metafunction SAValue:

---

[4] There are customisation points in the standard library that involve specialising a type trait, e.g.
`std::tuple_size`, but they are very few and clearly marked as such. It is also explicitly
stated that such specialisations may only affect newly defined types and not manipulate the traits
of existing types (ISO/IEC 14882:2017, 20.5.4.2.1).

```
  template<>
2 struct SAValue<String<Dna> >
  {
4   typedef unsigned Type;
  };
```

https://seqan.readthedocs.io/en/master/Tutorial/DataStructures/Indices/StringIndices.html

The implications of this are similar to the implications of being able to overload functions that manipulate the behaviour of existing types (see the previous subsection). Another non-obvious implication of the "global type trait modifiers" is that they are indeed "global": once one overrides the `SAValue` type, it affects all indexes over the respective text type and one cannot create indexes over the same text type with different traits—as would be possible if `SAValue` were a template parameter of the index.[5]

2.4   Discussion

Measuring the impact of the SeqAn library accurately is not easy. In general, research software has a hard time being properly attributed in many domains of science (Soito & Hwang, 2016). Even though citable publications have always existed for SeqAn, many instances have become known were software that uses SeqAn does not properly cite it, instead placing only link to the project homepage (Dröge et al., 2014) or not even that.

I also assume that the number of instances not known is far greater, since being a software library (and not an actual application) makes the contribution to research even less visible for many biologists and bioinformaticians. There are neither clear guidelines for citing software libraries nor enforcement of such practices by major journals (Soito & Hwang, 2016).

I would still maintain that the SeqAn library has been a big success. Some of the most highly cited bioinformatics applications released in the last decade make use of SeqAn, among them are Bowtie (Langmead et al., 2009), Tophat (Trapnell et al., 2009) and DELLY (Rausch et al., 2012). Furthermore, the team around SeqAn published applications based entirely on the SeqAn library that outperformed state-of-the-art competitors, often by multiple factors, e.g. RazerS (Weese et al., 2012), Masai/Yara (Siragusa et al., 2013; Siragusa, 2015) and Lambda (Hauswedell et al., 2014). SeqAn has also been used outside the domain of bioinformatics and computational biology, e.g. in image processing/text recognition (Yoon et al., 2016).

Gogol-Döring (2009) analysed existing C++ sequence analysis libraries, including BATS (Giancarlo et al., 2007), Bio++ (Guéguen et al., 2013), BTL (Pitt et al.,

---

[5] In practice, it is possible to workaround this limitation by defining different text type specialisations and then defining different `SAValue` specialisations for each. This implies substantial changes to the application code.

2001), libsequence (Thornton, 2003), the NCBI C++ Toolkit (Vakatov et al., 2003) and SCL (Vahrson et al., 1996). Out of these, only libsequence and Bio++ have had bug-fix releases in the last two years and only libsequence received new features. Development of the remaining libraries seems to have stalled. In the meantime, some important new libraries have been published, most of which are specialised and perform only a subset of SeqAn's features. A popular example is htslib, a library factored out from Samtools (Li et al., 2009), more on how SeqAn compares to htslib below. One of the few libraries aiming at a broader feature set is SeqLib (Wala & Beroukhim, 2017). It compared favourably against SeqAn in some published benchmarks, and however, it was later shown that the authors had built SeqAn in Debug mode, skewing the results in their favour.[6] There is notably less development lately and usage by other projects is insignificant compared with SeqAn. libgenometools is a C library developed together with the GenomeTools application (Gremme et al., 2013). Its feature set overlaps with SeqAn to a certain degree and it has some unique features (e.g. for data visualisation), but it has seen no release and almost no commits in 2018 and 2019.

On the other hand, SeqAn has had a continuous stream in contributions and a notable increase of contributors over the years. Contributions have come not only from labs closely associated with SeqAn like the Reinert lab but also from external researchers and developers all over the world. SeqAn picked up an (optional) update notification system with version 2.3.0. By aggregating and evaluating the requests received from applications, one can now get rough estimates of library usage. Plotting the approximate locations of the requesting IP addresses (resolved via geolocation) yields a map as in Fig. 2.1. It should be noted that this service is

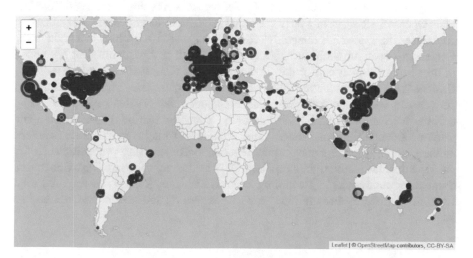

Fig. 2.1 Locations of SeqAn-based applications that performed update requests. Automatically generated image, which includes content licensed under ©①◎ by OpenStreetMap contributors

---

[6] https://github.com/walaj/SeqLib/issues/12.

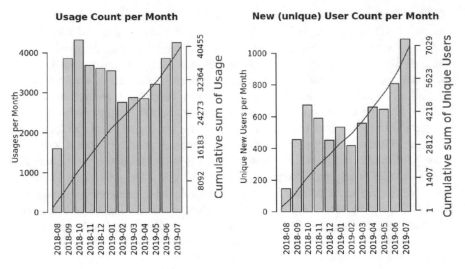

Fig. 2.2   Usage and user numbers reported during one year. Automatically generated image

entirely optional, many SeqAn-based applications do not make use of the argument parser (which is the component that triggers the request), and major operating system vendors like Debian GNU/Linux and derivates like Ubuntu deactivate the respective feature by default. So the data always only displays a subset of SeqAn use-cases, but it is still impressive to see the number of unique new users climb over time (Fig. 2.2).

SeqAn as a project is also part of multiple networks and initiatives. Together with OpenMS (Röst et al., 2016), KNIME (Berthold et al., 2007) and others, it constitutes the Center for Integrative BioInformatics (CIBI), which is a node in the German network for Bioinformatics (de.NBI, Tauch & Al-Dilaimi (2017)), which in turn is the German part of the ELIXIR network (Crosswell & Thornton, 2012). Besides these publicly funded initiatives, SeqAn has had research and development cooperation with important (hardware) companies like NVIDIA and Intel, being at times both an NVIDIA CUDA Research Center[7] and an Intel Parallel Compute Center.[8] Kristina Kermanshahche, Chief Architect of Intel Health & Life Sciences, announced the latter by saying " Intel regards SeqAn as a very promising software package that has all the right ingredients to considerably speed up Next Generation Sequencing analysis[.]".[9] According to Prof. Dr. Knut Reinert, he acquired total funding of close to three million euros in SeqAn-related grants over the last 10 years.

---

[7] https://developer.nvidia.com/academia.

[8] https://software.intel.com/en-us/ipcc.

[9] https://www.fu-berlin.de/en/presse/informationen/fup/2015/fup_15_285-professor-reinert-leitet-intel-parallel-computer-center/index.html.

All in all these facts add up to SeqAn being a success story and the involved researchers have ample reason to be proud. I would still like to reflect self-critically on the original design goals and decisions in the next sections. If some criticism reads as overly harsh, this is not to diminish the achievements of SeqAn1/2 but to raise the awareness of the reader for areas of potential improvement.

2.4.1 Performance

Performance, usually measured in execution speed but sometimes also in memory usage, has always been the stated primary goal of SeqAn (Gogol-Döring, 2009). Considering the challenges discussed in Chap. 1, this focus is and remains completely valid. And in fact the performance of SeqAn has been excellent in all important areas including Input/Output, Indexed Search and Alignment.

SeqAn supports many typical bioinformatics file formats for Input/Output, including FASTA, FASTQ, VCF, SAM and BAM. The performance of I/O is frequently cited as a main bottleneck in many data evaluation pipelines (Buffalo, 2015; Kosar, 2012). Routinely comparisons performed by third parties confirm that SeqAn performs very well, often better than the reference implementations, see Table 2.2.

Another core part of SeqAn is full text indexing, including q-gram/k-mer indexing, suffix arrays and FM-indexes. It allows for efficient searching of large databases and is a core part of read mappers and aligners alike. After the first release, significant contributions to this part of the library were made by Weese (2013), Siragusa (2015) and Pockrandt et al. (2017). As shown in Table 2.3, SeqAn's wavelet-tree-based FM-indexes are already very competitive. However, EPR-dictionaries (an FM-index type first available in SeqAn) deliver even higher speed-ups (more on this in Sect. 9.1).

The third pillar of SeqAn for which performance is crucial is sequence alignment. Sequence alignment is a part of almost all traditional sequence analysis tools, and SeqAn can perform all manner of different alignment algorithms (Needleman & Wunsch, 1970; Smith & Waterman, 1981 and many variations thereof) via its generic alignment module (Rahn et al., 2018). It also offers an implementation of a more specialised algorithm for edit-distance alignments (Myers, 1999). After the significant structural work by Rahn et al. (2018), this module displayed huge performance gains (see below).

Table 2.2 Parsing 2.2GiB of simulated reads in the BAM format

	Bamtools	htslib	SeqAn	PySAM
Time [s]	39 s	31 s	17 s	59 s

These results are taken from a third party benchmark performed on the current program versions in 2016: https://github.com/wilzbach/bam-perf-test

Table 2.3 Performance of different FM-indexes. This is part of Table 1 from Pockrandt et al. (2017) and only given here to illustrate the speed-up of SeqAn's new implementation (2EPR, in bold) over the original implementation (2WT) and competitors (2SDSL and 2SCH)

	DNA		Murphy10		IUPAC		Protein	
Index	Time	Factor	Time	Factor	Time	Factor	Time	Factor
2WT	9.32s	1.00	19.15s	1.00	23.44s	1.00	28.83s	1.00
2EPR	**4.69 s**	**1.99**	**5.78 s**	**3.31**	**5.67 s**	**4.13**	**6.21 s**	**4.64**
2SDSL	12.21 s	0.76	20.58 s	0.93	24.43 s	0.96	29.76 s	0.97
2SCH	14.08 s	0.66	22.18 s	0.86	26.11 s	0.90	31.81 s	0.91

These results show that the general strategy and design decisions were the right ones to achieve a high performance. But a notable *dimension* of performance was not addressed by the original SeqAn release at all: *parallelism/concurrency*. In fact the term "parallel" appears in none of the original publications (Gogol-Döring, 2009; Döring et al., 2008; Reinert et al., 2017). Parallelism is important because the hardware that is being programmed for has changed dramatically in the last years. The observation called "Moore's law" describes the doubling of the number of transistors in dense integrated circuits every one or two years (Moore, 1965). It is often misunderstood to mean the doubling of "CPU speed" or even raw CPU clock speed, because this used to be strongly correlated. Since a few years now, this has not been the case as Sutter (2005) explains well:

> Over the past 30 years, CPU designers have achieved performance gains in three main areas [...]
>
> - clock speed
> - execution optimization
> - cache
>
>   [...] Speedups in any of these areas will directly lead to speedups in sequential (nonparallel, single-threaded, single-process) applications, as well as applications that do make use of concurrency. [...] **CPU performance growth as we have known it hit a wall[.] [...] Applications will increasingly need to be concurrent** if they want to fully exploit continuing exponential CPU throughput gains[.] (Sutter, 2005; emphasis is mine)

SeqAn1 not offering parallelised algorithms does not mean that one could not have parallelism in applications based on SeqAn, it was simply the philosophy of the library that any parallelism should be implemented application-side. This shifted slightly with the introduction of parallel BAM I/O during a later release of SeqAn1, and much later in with the release of SeqAn-2.3 where parallelised and vectorised alignment code was added (Rahn et al., 2018). This yielded impressive speed-ups as can be seen in Fig. 2.3.

Pivotal to this change of philosophy was the realisation that certain forms of desirable parallelism are impossible or too difficult to achieve by SeqAn's users. And with growing relevance, it could not be left up to the individual application developer. Instead, important interfaces should directly offer access to high-level parallelisation. These changes were important to preserve SeqAn's status as a widely

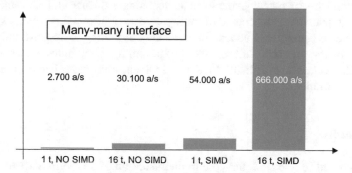

Fig. 2.3 Speed-up of alignment computation with threads and SIMD. Alignments per second (a/s) given for 2M 150bp Illumina reads at different threads and with/without AVX2. Image kindly provided by René Rahn

recognised performance-oriented bioinformatics library, but they were applied ex-post, and there was no clear strategy of implementation (some parts of SeqAn relied on OpenMP (Dagum & Menon, 1998), others on C++11 threads and others on Intel TBB (Pheatt, 2008)). Furthermore, the user-visible interfaces to parallelised features were not uniform: some aspects were controllable by runtime parameters, others by tags and others only via C macros or even shell environment variables.

It is clear that a successor to SeqAn1/2 would need to address parallelisation head-on and provide clear interfaces that enable users to easily choose between different levels of parallelisation.

In this context it should be mentioned that in the quest for even better perfor-mance, SeqAn developers put significant effort into targeting high-performance processors other than the CPU. Enrico Siragusa developed support for CUDA, which targets NVIDIA graphics processors (Nickolls et al., 2008), and Marcel Ehrhardt developed support for the Intel Xeon Phi Co-processor. For different reasons, none of these approaches ultimately led to usable applications. It remains to be seen whether it is feasible for a generic library to support such specialised devices.

2.4.2 Simplicity

The second goal formulated for SeqAn is *Simplicity*. This refers to both learning how to use the library and the ability to contribute to and maintain it continuously. While one might argue that SeqAn has been as simple as possible (under the primacy of performance and the constraints of C++ 98), I would argue that it was everything but simple.

A very steep learning curve is one of the criticisms heard most often about SeqAn, and my personal experience in teaching students and new members of the SeqAn team over the course of multiple years confirms this. Even experienced C++ developers struggle in understanding and contributing to SeqAn1 and SeqAn2.

This is the direct result of the programming techniques described above (Sect. 2.3). Some are difficult to apply in their own regard and some lead to secondary problems.

Non-locality

As mentioned in Sect. 2.3, the core implementation of a type in SeqAn1/2 is often not part of the type itself but implemented as free functions. These are not defined in the same header file as the type if a less specialised template/overload provides the functionality (which is the design for avoiding code duplication). The result is strong fragmentation of the implementation that is very difficult to track. This is reinforced through complex specialisation hierarchies that are not obvious from the code or the documentation and many intermediate layers of function wrappers and shims that obscure the call-graph. To add more complexity to the matter, header files in SeqAn1/2 do not include those headers that they require—which would give a hint on where to look for "inherited" functionality. Instead, there are singular "meta-includes" for every module and the headers inside the modules have no includes.

When attempting to understand the mechanics of a specific type in SeqAn1/2, one routinely has to open a debugger and step through the called functions, often jumping between multiple files. Understanding the path of template type instantiation (e.g. answering the question "which specialisation of metafunction X is selected for my type Y?") is even more difficult, because the "trick" using the debugger is not available for metafunctions.

Code Complexity and Feature Creep

Feature creep describes the continuous and excessive growth of features in a piece of software or hardware resulting in it becoming more difficult to use and/or less stable (Sullivan, 2005). Since SeqAn1/2 was developed in a single repository together with custom tooling and many applications,[10] the policy was that any code that *might* be useful to more than one application should become part of the library. Furthermore, the process of integrating a new application was very liberal, some applications being the results of small student projects or proof-of-concepts. At its height, the repository contained close to 40 applications (it has since been reduced to 28).

This combination led to a strong growth of the library code base and the incorporation of many features with little relevance to the general user base. The

---

[10] This is a problem in its own right.

number of modules in SeqAn2 increased to currently 48, containing a total of 706 header files and 181,000 lines of code.[11]

In the absence of clear policies and without project members dedicated to maintenance, modernisation and code quality, the general complexity of the code base increased significantly. This reflects the "second law of software evolution" formulated by Lehman (1980): "As an evolving program is continuously changed, its complexity, reflecting deteriorating structure, increases unless work is done to maintain it or reduce it.".

Unconstrained Templates

I elaborated on template subclassing in Sect. 2.3, and while in general this has been the type of polymorphism in SeqAn1/2, there are in fact also higher abstraction levels. One example is that specialisations of `seqan::String<>`, `seqan::StringSet<>` and `seqan::Segment<>` are all considered "sequences". Since they do not share a common base template, one cannot easily create a generic function that accepts *exactly* the specialisations of all three. The easiest way to write a function that accepts *at least* the specialisations of all three is to write an entirely unconstrained template (that formally accepts any type). This can be seen for `begin()` and `end()` defined in `sequence_interface.h`, but there are many more unconstrained templates in SeqAn1/2.

An effect of unconstrained templates is that misuse of the interface is not reported immediately. Instead, a compiler-error happens much further down the call-graph when an unsupported operation is called on the falsely given type (or possibly even a dependent type of that type). These kinds of error messages tend to be very long (often spanning multiple pages) and hard to understand (the error highlighted by the compiler seems to be entirely unrelated to the problem).

Another issue is that unconstrained templates increase the non-locality described above and make it harder to search for the relevant overloads in the code base. They also interact with implicit conversion in ways that are unexpected to many users which is why the CPP Core Guidelines have a rule against them (see the reference for an example).[12]

---

[11] SeqAn1 was split into `core` and `extra` with the intent being that `core` should be held to higher quality standards and `extra` be more of testbed for the actual library. But when I became involved more strongly with the library, this separation had already weakened significantly and `core` had strong dependencies on `extra` rendering the differentiation meaningless. They were merged for SeqAn2.

[12] "T.47: Avoid highly visible unconstrained templates with common names". http://isocpp.github.io/CppCoreGuidelines/CppCoreGuidelines#Rt-visible.

Documentation

The documentation of a software library is an integral factor of its maintainability and its ease of use (Geiger et al., 2018). Documentation includes API documentation (documents describing the interfaces of classes, functions, etc.), Tutorials, ReadMes, Wikis and possibly other resources that help in using the software. For libraries, API documentation is the most important aspect of documentation as it is the primary way users learn about features of the library and interact with the individual components. It should not be necessary for users to look at the source code of a library to use it, and the API documentation should provide all necessary information.

API documentation is typically written inside the source code as comments (in a certain style or markup language). These comments usually precede the entity that they document or are found in its proximity. Third party software then generates readable documentation (e.g. in HTML or PDF format) from the comments, often also performing rudimentary parsing of the source code and enforcing that the documentation matches the actual interfaces defined by the code.

The most common documentation generator for C++ software is Doxygen (van Heesch, 2008) which uses a syntax similar to Javadoc (Kramer, 1999), one of the earliest documentation generators. When SeqAn was first developed, the authors came to the conclusion that Doxygen would perform poorly on SeqAn (due to the unorthodox programming techniques) and decided to develop their own system: *dddoc*. It was part of the first SeqAn release and is briefly described in Gogol-Döring's dissertation (Gogol-Döring, 2009). I cannot judge whether developing a custom documentation generator was the most sensible option at the time, but it did increase the burden to contribute to SeqAn, especially since the syntax was very different from the well-known examples Doxygen and Javadoc. The code generator also performed no parsing of the source code; documentation entries were parsed completely independent of context. This increased the chance of documentation error and contributed to non-locality (documentation of an entity could be in an entirely different place than the entity itself). Furthermore, there was no method of enforcing that an entity be documented at all and casual examinations of the SeqAn-1.0 source code show that many were not.[13]

During the development of SeqAn2 an entirely new, stand-alone documentation generator, called *dox*, was created (Kahlert, 2015, see Fig. 2.4). This improved over dddoc in that its syntax was modelled after Doxygen and the visual appearance of the generated documentation was much more modern. However, the core problems mentioned above, the independence of the documentation and code as well as the lack of policy (enforcement) in regard to the completeness of the documentation, were not solved. The documentation as a whole was not able to explain the techniques of SeqAn well enough to make it appear like more traditional C++. For

---

[13] It is not clear whether this was a lack of "enforcement" or a general lack of policy in this regard.

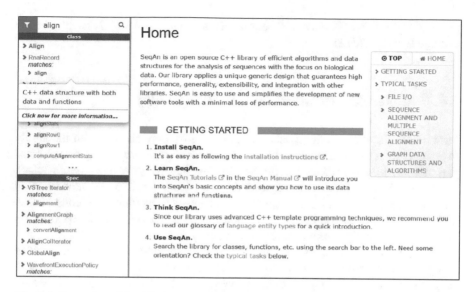

Fig. 2.4 Screenshot of the API documentation of SeqAn-2.4 (built with *dox*). Screenshot taken by me; content is part of SeqAn documentation, see Sect. A.3.1

Table 2.4
Source-lines-of-code and comment-lines-of-code in different SeqAn releases

SeqAn release	SLOC	CLOC	CLOC in %
SeqAn-1.0	88,332	36,578	29.28
SeqAn-2.0	168,488	94,635	35.97

example, template subclassing was explained similarly to inheritance, but typical documentation of the latter, like inheritance graphs, was notably absent.

To put the matter of *completeness* of documentation into perspective, I have given the source-lines-of-code and the comment-lines-of-code for the respective .0-releases in Table 2.4. These were measured with the *cloc* tool and only the library folder was considered.[14] Care should be taken when using these numbers to compare different projects, but considering that certain style decisions (e.g. the maximum line width and when/where to break lines) have remained constant from SeqAn1 until SeqAn3, they do have some descriptive value. The numbers are discussed and compared with SeqAn3's in Sect. 4.4.2.

Later criticism notwithstanding, it should be noted that the relative amount of comments in all SeqAn releases is well above average. The OpenHUB platform, which performs statistics and analytics of open source software projects and covers almost 500,000 projects, shows an average of 22% comment-lines-of-code for C++

---

[14] https://github.com/AlDanial/cloc/.

projects.[15] And there is reason to believe that academic software is usually below average (Lemire, 2012).

2.4.3   Generality, Refineability and Extensibility

I am discussing these design goals together, because they all deal with the ability to adapt SeqAn to one's needs (with some aspects of *generality* being discussed as part of *integration* below). In general, SeqAn1 and SeqAn2 offer a maximum degree of freedom in regard to their adaptability. The global function and metafunction interfaces described previously (especially when used as *modifiers*), in combination with a lack of the classic C/C++ protection model, place only few restrictions on how a user can apply, refine or extend the existing code.

While there are cases where this degree of freedom is useful, the added complexity should not be underestimated. The core problem for a user wishing to adapt the behaviour of the code is not knowing which entity to customise, because *any* entity *can* be customised. An example should explain this: given is an `int` property/member of a type that shall be refined to appear `1` larger than the actual value. One would typically specialise the accessor function to just add `1` when returning the value. But with multiple global shim functions, it may not be clear which function is the "accessor" (see Snippet 2.4.2), and specialising any function in the call-graph will likely yield the desired result. However, in a different context, the call-graph may look slightly different and the specialisation might be skipped, resulting in faulty behaviour.[16]

Users may also be tempted to not specialise the accessor function at all, and instead manipulate the private state of the object after creation—since the member is `public`, the user can change the value instead of overriding access functions. Gogol-Döring anticipates criticism of giving up the classic C/C++ protection model but explains:

> Global functions lack a protection model: They cannot be private nor [sic] protected, and they cannot access private and protected members of a class. [...] The main reason for a protection model is to prevent the programmer from accessing functions or data members that are intended for internal use only. A simple substitution for this feature is to establish clean naming conventions: We state that a '_'-character within an identifier indicates that it is for internal use only. [...] [We] decided to declare data members to be public, but only functions that belong to the core implementation of [...] [a class] are allowed to access them by convention. (Gogol-Döring, 2009)

---

[15] https://www.openhub.net/p/seqan/factoids (note that these statistics cover the entire repository, not just the library).

[16] The obvious solution to this problem is to specialise as close as possible to the type, but in absence of a language mechanism enforcing this, errors are easy to make—especially since the function most visible to the user is the one "furthest" from the type.

Having a convention is better than not regulating access at all, and however, a convention is a poor replacement for a language feature. Research has shown repeatedly that programming conventions are violated if they are not enforced via technical measures (Hedin, 1996; Prause & Jarke, 2015). A cursory examination of the most popular SeqAn applications shows that all of them make use of at least some "private" library functions or access "private" data members of library types. This is made easier by the fact that they are distributed in the same repository (see Sect. 2.4.4) and breaking changes to "private" library interfaces are visible in continuous integration (placing the burden of keeping the application in a functional state on the library maintainers).

I conclude that the chosen approach to customisation may be the most liberal, but neither the most user-friendly nor the one that guarantees the highest quality of code. Best practice guides for library design recommend limiting customisability to clearly specified *customisation points* and taking extra care in designing those (see Sect. 3.7).

2.4.4   Integration

Integration covers the ability to use the library with existing projects, both on source code level, i.e. the interaction with existing C++ types and functions, and on a project level, i.e. the interplay between repositories, build systems and packaging frameworks.

Source-Code Level Integration

Gogol-Döring mostly defines integration in terms of applying the extensibility discussed above to many or all types of the standard library or a third party library:

> The idea of *global Interfaces* imply the possibility of using *shims*, which make the library adaptable both for additional external data structures and for built-in types. We demonstrated in Sect. 6.1, that algorithms in SeqAn may be generic to an extend [sic] that we called 'library spanning programming', because they can be used for data structures from arbitrary sources, as soon as the necessary shims are available. SeqAn comes with an adaptor for `basic_string` of the Standard library (and its iterators), as well as for C-style strings, i.e. for zero-terminated char arrays. However, it is also quite possible to integrate other third party libraries easily into SeqAn. (Gogol-Döring, 2009)

```
  template <typename TChar, typename TTraits, typename TAlloc>
2 inline typename Size< basic_string<TChar, TTraits, TAlloc> >::Type
  length(basic_string<TChar, TTraits, TAlloc> const & str)
4 {
      return str.length();
6 }
```

Code snippet 2.3: Overload for `length()` and `std::basic_string` from Gogol-Döring (2009)

This approach works well for integrating a single type, but it scales very poorly to the size of a library. Consider the example of adapting a `std::basic_string` to work like a `Sequence` in SeqAn1/2 (Snippet 2.3). The interface consists of over 30 functions and over 15 metafunctions (the exact number depends on some special cases). If one were to add overloads/specialisations for all containers from the standard library (`std::basic_string`, `std::array`, `std::vector`, `std::deque`, `std::list`, `std::forward_list`), that amounts to over 270 functions/metafunctions and thousands of lines of "copy'n'paste" code. This is the opposite of *generic programming* and prone to errors.

To complicate matters further, adding a specialisation for a type is entirely orthogonal to any existing forms of refinement based on template subclassing (Sect. 2.3.2). For example, if an algorithm behaves in a generic way for `String<TAlph, TSpec>` and in a refined way for `String<TAlph, Alloc<TSpec>>`, one cannot cleanly express that the overload for `std::basic_string<TChar, TTraits, TAlloc>` should behave like one or the other; one needs to "copy'n'paste" code or change the library code by inserting another delegation layer that can be called from library code and the new overload.

As a result, SeqAn relied even more heavily on its own types and did not use standard library types. In fact, support for standard library types was very poor for a long time, something users often criticised.

The clean solution to these problems is using *C++ concepts*, a language feature which I will introduce in Sect. 3.4. Within the limits of C++ 98, *SFINAE* could have been used more often to facilitate refined overload resolution of functions (and specialisation of type templates). SFINAE stands for *substitution-failure-is-not-an-error* and describes how a failed template substitution does not result in a compiler-error, but only in not considering that function template in the set of possible overloads.[17] This effect can be used to craft overloads specifically for certain groups of types or based on certain conditions. Care needs to be taken, though, because no two such overloads should remain in the valid set to prevent ambiguity (there is no intrinsic notion of refinement/specialisation after the resolution of SFINAE). Järvi et al. (2003) performed early research on this and provide guidance on using SFINAE for controlling overloads.

---

[17] https://en.cppreference.com/w/cpp/language/sfinae.

For SeqAn's 2.1-release I, added support functions to have SeqAn recognise all standard library containers using a combination of SFINAE and C macros. Due to the abundance of unconstrained primary function templates (that always collide with overloads that do not use template subclassing), this was not possible without many library code changes.

In effect, I would argue that SeqAn1/2 was able to facilitate ad hoc specialisations of single third party types sufficiently well but was not able to properly handle third party libraries as a whole. Its reliance on self-provided types over standard library types and its poor handling of the latter underlines this weakness.

Project-Level Integration

A dimension of *integration* that played a much smaller role in Gogol-Döring (2009) is the integration on project level, and this includes the practical and legal implications of (re-)distributing the library and the administrative overhead of including it as a dependency and maintaining updates.

Legal Terms

The licence of the SeqAn library was originally the GNU Lesser General Public License (Free Software Foundation, 2002). It was changed to the 3-clause BSD License in SeqA-1.3.[18] Neither of the two licences requires that other software integrated with SeqAn have the same licensing terms (no *strong copyleft*), but the LGPL imposes some obligations regarding changes to the library itself. The BSD licence, on the other hand, is considered as one of the most permissive Free and Open Source Software licences and requires only attribution.[19]

Project Hosting

When SeqAn1 was released, public source code hosting was not yet popular for academic software. However, with the release of SeqAn-2.0, the project moved to GitHub (see Fig. 2.5).[20] Beyond the technical benefits of *git* as a version control system, having SeqAn hosted, there has increased the visibility of the project, the amount of external contributors and the ease with which it can be integrated in other repositories (e.g. via *git submodules*). While these aspects are not crucial to integration of the project, they most certainly help our users. Research suggests

---

[18] https://www.freebsd.org/internal/software-license.html.

[19] https://opensource.org/faq#permissive.

[20] https://github.com.

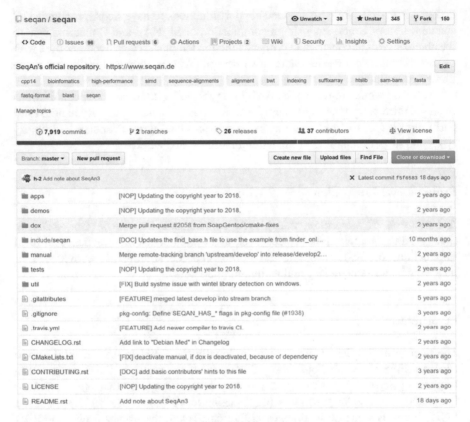

Fig. 2.5 Screenshot of the SeqAn project page on GitHub. Screenshot taken by me; the look and feel of the website and service is copyright © GitHub, Inc. All rights reserved

that most academic projects would benefit from being publicly hosted in a similar manner (Blischak et al., 2016).

Build Systems

Since SeqAn is a header-only library, it is distributed as source code and cannot be prebuilt as a shared object. This model is a direct consequence of the strong reliance on templates. While it implies longer compilation times, it has the added benefit of easy distribution and integration. In principle, it is sufficient to add SeqAn's include folder to the compiler's include path to start using SeqAn; special build steps are not required for the library. However, in practice, a few compiler flags do need to be set to enable threading support, raise the C++ standard level of the compiler and detect (optional) dependencies like ZLIB or BZip2. It is therefore advisable to use a build system that analyses the environment and sets respective flags automatically. As of version 1.2, SeqAn supports CMake,[21] the de facto standard for cross platform C and C++ projects (Wojtczyk & Knoll, 2008).

Semantic Versioning

An application that decides to add a dependency on a library needs to consider the stability of the library, i.e. the costs and risks associated with an upgrade of the library. Upgrades may seem optional, but often they are not, because new updates provide necessary fixes or security patches (Raemaekers et al., 2014). One way to clearly define the costs and risks associated with an upgrade is to follow *semantic versioning* and assign version numbers accordingly (Preston-Werner, 2013). Two core aspects of semantic versioning are having a clearly defined public API and promising to the user that no breaking changes to that API will be introduced within one *major* release (versioning is `major.minor.patch`). SeqAn1 already failed in regard to the first requirement.[22] The second aspect is the central paradigm of semantic versioning. This is a notable restriction on the changes developers can make to the project, but it provides a very strong guarantee for safe upgrades to the user. It was introduced for all SeqAn2 versions, beginning with SeqAn-2.1, by declaring all documented interfaces as part of the API.

An argument brought forth against the necessity of semantic versioning by previous maintainers of SeqAn is that header-only libraries (see above) can be shipped together with the application and that there is no "forced upgrade". This ignores the possibility of depending on a new feature or security update as well as the interdependencies of components in complex modern software; e.g. an application might depend on two third party components that each depend on SeqAn—if one updates its SeqAn requirement but not the other, the application can become unbuildable. To underline the disruptive nature of breaking updates, one should consider that OpenMS (Röst et al., 2016), a project closely affiliated with SeqAn, still uses SeqAn-1.4.1, because upgrading was seen as too expensive by the maintainers. The importance of semantic versioning for the long-term health of a software library cannot be underestimated (Raemaekers et al., 2014) and lack of semantic versioning in previous versions of SeqAn was a major problem.

Framework-Style Repository

An unfortunate development that began with SeqAn-1.1 was bundling applications together with SeqAn and later also the custom testing and documentation infrastruc-

---

[21] https://www.cmake.org.

[22] One could derive from the rules quoted in Sect. 2.4.3 that any name not containing _ be part of the API, but, since many such names are also not documented, it is not clear what the user should rely on.

ture. This was likely one of the results of the lack of semantic versioning, because to prevent regularly breaking the applications through changes in the library interface, they were simply tried and tested together. Many drawbacks of this approach have been discussed already. The relevant impact on *integration* was that developers who wanted to build an application with the SeqAn-**library** needed to checkout an entire ecosystem of library + applications + infrastructure. This conflation on repository level was mirrored also in the documentation where instructions on building "SeqAn applications" were mixed with the "first steps guide" to programming with SeqAn.[23] This first steps guide involved preparing a directory *inside* the repository by running a provided Python script and then editing a file in that directory. There were no instructions on adding SeqAn to an application with existing infrastructure and the provided CMake module failed when used individually. Other implications of the repository style are a confusing licensing situation (the applications each have individual licence files with different terms) and difficulties in packaging (see below). Some problems were fixed by myself and René Rahn after becoming the responsible developers, but the general structure of the repository remained largely the same.

Package Managers

Many users do not install their applications by downloading a package from the author/vendor but by utilising a *package manager*. Package managers automate the install, upgrade and removal of software packages, keep track of dependencies between packages and help maintain a consistent and up-to-date state of the entire set of installed software (Spinellis, 2012). Some operating systems provide package managers by default, typically GNU/Linux distributions or BSD-based operating systems (e.g. APT on Debian GNU/Linux[24]), but there are also stand-alone managers that can be installed individually (e.g. Homebrew (Jackman et al., 2016) on macOS or Conda (Grüning et al., 2018) which is popular with data scientists). Being present in such managers has the advantage that developers can easily get access to new SeqAn releases, but, most importantly, it is usually required so that application developers can add their SeqAn-based applications to said package managers. It is a quality indicator, because application developers know if they can easily integrate a library with their application when they ship it. Initially, SeqAn was only available in few managers and often only as part of an "application bundle". The lack of semantic versioning (see above) made packagers reluctant to add a dependency on SeqAn, because it meant unforeseen breakage could happen.

---

[23] This conflation also happened on a project level: events were hosted for "application users", e.g. someone wanting to perform read-mapping with RazerS, and "library users", i.e. developers interested in creating a new application, at the same time and place. In my opinion this needlessly complicated the situation for both audiences and greatly increased the notion that "SeqAn is difficult to use".

[24] https://wiki.debian.org/PackageManagement.

After this was fixed in SeqAn-2.1 and the CMake support was brought up to shape, SeqAn2 was packaged for all major GNU/Linux distributions, FreeBSD, the two macOS package managers Homebrew (Jackman et al., 2016) and MacPorts[25] as well as the domain-specific package managers Conda (Grüning et al., 2018) and Easybuild (Hoste et al., 2012).

Workflows

Bioinformatics applications are increasingly deployed as part of pipelines or workflows (Curcin & Ghanem, 2008). Workflows allow researchers that are not programmers to combine different applications and perform integrated analyses. They can help improve structure and reproducibility, and they replace many previous uses of shell-scripts and Makefiles (Leipzig, 2017). It is arguably the obligation of the application developer to ensure that their programs run in a workflow system and not the responsibility of a software library. However, the SeqAn project anticipated that many of the developers using SeqAn would welcome help in targeting workflow systems. Since SeqAn already comes with an argument parser that handles command line arguments to the application and can also generate help and manual pages, this was expanded to also generate descriptor files for the KNIME workflow system (Berthold et al., 2007). An example workflow is shown in Fig. 2.6. SeqAn chose to support KNIME, because the projects have a long-standing history of cooperation and KNIME is a promising workflow system with large industry support. It should, however, be noted that KNIME's largest user groups are in chemistry/cheminformatics, "business intelligence" and "predictive analytics" (Warr, 2012). It is not unpopular among bioinformaticians, but most comparative studies of workflow systems in bioinformatics and sequence analysis focus on other workflow systems (Curcin & Ghanem, 2008; Leipzig, 2017). Support for the Galaxy system (Afgan et al., 2018) was a frequently requested feature. More recently, the Nextflow platform (Di Tommaso et al., 2017) has gained popularity and there have been attempts to standardise workflow languages and the description of applications/nodes within them as CommonWL (Amstutz et al., 2016). Future versions of SeqAn should evaluate if more workflow systems can be targeted via specialised description generators or an open standard like CommonWL.

2.4.5   Summary

SeqAn1 and SeqAn2 were successful and influential C++ libraries in the domain of sequence analysis. Not only groundbreaking applications were built with the help of SeqAn, but also prototypes and small applets for the use in workflows.

---

[25] https://www.macports.org.

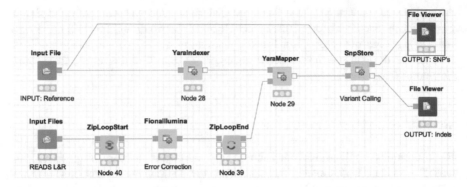

Fig. 2.6 A KNIME workflow that includes SeqAn applications. Image is part of the SeqAn1/2's documentation, see Sect. A.3.1

The performance of SeqAn was superb, although there was no coherent strategy for attaining the best possible speed in the context of an increasingly parallel execution environment.

The library strove to be as simple as possible, but the use of exotic programming techniques led to a very steep learning curve. The academic nature of the project and regular changes in its technical leadership led to a lack of consistent policy (enforcement) and direction. This in turn led to an ever-increasing size and complexity of the code base, further raising the bar for understanding and contributing to the library. In addition to having to understand the code base itself, SeqAn forced contributors to learn custom tooling, because it did/could not rely on industry standard tooling.

SeqAn anticipated many developments in the C++ language but had to rely on the now-old C++ 98 standard. It later adopted certain convenience features from C++ 11 and C++ 14, but the general design still reflected C++ 98 strongly and did not take the many structural advantages of Modern C++ into account.

Documentation of the library was always above average, especially for an academic project. However, there was no policy (enforcement) that ensured that (at least) all *public* entities were documented. Considering the extraordinary complexity of the library, better documentation would have certainly been helpful.

SeqAn allowed for a high degree of customisation in regard to small changes, adapting single type and overriding the behaviour of almost any library routine ("hacks"). However, the manner of customisation was obscure and the potential for error high. It had poor support for adapting large number of types from third party libraries, and standard library types were always second-class citizens. Not relying on standard library types and functions implies a lot of code/logic duplication and the additional overhead for users to re-learn.

On project level, there was a lot of conflation between application development and library development, needlessly complicating the maintenance, distribution and packaging of the library. Many best practices in software development were introduced by myself and René Rahn in the last versions of SeqAn2. These improved the quality of changes and new additions to the library, but ultimately it was decided that the *technical debt* was too large to continue improving on the library incrementally and that a more radical re-design was necessary.

Chapter 3
Modern C++

The C++ programming language is a general purpose programming language created by Bjarne Stroustrup in the early 1980s. The original intent was to extend the C programming language by features for object-orientation similar to the programming language Simula that was popular at the time (Stroustrup, 1993). Current versions of C++ combine elements of procedural, functional, generic and object-oriented programming.

Table 3.1 shows important milestones in the history of C++. Since 1998, C++ is an international standard, governed by the International Organization for Standardization (ISO). Inside ISO, Workgroup 21 (WG21) of subcommittee 22 (SC22—Programming Languages) is responsible for C++. Together with C and ECMAScript/JavaScript, it is one of the few general purpose programming languages that are standardised and have multiple competing, standards-conform implementations. All major IT companies, among them the top five most valuable brands on the planet (Apple, Google, Microsoft, Amazon, Facebook),[1] are involved in the standardisation process.[2] Other attendees include members of research institutes, universities and companies from such diverse fields as finance, graphics, video gaming and embedded computing. These facts highlight C++'s central role as an industry language and its importance worldwide.

There are numerous comparisons of programming languages in the literature and it is not my aim to present C++ as the solution to all programming problems. However, when *performance* and/or *stability* (as in the "longevity" of software) are a concern, C++ is a very good choice.

Performance is not always a concern in Bioinformatics, and Python, Perl and R are established languages to solve problems that are not performance-critical and/or more geared towards statistics (Nattestad, 2017). As I elaborated in Chap. 1,

---

[1] https://www.forbes.com/powerful-brands/list/.

[2] Attendance lists are public, e.g. http://www.open-std.org/jtc1/sc22/wg21/docs/papers/2019/n4826.pdf.

Table 3.1 History of C++
releases/revisions

Year	Version	Codename	ISO standard
1985	C++ 1.0		
1989	C++ 2.0		
1991	C++ 3.0		
1998	C++ 98		ISO/IEC 14882:1998
2003	C++ 03		ISO/IEC 14882:2003
2011	C++ 11	C++ 0x	ISO/IEC 14882:2011
2014	C++ 14	C++ 1y	ISO/IEC 14882:2014
2017	C++ 17	C++ 1z	ISO/IEC 14882:2017
2020	C++ 20	C++ 2a	ISO/IEC 14882:draft

however, many fundamental problems in sequence analysis are only solvable with the most efficient engineering in which the programming language plays a crucial role. While languages like Python and Perl were never designed for this domain, other languages like Java, C#, Go and (more recently) Rust also advertise a high performance. Most empiric comparisons still conclude that C++ is superior to the others or at least among the best (Prechelt, 2000; Fourment & Gillings, 2008; Aruoba & Fernández-Villaverde, 2014),[3] although under very specific circumstances other programming languages also take the lead (Costanza et al., 2019).

Young programming languages often evolve quickly which has the benefit of also being able to fix design mistakes quickly, but the down-side is that old code often breaks and different pieces of software become incompatible to each other (Malloy & Power, 2017). C++ on the other hand is older than most people who program today, and although this chapter will present radical changes in the language, almost all C and C++ code ever written is still valid C++ code today. In contrast to languages like Fortran and to some degree C, this backwards-compatibility does not express stagnation, but the joint interest (and hard work) of the involved parties to evolve the language in a non-breaking manner. This has resulted in a wealth of software libraries and in a stability of the language that is unrivalled. Associated with backwards-compatibility is also the promise for a stable future, i.e. a company or project that decides to commit to a new piece of infrastructural software can expect this piece of software to be usable many years after the conception.

This does not preclude rewriting software to improve the design, decrease the technical debt or make use of new language features—which this book is all about! But this decision is not a forced one, and it does not break compatibility with other parts of the ecosystem (as, for example, moving from Python2 to Python3 does). With long-term reproducibility becoming more of a focus in science in general (Baker, 2016), this kind of backwards-compatibility might also become more important for research.

---

[3] Admittedly, not all of these comparisons are recent and more research into this area would be very welcome—especially considering the advances of C++ also discussed in this chapter.

The formal name of C++ is *"International Standard ISO/IEC 14882:xxxx— Programming Language C++"* (where *xxxx* refers to the year of publication). For readability, I will abbreviate this to "C++ 11", "C++ 17", etc. in regular text and to *"ISO/IEC 14882:xxxx"* when citing certain sections or paragraphs. At the time of writing, the current working draft of the standard has reached the level of *committee draft* indicating a state that is almost final. I will refer to this draft as "C++ 20" although it is not the formally published C++ 20 standard. The primary document for citations is ISO/IEC 14882:2017 as it is the currently valid standard and section numbering is guaranteed not to change. I only refer to the draft standard (denoted by ISO/IEC 14882:draft) when discussing changes after C++ 17. Section references for this document pertain to the aforementioned *committee draft*, also known by the paper number N4830.

Proposals to the standard (denoted by P-numbers) are cited as regular sources. Although ISO publications themselves are not public, all P-number papers are, and versions of the standard published immediately before and after official standard releases are, as well. (They have N-numbers.) All of these can be retrieved via https://wg21.link/NUMBER, where NUMBER can e.g. be a P-number like P1739 or an abbreviation for the standard itself (std11 for C++ 11, std17 for C++ 17, std for the current draft, ...).

This chapter will be the most technical as it needs to lay the foundation for understanding the design decisions and the implementation in later chapters. I do not attempt to cover all changes in the standard since C++ 98, instead I focus on those aspects that I deem fundamental to SeqAn3. Smaller changes and improvements that do not warrant their own section in this chapter will be explained when and where they are first used. Beyond actual C++ standard changes I will also describe certain best-practices and techniques that have become widely accepted in the C++ community during the last decade. Please note that I try to be *accurate* in the sense that I do not make statements that contradict the standard or betray the intention of the authors. However, I make no claim to completeness or formal correctness, i.e. I will not cover edge-cases, exceptions or detailed standard wording. I hope readers understand the motivation for the discussed changes, see how they are applied in a very basic fashion and comprehend why they are relevant for SeqAn3. This book does not replace programming guides, tutorials, etc. Readers interested in details are invited to read the respective sections in the standard; I provide references wherever this is reasonable.

3.1 Type Deduction

This section introduces some places where *type deduction* happens in Modern C++. Type deduction means that the developer uses a placeholder instead of a type and that the compiler then deduces the concrete type (and replaces the placeholder) from the context. All of this is still part of *static typing*, i.e. the actual type must be unambiguously deducible at compile-time.

3.1.1   The auto Specifier

The `auto` specifier[4] can be used as a placeholder in the following circumstances
where one would otherwise specify a type:

1. As the type of a locally declared variable, deduction happens from the initialiser
   (ISO/IEC 14882:2017, 10.1.7.4).
2. As the return type of a function or function template, deduction happens from a
   "trailing return type" (ISO/IEC 14882:2017, 11.3.5).
3. As the return type of a function or function template; deduction happens from
   the return statement(s) in the function body (ISO/IEC 14882:2017, 9.4.1).
4. As the type of a parameter in a lambda expression (making it a *generic lambda*),
   deduction happens from the type of the argument (ISO/IEC 14882:2017, 8.1.5.1).
5. As the type of a parameter in a function definition (making it an abbreviated
   function template), deduction happens from the type of the argument (ISO/IEC
   14882:draft, 9.2.3.5).

A use-case of `auto` where it cannot be substituted by a concrete type is in the
declaration of *structured bindings* (ISO/IEC 14882:2017, 11.5).

```
  auto i1 = 3;          // the type of i1 is deduced to int
2 auto i2(5);           // same; but different syntax
  auto i3{7};           // same; but different syntax (C++11)
4
  auto j1 = i1;         // the type of j is deduced to i1's
6 auto j2 = foobar();   // the type of j is deduced to the return type of foobar()

8 //auto k;             // not valid code, because there is no initialiser
```

Code snippet 3.1: The `auto` specifier in variable declarations

```
  template <typename T>                      template <typename T>
2 void swapvalues(T & container)           2 void swapvalues(T & container)
  {                                          {
4     typedef typename Value<T>::Type TValue; 4
      TValue help = container[0];                auto help = container[0];
6     container[0] = container[1];           6     container[0] = container[1];
      container[1] = help;                       container[1] = help;
8 }                                        8 }
```

Code snippet 3.2: Use of `auto` in generic programming: Excerpt of "Listing 4"
from Gogol-Döring (2009) vs a version written in Modern C++

---

[4] There is also a `decltype(auto)` specifier with slightly different deduction rules, but the
differences are not discussed here, see ISO/IEC 14882:2017 (10.1.7.4) for details.

Snippet 3.1 shows examples for deduced variable declaration (case 1.). Type deduction has the potential to increase readability, but the reduced verbosity can also make the code more difficult to understand for other programmers. In Snippet 3.1 the first examples do not improve readability (over using `int`), but I would argue that the change presented in Snippet 3.2 does.

For these uses of `auto`, I agree with the rule stated in the Google C++ style guide, to not use it arbitrarily, but only "to avoid type names that are noisy, obvious, or unimportant".[5] We will later see standard library types that become complex very quickly (Sect. 3.6.4) and that profit strongly from `auto`. Other expressions, like lambdas, do not have a named type and need to be stored in variables of deduced type (see the excursus on p. 38).

```
 2   auto foo() -> int       // C++11          12   template <typename T1, typename T2>
     {                                              auto bax(T1 a, T2 b)      // C++14
         return 3;                                  {
 4   }                                         14       return a + b;
                                                    }
 6                                             16
     auto bar(int i)         // C++14               auto baz(auto a, auto b)    // C++20
 8   {                                         18   {
         return i + 1;                                  return a + b;
10   }                                         20   }
```

Code snippet 3.3: Deduction of function return and parameter types

Snippet 3.3 shows two functions (left) and two function templates (right) with deduced return type. `foo()` is an example of the trailing return type syntax introduced in C++ 11 (case 2 above). Since C++ 14 the type can be deduced from the return statements inside the function body (case 3 above) and the `-> type` notation is not required any more. As the function `bar()` illustrates, the return type can also be deduced from expressions, not just concrete variables or values. It is debatable whether any of the examples on the left in Snippet 3.3 improve readability, and I would suggest to just use `int` in these cases (similar to the rules defined above).

For `bax()` on the right of Snippet 3.3, however, one can see an actual advantage of return type deduction as it is non-trivial to determine the result type of the expression `a + b` in C++ 98 for two objects of different arbitrary types `T1` and `T2`. Finally, `baz()` is an example of a function template introduced via the `auto` keyword (case 5 above). The semantics of `bax()` and `baz()` are identical and I would argue that the syntax is much easier to read and can reduce the (perceived) complexity of templates. The latter is only available with C++ 20 or the Concepts TS (see also Sect. 3.4).

---

[5] https://google.github.io/styleguide/cppguide.html.

Excursus: Lambda Expressions

Lambda expressions are a way to create simple function objects (ISO/IEC 14882:2017, 8.1.5.1). They are *anonymous*, i.e. do not have a name, and they also have a unique and unnamed type—even two lambdas with identical signature and body! This means they can only be bound to a variable of deduced type (`auto` or a template parameter).[6]

```
   struct abs_comparator_t
2  {                                       2
      bool operator()(int lhs, int rhs) const    auto abs_comparator = [] (int lhs, int rhs)
4  {                                       4  {
         return std::abs(lhs) < std::abs(rhs);      return std::abs(lhs) < std::abs(rhs);
6  }                                       6  };
   };
8                                          8
   abs_comparator_t abs_comparator;
```

```
    int arr[] = {10, -3, 1, 2, -7};
11  std::sort(arr, arr + 5, abs_comparator);    // arr == {1, 2, -3, -7, 10}
```

Code snippet 3.4: Sorting an array of `int`s by absolute value. On the top left via a full function object (C++ 98) and in the top right via a lambda expression (C++ 11). The lambda expression could have been placed entirely in the call of `std::sort()` instead of binding it to a variable

Lambda expressions are a core part of functional programming languages (Thompson, 1991) and their introduction in C++ 11 strengthens C++ in this domain. They are typically used as arguments to higher-order functions and can reduce the complexity of code. Snippet 3.4 shows such an example where the lines of code are reduced through the use of a lambda expression. `[]` is the so-called *capture* of a lambda (in this case empty). It can be used to enable access to objects outside the lambda's scope. (The syntactical details are not important here.) The rest of a lambda expression's syntax is very similar to that of a function definition: the `()` hold the parameters and the `{}` hold the body. An advantage of lambda expressions beyond code reduction is that no auxiliary definitions outside the local scope are needed which improves the readability of the code.

As previously mentioned, lambda expressions are bound to variables introduced by `auto`. (If they are bound at all.) Their return type is also always deduced, although it can be forced to a type via trailing return type syntax (similar to that of functions). In contrast to functions/function templates, lambda expressions can have parameters of deduced type (denoted by `auto`) already since C++ 14—while for function templates this is only valid since C++ 20. This is case 4 of the enumeration at the start of Sect. 3.1.1. Lambdas with deduced parameter types are called *generic*

---

[6] They can also be bound to objects of types that perform *type erasure*, like `std::function`, but this incurs a performance overhead when invoking.

lambdas. They behave like function templates and thus provide a significant benefit over other approaches. (It is not possible to define class templates or function templates inside the body of a function but defining generic lambdas is legal.)

3.1.2 Class Template Argument Deduction (CTAD)

```
std::vector vec{1, 2, 3, 4};    // type of vec is deduced to std::vector<int>
std::tuple tup{vec, 3.3};       // type of tup is deduced to std::tuple<std::vector<int>, float>
```

Code snippet 3.5: Deduction of class template arguments from the constructor

Under certain circumstances, one can also use the name of a class template (without template arguments and without `<>`) as a placeholder in type deduction (ISO/IEC 14882:2017, 10.1.7.5). This is possible since C++ 17 if the compiler can unambiguously deduce the template arguments from the initialising expression. An example can be seen in Snippet 3.5.

Deduction happens either via implicitly generated deduction guides based on constructors or via deduction guides provided by the developer. The specifics of this are not essential here, but the impact of this feature on users of generic code is fundamental: well-designed class templates can now be used entirely without `<>` in most cases. Although this kind of deduction is not usable in all situations where `auto` is currently usable,[7] it plays an important role in making generic (template-rich) code more accessible.

3.2 Move Semantics and Perfect Forwarding

Expressions in C++ are either *lvalues* or *rvalues*.[8] *lvalues* are values that determine the identity of an object or function, they represent a fixed area of memory and their address can be taken. After a variable has been defined (e.g. `int i;`), the name of the variable (e.g. `i`) is an lvalue. Notably, lvalues can be on the left-hand side of assignments.

Those expressions that are not lvalues are *rvalues*. This includes expressions resulting in a new object not bound to a variable, often called *temporaries*.

---

[7] It is only valid when declaring variables or performing casts and not in the signature of function templates or lambda expressions (although this might change in a future version of C++).

[8] The value categories used here are simplified and based more closely on their C++ 98 definitions. However, I think these are better at explaining the phenomena in such a compressed form. See ISO/IEC 14882:2017 (6.10) for C++ 17's formal definitions.

Examples are string literals (``foo''), numeric literals (42) or values returned from functions. Historically, rvalues could only be on the right-hand side of an assignment.

3.2.1   Move Semantics

One of the very fundamental features of Modern C++ are *move semantics*. They were introduced with C++11 and help avoid copying dynamically allocated memory when it can instead be *moved*. To *move* in this context does not mean that data changes its place in memory, instead its ownership is transferred from one object to another.

```
   class Foo
2  {
       std::vector<int> s;
4  public:
       void set_s(std::vector<int> const & t)
6      {
           s = t;
8      }
   };
```

```
   Foo f;
11
   std::vector s1{1, 2, 3};
13 f.set_s(s1);                         // [1]
   std::vector const s2{42, 23};
15 f.set_s(s2);                         // [2]
17
   f.set_s(std::vector{7, 9});          // [3]
```

Code snippet 3.6: A simple "set()-er"-function in C++98 and ways to invoke it

Typically, functions that take an argument that they do not modify qualify the parameter as `const &`. An example of this can be seen in Snippet 3.6. This prevents needless copying at the time the function is invoked and it can handle any kind of input, also temporaries created at the time of invocation ([3] in Snippet 3.6). In the latter case the temporary is *materialised* and then bound to the reference; its lifetime ends at the end of the function call.

This is convenient when the origin of the input is irrelevant to the function and a cursory analysis of the function in Snippet 3.6 might come to the conclusion, "after all we need to copy the data during assignment anyway". This is, however, not true. Data has to be copied in the cases [1] and [2], because an external object is referenced that shall not be modified. But in case [3] an object is copied that has just been created and that will go out of scope immediately after the assignment (in line 7). In many cases copying the data is also associated with dynamic memory allocation making this even more expensive.

The point of *move semantics* is to reuse the memory of the materialised temporary instead of copying it (ISO/IEC 14882:2017, 15.8), i.e. the goal is to prevent a copy for case [3]. An example of this can be seen on the left side in Snippet 3.7:

- There is a second function overload denoted by a `&&`-qualified parameter (line 10); it is preferred by overload resolution when rvalues are passed to the function, i.e. [1] and [2] will trigger the first overload and [3] will trigger the second. The exact meaning of `&&` will be discussed in Sect. 3.2.2.

```
   class Foo                              class Foo
2  {                                   2  {
       std::vector<int> s;                    std::vector<int> s;
4  public:                             4  public:
       void set_s(std::vector<int> const & t)    void set_s(std::vector<int> t)
6      {                               6      {
           s = t;                                  std::swap(s, t);
8      }                               8      }
                                          };
10     void set_s(std::vector<int> && t)  10
       {
12         s = std::move(t);           12
       }
14 };                                  14
```

Code snippet 3.7: Overload resolution with `&&` and the "copy and swap idiom"

- Instead of doing assignment (now called *copy assignment*), the assignment operator in line 12 performs *move assignment*.
- Perhaps surprisingly, it is not `std::move()` that manipulates ownership itself; `std::move()` simply makes its argument *appear* like a temporary (by the means of a cast). This then leads to a different function overload of the assignment operator being called on the vector, namely the move assignment operator.
- This move assignment operator of `s` then "steals" the memory from `t` which happens in constant time.[9]

Using this method, unnecessary copy operations can be avoided which is an important improvement for C++ as a performance-oriented programming language. It is, however, cumbersome to add these additional overloads everywhere, and for most cases there is an elegant alternative which can be seen on the right side of Snippet 3.7:

- There is only one function, it takes the argument by value, leading to a new object being created for every function invocation.
- Inside the function there is a swap instead of an assignment; this exchanges the contents of two vectors without copying and in constant time (e.g. exchanging pointers to dynamic memory).
- Invoking the function with existing objects ([1] and [2] in Snippet 3.6) will lead to exactly one copy operation (as before), but as part of a copy constructor call at the beginning of the member function invocation—not as part of an assignment.
- Invoking the function with a temporary ([3] in Snippet 3.6) will lead to no copy operation, because the *move constructor* of `std::vector` is called which takes over the dynamic data from the temporary (also possibly implemented via swap).

---

[9] Transfer of ownership could be implemented as exchanging the pointers to dynamically allocated memory so that `t` destructs the old data of `s` and `s` holds the old data of `t` (which was allocated at the beginning of the function call through above-mentioned materialisation).

This is called the "copy and swap idiom" (Mansfield, 2017).[10] It shows that one can often delegate move semantics to the constructors and assignment operators of a class and that one does not need to add custom overloads to all of one's functions. Implementing correct (move) constructors and assignment operators is an essential part of Modern C++, but in many cases the compiler generates correct default implementations.

3.2.2   Reference Types and Perfect Forwarding

For a concrete non-reference type `T`, the expression `T &` denotes the *lvalue reference* of that type and `T &&` denotes the *rvalue reference* (ISO/IEC 14882:2017, 11.3.2). As the previous examples illustrated, rvalue references and lvalue references-to-const can both bind rvalues. But in contrast to the latter, rvalue references **only** bind rvalues. See the left side of Snippet 3.8 for an example.

```
   size_t         i1 = 1;
 2 size_t const   i2 = 1;


   size_t       & i1a = i1;                    auto         & a1a = i1;
 4 //size_t      & i1b = i2;               15  auto         & a1b = i2;
   //size_t      & i1c = 1;                    //auto        & a1c = 1;
 6                                         17
   size_t const & i2a = i1;                    auto   const & a2a = i1;
 8 size_t const & i2b = i2;               19   auto   const & a2b = i2;
   size_t const & i2c = 1;                     auto   const & a2c = 1;
10                                         21
   //size_t      && i3a = i1;                  auto         && a3a = i1;
12 //size_t      && i3b = i2;             23   auto         && a3b = i2;
   size_t       && i3c = 1;                    auto         && a3c = 1;
```

Code snippet 3.8: Rvalue references vs forwarding references. Code that is commented out is invalid. `const &&` is omitted, because it is used very rarely. `i1` and `i2` are lvalues; `1` is an rvalue

When overloads with differently qualified parameters of some concrete type `T` are competing, the overload with `T const &` has the lowest priority, i.e. `T &` is preferred for non-constant lvalues and `T &&` is preferred for non-constant rvalues. Thus, there is never any ambiguity when such overloads are defined together. This specific order in overload resolution is the key to move semantics, it allows differentiating between lvalues and rvalues and providing specialised behaviour for the latter (e.g. the aforementioned "stealing" of memory).

In addition to *rvalue references* which mostly appear in constructor and assignment operator definitions, there are also *forwarding references*. Forwarding refer-

---

[10] It has the added benefit of stronger exception safety, but this comes at the cost of always allocating the memory anew—even when the new data fits into the old memory region.

ences appear like rvalue references (denoted by `&&`) but behave slightly different: they also bind lvalues, even lvalue constants (see Snippet 3.8).

`T &&` is a forwarding reference if and only if the type is part of type deduction, i.e. if `T` is `auto` or a template parameter (ISO/IEC 14882:2017, 17.8.2.1). If, on the other hand, `T` is a concrete type (e.g. `int` or a user-defined class type), `T &&` is an rvalue reference.[11]

```
  template <typename T>
2 void bar(T && v) { /*_*/ }

4 template <typename T>
  void foo(T && v)
6 {
      /*_*/
8     bar(std::forward<T>(v));
  }
```

Code snippet 3.9: Perfect forwarding

Forwarding references are used often in generic programming as they can capture and forward (hence the name) the original input type fully (lvalues and rvalues, `const` and non-`const`). This is called *perfect forwarding*, an example can be seen in Snippet 3.9. `std::forward()` preserves lvalue references and moves temporaries. In Modern C++, this is the recommended way of taking and passing most generic parameters (instead of `T const &`) even if the function does not modify the argument. This allows for functions further down the call-graph to differentiate between `const` and non-`const` via different overloads.[12] It also prevents certain non-`const` objects from being forced into a `const`-context in which they might exhibit different (potentially slower) behaviour.

3.2.3 Out-Parameters and Returning by Value

The previous subsections dealt with move semantics in the context of function parameters. I will now discuss return values. In C++98 returning complex types by value from a function was strongly discouraged, because it could lead to multiple copies being created in the process of returning the value. While the language explicitly allowed for compilers to optimise away some of these copies, there was no guarantee. Performance-focused and portable code, like SeqAn1/2, thus never returned non-built-in types from functions and relied on the so-called *out-parameters*.

---

[11] See also https://en.cppreference.com/w/cpp/language/reference.

[12] F.19: For 'forward' parameters, pass by `TP&&` and only `std::forward` the parameter. https://isocpp.github.io/CppCoreGuidelines/CppCoreGuidelines#Rf-forward.

```
   void generate_vec(size_t i,
2                    std::vector<size_t> & out)
   {
4      out.resize(i);
       for (size_t j = 0; j < i; ++j)
6          out[i] = i;
   }
8
   int main()
10 {
       std::vector<size_t> out;
12     generate_vec(5, out);
   }
```

```
   std::vector<size_t> generate_vec(size_t i)
2  {
       std::vector<size_t> out;
4      out.resize(i);
       for (size_t j = 0; j < i; ++j)
6          out[i] = i;
       return out;
8  }

10 int main()
   {
12     std::vector<size_t> out = generate_vec(5);
   }
```

Code snippet 3.10: "Out-parameters" in C++ 98 vs returning by value in C++ ≥ 11. Both code snippets show a function that generates a vector of size i that contains the elements [0..i-1]

Out-parameters are a way of passing to a function by reference one or multiple objects that are then "filled" by the function. In contrast to "in-out-parameters" that are expected to have a previous value that is acted upon, out-parameters simply act as a way of returning values from the function. They prevent any form of unintended copy but lead to function interfaces that are harder to read and use. The sheer number of parameters increases and conventions need to be established (e.g. "out-parameters before in-parameters" as in SeqAn1/2) so that users do not accidentally switch the input and output (which can still happen if both have the same type). Furthermore, out-parameters lead to uncertainties for the programmer, e.g. whether a function should assume that out-parameters are in a default-constructed (empty or "null") state or whether the functions need to establish this itself by clearing the arguments.

Since C++ 11, an object that is returned from function is moved instead of copied. (If a move constructor is available.) This happens automatically and does not require the use of `std::move()`. The move operation may be even by *elided* completely which means the function's local variable is directly created in the place of the variable that stores the return value in the surrounding scope. (This even avoids calling the move constructor.)

A comparison of both approaches can be seen in Snippet 3.10. To return multiple values from a function, one can wrap them in a `std::tuple` or a custom type. Since C++ 17 some forms of *copy elision* are mandatory, but copy elision in Snippet 3.10 would still be an optional optimisation of the compiler.[13]

Independent of copy elision, the implicit fallback to move semantics means library designers can make function interfaces much more readable by avoiding out-parameters. While the other aspects of move semantics are more relevant to library and "systems" developers, this aspect has very visible consequences for the users of a library like SeqAn. And, generally speaking, it should improve the learnability of C++ for programmers coming from other programming languages.

---

[13] See https://en.cppreference.com/w/cpp/language/copy_elision for helpful examples.

3.3 Metaprogramming and Compile-Time Computations

3.3.1 Metafunctions and Type Traits

As mentioned in Sect. 2.3.4 the term *metafunction* is not clearly defined and its synonymity with *type trait* is also not widely accepted.[14] I prefer the term *type trait*, because it is defined in ISO/IEC 14882:2017 (13.15.1) and it is more descriptive. ("A type trait is a property of a type".)

In contrast to Table 3.2, Gogol-Döring always uses the term *metafunction*; or more precisely the term *value metafunction* whenever a value is "returned" and *type metafunction* whenever a type is "returned". Thus, the term *metafunction* better describes the "how" than the "what".

Historically type traits were always implemented as class templates that expose a member value or a member type (in the case of transformation traits). Since the introduction of variable templates in C++ 14 (ISO/IEC 14882:2017, 17.0) a less verbose syntax with the same power is available for type traits that expose values (see Snippet 3.11).

```
   template <typename T>
2  struct num_bytes
   {
4      static size_t const value = sizeof(T);
   };

6
   template <>
8  struct num_bytes<my_type>
   {
10     static size_t const value = 7;
   };
12
   // values can be checked at compile time
14 static_assert(num_bytes<int32_t>::value == 4);
   static_assert(num_bytes<my_type>::value == 7);
```

```
   template <typename T>
2  constexpr size_t num_bytes = sizeof(T);

4

6
   template <>
8  constexpr size_t num_bytes<my_type> = 7;

10

12
   // values can be checked at compile time
14 static_assert(num_bytes<int32_t> == 4);
   static_assert(num_bytes<my_type> == 7);
```

Code snippet 3.11: A type trait example in C++98 and C++ >=14. It exposes the number of bytes a type occupies in memory by delegating to the `sizeof()`-built-in by default. I have added a specialisation for a fictitious type `my_type`. Note that the `static_assert()` is itself a feature of C++ 11 (it halts compilation if the argument evaluates to false)

Table 3.2 Terminology surrounding *type traits*

Input/Output	One type	A value
One type	TransformationTrait	UnaryTypeTrait
Two types	–	BinaryTypeTrait
Value(s)	–	"Compile-time computation"

---

[14] https://stackoverflow.com/questions/32471222/c-are-trait-and-meta-function-synonymous.

Similar in appearance to this change of the language, there are also *alias templates* in C++ 11 which are a "templatised" version of a `using` or `typedef` declaration (ISO/IEC 14882:2017, 17.5.7). These can make transformation traits more accessible by serving as shortcuts to the exposed type of a class template, but they cannot replace the class template (like variable templates for value type traits), because they cannot be partially specialised ("overloaded" in metafunction terminology).

3.3.2   Traits Classes

Traits classes, sometimes also called *traits types*—not to be confused with *type traits*—are types that provide another type's traits as members (member types, member constants, etc). They can be used in place of many individual template parameters to simplify the interface of a template. A "traits class" is defined by the standard as "a class that encapsulates a set of types and functions necessary for class templates and function templates to manipulate objects of types for which they are instantiated" ISO/IEC 14882:2017, 20.3.24.

```
   template <typename param1_t = int,               struct default_traits
2            typename param2_t = bool,            2  {
            typename param3_t = double,                using param1_t = int;
4            size_t constant1  = 42,              4      using param2_t = bool;
            size_t constant2  = 23>                    using param3_t = double;
6  struct my_type                               6      static constexpr size_t constant1 = 42;
   {                                                   static constexpr size_t constant2 = 23;
8      //...                                     8  };
   };
10                                               10 template <typename traits_t = default_traits>
                                                    struct my_type
12                                               12 {
                                                        //...
14                                               14 };

16                                               16 // new traits class with one parameter changed
                                                    struct adapted_traits : default_traits
18                                               18 {
                                                        static constexpr size_t constant1 = 1000;
20                                               20 };

   // only change one parameter ↓
22 my_type<int, bool, double, 1000, 23> m;      22 my_type<adapted_traits> m;
```

Code snippet 3.12: Multiple template parameters vs a single traits class

An example for such a traits class can be seen in Snippet 3.12. Evidently, using the traits class results in more lines of code and added complexity at the time of specialisation. However, the approach offers some distinct benefits:

- The interface of `my_type` becomes simpler.
- The code is more expressive, because only the parameter that is supposed to be changed needs to be changed for the specialisation.

- This makes the code more resilient to a change of defaults (in the example without traits class also those parameters are hard-coded that are meant to be defaulted).
- It is possible to provide different traits classes as "sets of default parameters" for an algorithm, e.g. a configuration that favours speed over memory usage and another one that does the opposite.

Typically, traits classes are used when the number of template parameters is high, there are inter-dependencies between the template parameters, pre-defined combinations are desirable (see above) and/or the template parameters are not specialised frequently at all (reducing the added complexity of defining more and more traits classes). It is also possible to have "frequently used" template parameters preserved as such, and less frequently used template parameters subsumed into a traits class. Using traits classes predates "Modern C++", an early example of this design in the standard library is `std::basic_string` whose signature is shown in Snippet 3.13.

```
   template <typename CharT,
2            typename Traits = std::char_traits<CharT>,
             typename Allocator = std::allocator<CharT>>
4  class basic_string
   {
6      /* ... */
   };
8
   template <typename CharT>
10 struct char_traits
   {
12     using char_type  = CharT;
       using int_type   = unsigned long;
14     using pos_type   = std::streampos;
       using off_type   = std::streamoff;
16     using state_type = std::mbstate_t;

18     static constexpr void assign(char_type & c1, char_type const & c2) { c1 = c2; }
       static constexpr bool eq(char_type const & c1, char_type const & c2) { return c1 == c2; }
20
       /* ... */
22 };
```

Code snippet 3.13: The signature of `std::basic_string` and parts of its traits class

Here the character type is considered the essential parameter that is set directly and the remaining parameters (minus the allocator) are part of the traits class. The default traits class is in turn a template specialised over the character type, because the defaults (can) depend on the character type. This design highlights to the user of the interface that for many situations it will suffice to change the first parameter while enabling more sophisticated specialisations via the traits class.

In SeqAn1/2 traits classes were used rarely, although they could arguably have replaced globally modifiable metafunctions (see Sect. 2.3.4). Some SeqAn-based applications, like Yara, used traits classes (called "configs") and SeqAn3 makes use of them in several places.

3.3.3   Compile-Time Computations

It has long been known that C++'s template preprocessor is Turing complete (Veldhuizen, 2003). It can thus be used to not only resolve types and their traits but also perform actual computations that return values at compile-time. This technique is often referred to as metaprogramming. Gogol-Döring (2009) gives the example shown in Snippet 3.14.

```
  template < int numerus >
2 struct Log2
  {
4     enum { VALUE = Log2<(numerus+1)/2 >::VALUE + 1 };
  };
6
  template < > struct Log2<1> { enum { VALUE = 0 }; };
8 template < > struct Log2<0> { enum { VALUE = 0 }; };
```

Code snippet 3.14: "Listing 5: **Metaprogram Example.** This metaprogram computes the rounded up logarithm to base 2. Call `Log2<c>::VALUE` to compute $\lceil log_2(c) \rceil$ for a constant value c" from Gogol-Döring (2009)

While the technique is powerful, Gogol-Döring already admits that "[it] is rather complicated and hard to maintain, we decided to use it only in limited circumstances."

Fortunately, C++ 11 introduced the notion of *constant expressions* (ISO/IEC 14882:2017, 8.20). A variable declared `constexpr` holds a compile-time constant and is required to be initialised by such a constant expression. A function declared `constexpr` can be used in a constant expression, i.e. evaluated at compile-time. Note that it can also be evaluated at runtime if called in a non-constant expression.

Since working with `constexpr` functions is very much part of the regular language now, I prefer to use the term "compile-time computation"; it is also used by Stroustrup (2012).

```
  constexpr size_t log2(size_t const c)
2 {
4     return (c <= 1) ? 0 : log2((c + 1) / 2) + 1;
  }
6
  constexpr size_t i = log2(8);    // == 3, always evaluated at compile-time
  size_t j = log2(8);              // == 3, may or may not be evaluated at compile-time
```

Code snippet 3.15: `constexpr` variables and functions. There is a single function to compute $\lceil log_2(c) \rceil$ for runtime and compile-time contexts

An example can be seen in Snippet 3.15. Some things to note:

- It would not be possible to initialise `i` from the return value of a function that is not qualified as `constexpr`.

- The function can only be evaluated in a constant expression if its arguments are also constant expressions, i.e. if one passes a non-`constexpr` variable to `log2()` one can use the function to initialise `j`, but not `i`.

Not all functions can be declared `constexpr`, but the number of restrictions has been reduced with every C++ standard since C++ 11. Notably with C++ 17 it is now possible to declare *lambda expressions* (see Sect. 3.1.1) as `constexpr`—in fact they are `constexpr` by default if possible. Combining both allows to initialise constants via immediately evaluated, `constexpr` lambda expressions.

```
2  struct foo
   {
4      static bool const data[256];
       /* ... */
   };
6
   bool const foo::data[256] =
8  {
       0, 0, 0, 0, 0, 0, 0, 0, 0, 0,
10     0, 0, 0, 0, 0, 0, 0, 0, 0, 0,
       /* ... */
12     1, 0, 0, 0, 0, 0, 0, 0, 0, 0,
       1, 0, 0, 0, 0, 0, 0, 0, 0,
14     /* ... */
   };
```

```
2  struct foo
   {
4      static constexpr std::array<bool, 256> data = [] ()
       {
           std::array<bool, 256> ret{}; // initialised to 0s
6
           ret[50] = true;
8          ret[60] = true;

10         return ret;
       } ();
12     /* ... */
   };
14
```

Code snippet 3.16: Both snippets show a static member array whose 50th and 60th element are set while the rest is not. The left is C++ 98-style "manual" initialisation, the right uses an immediately evaluated `constexpr` lambda function

Snippet 3.16 shows an example of this technique. The "manual" style shown on the left is used frequently in SeqAn1 and SeqAn2, e.g. in the character/ord-value conversion tables of alphabets. As can be seen in the example, it is also required to initialise the member variable outside of the class. In C++ 17, on the other hand, it is possible to initialise the member directly through the return value of a lambda expression that is defined ad hoc and evaluated immediately. (The final `()` invokes the previously defined function object.) The advantages of the approach should be obvious: instead of hard-coding magic numbers (or arrays thereof), the programmer expresses the intent (that bit number 50 and 60 are to be set) in the code itself. No extra member or detail functions need to be defined just for initialisation; all logic is local to the only place where it is used. This is much easier to read and maintain and the runtime behaviour is identical.

SeqAn3 uses this mechanism extensively (see e.g. Snippet A.3 on p. 331). Since the complete interface of a type can be marked `constexpr`, construction and manipulation of user-defined objects can even happen at compile-time. This allows for "reflecting" on types and defining new types based on another type's properties; examples for this are introduced later (Sect. 6.5).

3.3.4 Conditional Instantiation

When working with type traits and compile-time computations, it often happens that one wishes to choose one code block or another depending on a condition whose result is known at compile-time. This typically happens inside function templates where a large part of the code is shared between instantiations, but small parts differ. The intuitive approach is to use an `if`-statement to evaluate the condition and then either execute the block after the `if` or the `else` depending on the outcome. There are two issues with this approach:

1. The condition of an `if`-statement is evaluated at runtime; while the result will always be identical for one instantiation of the template, the check cannot be optimised away by the compiler leading to sub-optimal performance.[15]
2. It is not always possible to use an `if`-statement, because some expressions might only be valid in one instantiation and not the other; i.e. one "knows" that the first code block will only ever be executed for types where the first code block is valid, but the compiler rejects the code, because both code blocks need be formally valid for all instantiations.

```
   template <typename it_t, typename tag_t>
2  void advance_by_5(it_t & it, tag_t)
   {
4      it++; it++; it++; it++; it++;
   }
6
   template <typename iterator_t>
8  void advance_by_5(it_t & it, std::random_access_iterator_tag)
   {
10     it += 5; // jump in O(1)
   }
12
   template <typename it_t>
14 void algorithm(it_t & it)    // generic algorithm that takes an iterator
   {
16     /* ... */
       typedef typename std::iterator_traits<it_t>::iterator_category it_tag;
18     advance_by_5(it, it_tag{});
       /* ... */
20 }
```

Code snippet 3.17: Conditional instantiation in C++ 98. The `operator+=` is only defined on random access iterators so it would be invalid to switch between the two alternatives via `if` inside `algorithm()`. Instead, a tag is used to select the appropriate code at compile-time

The solution to the second problem is to move the code blocks into separate overloads of an extra function template and dispatch to these via some auxiliary type (usually a tag). Due to *inlining*, this dispatching typically comes with no

---

[15] *Branch prediction* in modern processors mitigates this to some degree (Mittal, 2019), but the behaviour is still dynamic where it could be static.

runtime overhead, thus also solving the first problem. But, as can be seen in Snippet 3.17, the solution is not very pretty. It should be noted that this example is "easy", because the `iterator_category` tags already exist to facilitate tag-based dispatching. Dispatching on other conditions leads to even more code.

SeqAn1/2 is littered with small functions that solve exactly these problems. However, even in SeqAn1/2 with its focus on performance, plain `if`-statements on compile-time decisions are just as common, leading to problem 1 and sub-optimal performance. This highlights that due the high code overhead, programmers tend to use tag-based dispatching (as in Snippet 3.17) only if they are forced to, i.e. when they suffer from problem 2.

```
   template <typename it_t>
2  void algorithm(it_t & it)    // generic algorithm that takes an iterator
   {
4      /* ... */
       using it_tag = typename std::it_traits<it_t>::iterator_category;
6      if constexpr (std::is_same_v<it_tag, std::random_access_iterator_tag>)
       {
8          it += 5; // jump in O(1)
       } else
10     {
           it++; it++; it++; it++; it++;
12     }
       /* ... */
14 }
```

Code snippet 3.18: Conditional instantiation in C++17 via `if constexpr`

In C++17, there is a simple and elegant solution: `if constexpr` (ISO/IEC 14882:2017, 8.4.1 §2). It behaves essentially like a regular `if`-statement, but the condition is always evaluated at compile-time. (This means it is required to be a constant expression.) Furthermore, inside a function template only the code block that is chosen based on the condition is instantiated and needs to be valid. This solves both discussed problems with minimal visual overhead.

See Snippet 3.18 for an example. While this feature might appear trivial compared to larger changes discussed in this chapter, I would argue that its impact on the readability and compactness of a codebase like SeqAn's would be tremendous. Persistent use of `if constexpr` instead of plain `if` also reduces the total amount of instantiated code thereby reducing compile-times and binary sizes.

3.3.5 Standard Library Traits

The standard library of Modern C++ provides a rich set of traits that can be used in metaprogramming. Most of these are Boolean type traits whose names begin with `is_`, but there are other type traits and many transformation traits, as well. Some examples are shown in Snippet 3.19.

All standard library traits are implemented as class templates with those representing values (unary and binary type traits) exposing that value as a static `::value` member. Transformation traits expose the transformed type as a member type called `::type`. However, there are "shortcuts" for these members implemented as variable templates for the values and alias templates for the member types. These have the names `std::is_foo_v<T>` (equivalent to `std::is_foo<T>::value`) and `std::bar_t<T>` (equivalent to `typename std::bar<T>::type`). This methodology has historic reasons (the traits were introduced in C++ 11, variable templates were introduced in C++ 14) and technical reasons. (Some metaprogramming is easier when always working on types.)

```
   // boolean unary type traits
 2 std::is_integral              // int, uint64_t...
   std::is_signed                // int, but not uint64_t...
 4 std::is_unsigned              // uint64_t, but not int...
   std::is_pointer               // any *
 6 std::is_lvalue_reference      // any &
   std::is_rvalue_reference      // any &&
 8 std::is_copy_constructible    // has copy constructor

10 // boolean binary type traits
   std::is_same                  // check identity
12 std::is_base_of               // check inheritance
   std::is_convertible           // check convertibility
14
   // transformation traits
16 std::make_signed              // "unsigned" to "int"
   std::make_unsigned            // "int64_t" to "uint64_t"
18 std::remove_pointer           // "int *" to "int"
   std::remove_reference         // "int &" to "int"
20 std::add_lvalue_reference     // "int" to "int &"

22 // miscellaneous traits
   std::conditional<b, T1, T2>   // if b then "T1", else "T2"
24 std::enable_if                // helper for SFINAE
```

Code snippet 3.19: A non-comprehensive list of traits available from the `<type_traits>` header

3.4   C++ Concepts

C++ Concepts are one of the Modern C++ features that have taken the longest to make it into the standard. They were originally scheduled for C++ 11, then for C++ 17, but only made it into C++ 20. In 2015 a stand-alone technical specification (TS) was published by ISO (ISO/IEC 19217:2015). The wording accepted into C++ 20 is a subset of the TS with some further changes. This long process was the result of various disagreements within the committee and the importance attributed to concepts. The inventor of C++ describes concepts as "a foundational feature that in the ideal world would have been present in the very first version of templates and the basis for all use" (Stroustrup, 2017).

3.4.1 Introduction

Generic programming in C++ is based on templates. This is a very versatile solution, but also very verbose, visually distinct from non-generic code and more difficult to debug (Stroustrup, 2017). Formally, a template parameter in C++ accepts any type. However, almost all function templates and the majority of class templates make some assumptions about the types that one specialises them with. A `sort()` function template would e.g. depend on the argument type being *sortable*, i.e. "have elements" and provide certain operators to access and compare them.

To communicate these assumptions to the user, the standard library has annotated its template parameters with the so-called requirements that specify in standard language which criteria the template expects from types (ISO/IEC 14882:2017, 20.4.1.3 & 20.4.2.1.1). Sometimes names are given to a set of requirements, e.g. *Iterator* encompasses the requirements all types need to satisfy to be accepted as iterators by standard library algorithms. These named requirements are historically called *concepts*, as Alexander Stepanov, the original creator of the C++ standard template library, points out:

> We call the set of axioms satisfied by a data type and a set of operations on it a *concept*. Examples of concepts might be an integer data type with an addition operation satisfying the usual axioms; or a list of data objects with a first element, an iterator for traversing the list, and a test for identifying the end of the list (Dehnert & Stepanov, 2000).

However, since these concepts are just part of the "documentation" and not the language itself, it is easy to misuse a template leading to a painful user experience. Snippet 3.20 illustrates this. Programmers familiar with C++ will know that the error message for the unconstrained template is a lot more complex in reality; it can cover multiple pages for nested function calls. This problem is also discussed in Sect. 2.4.2.

C++ 20 changes this by introducing concepts as a language feature. Developers can now *constrain* templates (ISO/IEC 14882:draft, 13.4). These constraints can be formulated ad hoc when defining a template, or they can be given a name, turning them into a *concept*. The concept can then later be used to constrain templates again.

As such the term "concept" has been dropped in C++ 20 for requirements expressed solely as part of the specification, they are now simply called *named requirements*. The word "concept" is now only used for named constraints written in code.

The goals of using concepts as a language feature can be summarised as (Stroustrup, 2017):

1. Reducing the verbosity of templates, visually bringing generic programming closer to object-oriented programming.

```
   // 1990s style generic code:                  // Generic code using a concept (Sortable):
 2 template <class T>                          2  void sort(Sortable auto & c)
   void sort(T& c) // C++98: accept any type T      // Concepts: accept any c that is Sortable
 4 {                                           4  {
       // code for sorting (depending on various       /* ... */
 6     // properties of T, such as having [] and 6
       // a value type with <)
 8 }                                           8  }

10 vector<string> vs = {"Good","old","templates"}; 10 vector<string> vs = { "Hello","new","World" };
   sort(vs);// fine: vs happens to have all the    sort(vs);// fine: vs is a Sortable container
12              // syntactic properties required by sort 12
                                                  double d = 7;
14 double d = 7;                               14 sort(d); // error: d is not Sortable
   sort(d); // error: d doesn't have a [] operator     // (double does not provide [], etc.)
```

Code snippet 3.20: Example snippets adapted from Stroustrup (2017). They illustrate the error diagnostics in pre- and post-concepts-C++; example has been changed to the final concept syntax of C++ 20

2. Improving the compiler diagnostics when (mis)using templates, thus also lowering the barrier to generic programming.
3. Moving more of the specification from documentation into the code and increasing the amount and quality of specification.

Another very important "by-product" is concepts-based polymorphism which I will introduce in Sect. 3.4.4. The syntax shown in the next subsections reflects Concepts as in C++ 20, but I will point out if shown code is not compatible with GCC7–9 (which only support the Concepts TS).

3.4.2 Defining Concepts

```
   template <typename T>
 2 concept not_useful = true;                      // all types satisfy this concept

 4 template <typename T>
   concept integral = std::is_integral_v<T>;       // delegate to type trait
 6
   template <typename T>
 8 concept addable = requires (T a, T b)           // initialised with requires expression
   {
10     a + b;                       // requires that "a + b" is valid
       { a + b } -> std::same_as<T>;// requires that result of "a + b" is of type T
12 };

14 template <typename T>
   concept addable_integral = integral<T> && addable<T>;  // can be combined with logical operators
```

Code snippet 3.21: Examples of short concept definitions

A *requires expression* is an expression of type `bool` introduced by the `requires` keyword. It contains one or many *requirements* and becomes `true` if and only if all requirements are satisfied. Requirements can be: evaluating whether a given expression is well-formed, whether a type definition is valid or whether an expression has certain properties. Typical requirements are the existence of a member function with a specific name, the existence of a member type or the validity of a free function call on an object of the respective type.[16] In simple terms, a requires expression can check whether a piece of code "would be valid".

Concepts appear like Boolean variable templates (they are templates!) except that they are introduced by the keyword `concept` and have no type specifier. (`bool` is implicit.)[17] They can be initialised with primary expressions of type `bool`, but importantly also with a requires expression.

Snippet 3.21 shows some examples of concept definitions. The concept `integral` simply invokes a type trait as primary expression, i.e. it will evaluate to true for built-in integer types like `int`, `long`, `unsigned`, etc. `addable` is defined by a requires expression that evaluates whether an `operator+` is defined for the type and whether the return type of that operator is again the type. This would be satisfied by e.g. `int` and `std::string`. Note that the requirement in line 10 is redundant, it is already implied by the requirement in line 11.

The last concept, `addable_integral`, is composed of two other concepts. Beyond the reuse of code, this form of composition also implies a hierarchy between the concepts: `addable_integral` is said to be *more refined* than either of `addable` and `integral`. This will be important later on.

3.4.3   Using Concepts

```
   template <typename T>
2  concept integral = std::is_integral_v<T>;

4  template <typename T>
   void foobar(T const & val)
6  {
       if constexpr (integral<T>)          // same as calling std::is_integral_v<T>
8          /* code specific to integers */
       else
10         /* generic code */
   }
```

Code snippet 3.22: Using concepts as type traits. In this case `integral` behaves just as if it had been defined as `constexpr bool` instead of `concept`

---

[16] https://en.cppreference.com/w/cpp/language/constraints.

[17] In the Concepts TS, concepts are introduced by `concept bool` instead of just `concept`. An implementation that targets, both, C++20 and the Concepts TS can use a macro to resolve this incompatibility.

Concepts can be used just like type traits/variable templates (see Sect. 3.3.1),[18] i.e. for a concept `C` and some type `T`, one can instantiate `C<T>` which will resolve to `true` if `T` satisfies `C` and `false` otherwise. This is particularly useful in metaprogramming and in combination with `if constexpr` (see Snippet 3.22).

```
   template <typename T>                    template <typename T>
2  concept integral = std::is_integral_v<T>;   18 concept c2 = /* ... */;

4  // terse syntax                          20   // terse and intermediate syntax combined
   void foobar0(integral auto i) { /* ... */ }  template <integral T>
6                                           22   void foobar3(c2 auto i)          { /* ... */ }

8  // intermediate syntax                   24   // verbose syntax can include expressions
   template <integral T>                         template <typename T>
10 void foobar1(T i)              { /* ... */ }  26   requires c2<T> && std::is_integral_v<T>
                                                  void foobar4(T i)               { /* ... */ }
12                                          28
   // verbose syntax                             // verbose syntax can include ad-hoc req. expr.
14 template <typename T>                     30   template <typename T>
     requires integral<T>                           requires (requires (T a) { a + a; })
16 void foobar2(T i)             { /* ... */ }  32   void foobar5(T i)            { /* ... */ }
```

Code snippet 3.23: The three equivalent syntaxes of constraining function templates are on the left. The right shows various combination possibilities and an ad hoc definition of a requires expression (see below for why `requires` appears twice)

But the important advantage of concepts is that they can be used to *constrain* templates. Different syntaxes are available to constrain function templates as can be seen in Snippet 3.23. The terse syntax (l. 5) is an extension of the abbreviated function template syntax introduced in Sect. 3.1.1. It is only available for function templates and not for class or variable templates and one can only use a single concept to constrain the type.[19] The combination of "concept name + `auto`" can also be used when declaring variables at block scope. This does not influence the type deduction process; it merely asserts after type deduction that the deduced type satisfies the given requirements. But it does increase readability of the code by narrowing what `auto` can mean.

The intermediate syntax (l. 9) is a little more verbose, it consists of replacing the `typename` keyword in the template definition with a concept name. It also enforces exactly one concept on a given type, but it is available for class and variable templates, too. Additionally, the type of `i` is now aliased as `T` whereas in the terse syntax there is no immediate access to the type.

The verbose syntax (l. 14) adds a *requires clause* (not be confused with the *requires expression*!) to the template signature. The requires clause can contain any primary expression that is a constant expression and convertible to `bool`. This means it is not required to use concepts to constrain a template, one can e.g. use

---

[18] With the exception that concepts themselves cannot be constrained or (partially) specialised.

[19] There is a terse form in the TS, but both the syntax and the semantics are slightly different from C++20. Code targeting C++20 and the TS should avoid the terse syntax.

type traits and/or combine multiple concepts and type traits (see `foobar4()` in Snippet 3.23). The requires clause can also contain a requires expression, i.e. it allows for the local specification of more complex requirements (see `foobar5()` in Snippet 3.23). Note that in this case the `requires` keyword indeed appears twice: once for introducing the requires clause and once for introducing the requires expression.

Syntaxes can be freely combined and it is generally recommended using the shorter forms for improved readability (Stroustrup, 2017).

```
 1  // base template
 2  template <typename TSpec>
    struct IntContainer
 4  {
        /* ... */
 6  };

 8  // most generic overload
    template <typename TSpec>
10  size_t find(IntContainer<TSpec> & c, int i)
    {
12      /* find index of i by linear scan */
    }
14
    // tag for derived type
16  struct MapSpec;
    // derived type via specialisation
18  template <>
    struct IntContainer<MapSpec>
20  {
        /* ... */
22  };

24  // refined overload
    size_t find(IntContainer<MapSpec> & c, int i)
26  {
        /* find index of i by binary search */
28  }

30  // polymorphic interface
    template <typename TSpec>
32  void print_idx_of(IntContainer<TSpec> & c,
                      int elem)
34  {
        std::cout << find(c, elem);
36  }
```

```
 1  // most generic concept
 2  template <typename T>
    concept int_container = /* e.g.:
 4  requires (T vec)
    {
 6      { vec[0] } -> std::same_as<int>;
    } */;
 8
    // most generic overload
10  size_t find(int_container auto & c, int i)
    {
12      /* find index of i by linear scan */
    }
14
16  // refined concept
    // (subsumes the generic concept)
18  template <typename T>
    concept int_map =
20      int_container<T> && /* e.g.:
    requires (T vec) { vec.smart_insert(0); } */;
22
24  // refined overload
    size_t find(int_map auto & c, int i)
26  {
        /* find index of i by binary search */
28  }

30
    // polymorphic interface
32  void print_idx_of(int_container auto & c,
                      int elem)
34  {
        std::cout << find(c, elem);
36  }
```

Code snippet 3.24: Template subclassing vs concepts-based polymorphism. See also Snippet 2.1

3.4.4 Concepts-Based Polymorphism

In Sect. 2.3.2 I introduced different forms of polymorphism:

1. polymorphism via `virtual` functions and inheritance (typically found in object-oriented C++ code); and
2. *template subclassing*, a form of *static polymorphism* used in SeqAn1/2.

I discussed why static polymorphism is necessary in performance-oriented codebases but also the problems associated with template subclassing. C++ concepts enable a new form of static polymorphism that can be seen in Snippet 3.24. This polymorphism is based on the previously mentioned notion that some concepts are considered more *refined* than others. In the example on the right side of Snippet 3.24, `int_container` is the generic concept and `int_map` is the refinement, because the latter subsumes the former. Note that `int_map` describes types that are intrinsically sorted and somehow expose this in their interface (it is not about the sorted/unsorted state of a type that can be either!). Simplified example requirements are given for `int_container` ("it shall have a random access operator that returns a type convertible to `int`") and `int_map` ("it shall model `int_container` and additionally provide a `smart_insert()` member that is assumed to perform an insert preserving the order").

An overload constrained with a concept that is more refined than that of a competing overload is considered *more constrained*. During overload resolution the most constrained overload that satisfies the requirements is chosen. The same is true for the instantiation of constrained type and variable templates.

This mechanism allows a "refinement hierarchy" similar to that buildable around virtual function lookup or partial template specialisation. But the very important difference is that it presupposes no derivation of the types, i.e. neither inheritance from a base type nor specialisation of a type template. It operates one abstraction level higher: on type requirements.

That aspect is the most important benefit of concepts-based polymorphism: templates constrained in this way take types that are designed for them but also third party types that satisfy the requirements. Depending on the exact requirements of a constrained template, third party types that do not meet the requirements by themselves can be *adapted* to do so. Typically, this involves adding a free function wrapper with the expected name, but more elaborate methods of adaptation exist, see the section on customisation points (Sect. 3.7).

This not only helps integrate third party types. For the types provided by the library, the choice of sharing code (e.g. via inheritance) is now independent of their role in polymorphic interfaces. Furthermore, constraining an algorithm does not preclude other forms of static polymorphism. It is still possible to add overloads/specialisations for concrete types and even partial specialisations, because resolution based on specialisation is resolved before constraints are compared.[20]

---

[20] Although it should not be necessary to mix partial specialisations and constraints in most cases—and the result is not very easy to read and interpret.

3.4.5 Standard Library Concepts

Analogous to the type traits provided by the standard library, it ships a set of concepts, as well. Some of these are very similar to traits or even defined in terms of them, but, as I discussed above, concepts are usable in many more contexts than traits.

While not formulated as part of the standard, WG21 agreed in their Cologne meeting in 2019 that standard library concepts (indeed most concepts) can be broadly divided into the following categories:

Abstractions: encompass the entire (relevant) interface of a type. The concept name should be a noun.

Capabilities: represent one aspect/property/capability of a type. The concept name should be an adjective, preferably ending in "-ible" or "-able."

Relations/misc. predicates: take two or more parameters. The concept name should end in a preposition.

```
   /* Abstractions */
2  std::integral                 // == std::is_integral_v
   std::signed_integral          // == std::is_signed_v (and refines integral)
4  std::input_iterator           // all input iterators
   std::forward_iterator         // all forward iterators (refines input_iterator)
6  std::predicate                // an invocable type (function obj, function *, …) returning bool

8  /* Capabilities */
   std::default_constructible    // ~= std::is_default_constructible_v
10 std::swappable                // std::ranges::swap can exchange two values of two objects of the type
   std::regular                  // default constructible + copyable + equality-comparable
12
   /* Relations / misc. predicates */
14 std::same_as                  // == std::is_same_v
   std::derived_from             // ~= std::is_base_of_v (with reversed arguments)
16 std::convertible_to           // == std::is_convertible_v
```

Code snippet 3.25: A non-comprehensive list of standard library concepts (most are available from the `<concepts>` header)

Snippet 3.25 shows examples for each of the categories. It also highlights that some concepts are extensions of type traits while others implement *named requirements*, e.g. `std::input_iterator`.

3.5 Code Reuse

In the previous section I decoupled derivation of types (inheritance, specialisation) from their use in polymorphic interfaces. This means types do not need to share code—the huge advantage of concepts. However, in many cases it is still desirable to share code between implementations to reduce redundancy in the codebase.

Reducing redundancy is a goal in and of itself, often described as the *DRY* ("don't-repeat-yourself") principle (Hunt & Thomas, 1999). In object-oriented programming, this is intrinsic to inheritance where a derived type only overrides those members that are meant to be different and inherits the rest from the base class. And similarly in template subclassing, additional free function overloads with stronger partial specialisation are added only for those functions that are designed to deviate while relying on existing overloads for rest.

The design (and problems) of template subclassing and only using free functions have been discussed extensively in Chap. 2. And since concepts-based polymorphism allows requirements on member functions, it is safe to assume that Modern C++ code will contain member functions and that developers will want to reuse code between types that haver member functions.

3.5.1   The Curiously Recurring Template Pattern (CRTP)

```
 2   struct Base
     {
 4       virtual void foo()
         {
 6           std::cout << "base";
         }
 8
         virtual void bar()
10       {
             /*...*/
12           foo();
         }
14   };

16   struct Derived : Base
     {
18       void foo()
         {
20           std::cout << "derived";
         }
22   };
```

```
     template <typename Spec>
 2   struct Base
     {
 4       void foo()
         {
 6           std::cout << "base";
         }
 8
         void bar()
10       {
             /*...*/
12           static_cast<Spec &>(*this).foo();
         }
14   };

16   struct Derived : Base<Derived>
     {
18       void foo()
         {
20           std::cout << "derived";
         }
22   };
```

Code snippet 3.26: Code reuse through inheritance and virtual functions versus CRTP

When ruling out pure free function interfaces and template subclassing, the obvious answer to sharing code between classes is using inheritance. Classic inheritance as used in OOP is displayed on the left side of Snippet 3.26. When invoking `bar()` on an object of type `Derived`, it will call the implementation of `bar()` in the base class `Base` which will then invoke the implementation `foo()` in `Derived`. The latter is crucial, any other behaviour would mean member functions cannot call each other without breaking the implicit assumptions of code sharing through inheritance. But it is important to note that this behaviour is only due to `foo()` being marked `virtual` in `Base`. Had it not been marked `virtual`, `foo()`'s implementation in the base class would have been called. Virtual function calls are associated with runtime overhead (Driesen & Hölzle, 1996) and if one decides to avoid them in selecting the best operations in an algorithm (where they influence semantics), it makes no sense to introduce them for syntactic reduction of the code.

A way of using inheritance without running into the previously mentioned problem is the *curiously recurring template pattern* (CRTP), named so by Coplien (1995). By the means of CRTP, one uses both inheritance and template specialisation to create a derived class from a base class template. It does not rely on any features of Modern C++ and has been used in generic programming with C++ extensively (Coplien, 1995; Duret-Lutz et al., 2001), but since it is not used in SeqAn1/2 and the technique is not trivial to understand, I want to briefly introduce it here.

As mentioned above, the core problem that virtual functions solve is selecting the override from the derived class when the function is invoked within the function body of another member of the base class. CRTP solves this problem by `static_cast`ing the object to its derived type before invoking the member.[21] The only difficulty of this cast is that the base type needs to *know* the name of the derived type. With regular inheritance this is impossible, but by making the base class a template and instantiating it with name of the derived type during the declaration of that derived type, it becomes valid C++. This is because the template argument does not need to be a complete type at the time of declaration.

The right side of Snippet 3.26 shows an example of CRTP. It looks very similar to the example with virtual function calls on the left, but it does not use the `virtual` keyword. The declaration of `Derived` now inherits a template specialised by its own name (line 16) and the invocation of `foo()` needs to be prefixed with the aforementioned cast (line 12).[22] When invoking `bar()` on an object of type `Derived`, it will call the implementation of `bar()` in the base class `Base<Derived>`, because no own implementation is provided by `Derived`. It then casts itself back to the derived type and invokes `foo()` on that. Because a (specialised) implementation of `foo()` is provided by `Derived`, that is chosen, otherwise the inherited `foo()` from base would have been selected. No runtime overhead is incurred and compiler optimisations like *inlining* can take place.

The attentive reader might have noticed that one can use CRTP to perform static polymorphism and not just code reuse. This is true and in a C++98 codebase it has the important advantage over template subclassing that it looks more similar to object-oriented programming. However, it shares the fundamental problem that refinement is performed by derivation and third party types are not easily

---

[21] A `static_cast` is always resolved at compile-time and casting pointers and references does not involve copying the object.

[22] Typical implementations provide a `to_derived()` member function that performs the cast and is easier to read.

integrated.[23] Thus, concepts-based polymorphism is preferred, and CRTP is only of interest for its ability to reduce duplication of logic and code.

3.5.2   *Metaclasses*

While CRTP works well in practice and the above example illustrates that simple use-cases are close in appearance to object-oriented programming, it should be noted that the technique is far from perfect:

- It is surprising to most programmers that it works at all (especially passing the incomplete type as template argument).
- It is easy to forget the `static_cast`, resulting in hard-to-detect errors. (The `virtual` keyword in OOP is only needed in the function signature, while the casts are necessary in every invocation.)
- It is possible to wrongly derive from a CRTP-base, e.g. by specialising over a different type than one is defining. There are techniques to detect these errors, but they increase the complexity of the code.

In the future a different proposal could help reduce duplication of logic and code in a much cleaner way: *Metaclasses* (Sutter, 2019). Metaclasses would be a very fundamental change, likely even more so than concepts. The proposal is based on two other major features that are not yet part of C++ :

Reflection   The ability to introspect into code from within the code, e.g. "iterate over the members of a class".

Injection   The ability to generate valid C++ code from within C++ , e.g. "generate a class type with *n* member variables via a compile-time `for`-loop".

The basic idea of metaclasses is to provide an abstraction on a level above types, but very different than what templates currently do. Similar to how the rules for `struct` and `class` are different in regard to which members are `public`/`private` by default, and similar to how there are rules for automatically generating constructors and assignment operators (depending on the presence/absence of any user-defined ones), metaclasses will enable the programmer to define their own type categories with custom rules for generating members or enforcing certain properties on derived types:

> Metaclass functions (provisional name) let programmers write a new kind of efficient abstraction: a user-defined named subset of classes that share common characteristics, typically (but not limited to): user-defined rules, defaults, and generated functions. [...] The goal is to elevate idiomatic conventions into the type system as compilable and testable code, and in particular to write all of the same diverse kinds of class types we already write today, but more cleanly and directly (Sutter, 2019).

---

[23] In fact, it is even more difficult to adapt third party types when polymorphism happens on members, because members cannot be overloaded at all.

Avoiding code duplication is not the primary aim of the proposal but a likely outcome. It would mean that one could avoid techniques like CRTP which appear more like "hacks" of the language in comparison. The proposal has gained lots of interest in WG21 but is currently far from being accepted, because the changes are very fundamental and even the prerequisites mentioned above are still in need of a lot of fine-tuning. A possible target is C++23 with C++26 being more likely, so it has no bearing on SeqAn3. It is, however, one of the things I recommend future developers keep track of when designing a successor to SeqAn3.

3.6   C++ Ranges

3.6.1   Introduction

Traditionally, generic algorithms in the standard library that work on a collection of items take a pair of iterators into that collection as arguments. Iterators are the objects returned by calling `begin()` and `end()` on e.g. a `std::vector`. So assuming that `vec` is a vector of integers, the way to sort that vector is calling `std::sort(vec.begin(), vec.end())` (or `std::sort(std::begin(vec), std::end(vec))` using the free function wrappers).

Surprising to some, one cannot call `std::sort(vec)`, i.e. the standard library algorithms do not work with the collection directly, only with its iterators. This has several reasons:

1. Internally the algorithms work with iterators anyway and prior to C++11 there was no generic way to get to the iterator from the collection.
2. Iterators have a well-defined hierarchy of capabilities (Table 3.3) and therefore provide a clean abstraction from the collection. Many algorithms behave differently based on those capabilities and in pre-concepts-C++ it would have been more difficult to implement this refinement on collection types.
3. Iterator-based interfaces provide a level of flexibility to the user that is not possible otherwise, e.g.:

Table 3.3 The traditional input iterator hierarchy. Each subsumes the capabilities of the previous one. Contiguous iterator was introduced in C++17

Name	Capabilities (simplified)
Input iterator	Reading (*), moving right (++), and eq. comparison (==)
Forward iterator	Multi-pass (incrementing one iterator does not break another)
Bidirectional iterator	Moving left (- -)
Random access iterator	Jumping in $O(1)$ ([i], +, -), ordered (<, <=,...)
Contiguous iterator	Elements pointed to by iterator are adjacent in memory

- Sort all elements after the fifth: `std::sort(vec.begin() + 5, vec.end())`
- Sort in reverse order (using reverse iterators): `std::sort(vec.rbegin(), vec.rend())`
- Combining both approaches: `std::sort(vec.rbegin() + 5, vec.rend())`

C++ Ranges address all the above issues by providing single argument algorithm interfaces and facilities that are even more flexible than working with iterators. Ranges did not make it into C++ 17, because they depend on C++ Concepts, but as with concepts ISO published a technical specification for ranges (ISO/IEC 21425:2017). The proposal that was finally merged and contains most of the content is P0896 (Niebler et al., 2018)—with 226 pages one of the largest changes ever made to the standard by a single proposal. It adds a new chapter to the C++ standard, the "Ranges library" and changes many other parts, mostly in the "Iterator library" and the "Algorithms library". All the changes are standard library changes (no new language features) and a stand-alone library that contains a super-set of the standardised contents has been widely used for many years.[24] I will use the notation based on the C++ 20 wording in this chapter; a way to use the stand-alone range-v3 library as a "placeholder" with a C++ 17 compiler is described on p. 109 in Sect. 4.4.1.

The first of the aforementioned problems has been addressed in C++ 11 already with the introduction of the `std::begin()` and `std::end()` free functions. C++ 20 introduces the namespace `std::ranges::` in which most of the ranges library resides, as well as the single argument interfaces for the algorithms library (so as not to conflict with their overloads in `std::`). It also provides `std::ranges::begin()` and `std::ranges::end()` which behave very similar to their counterparts in `std::` (ISO/IEC 14882:draft, 24.3).[25]

The second problem is solved by introducing C++ concepts (instead of just named requirements) for iterators and also for *ranges*. The iterator concepts are named like their named requirements/capabilities: `std::input_iterator`, `std::forward_iterator`, etc.[26] Then the concept `std::ranges::range` is defined which represents "anything iterable" and formally requires only that `std::ranges::begin()` and `std::ranges::end()` are defined for objects of that type (ISO/IEC 14882:draft, 24.3). In Sect. 3.6.2 I will discuss the different refinements of the range concept.

The last problem is solved by introducing *views*. A view is a kind of range that provides functionality like I used above ("skip first five elements", "reverse elements"). Two views can be composed into a new view resulting in a design at least as flexible as using iterators. The machinery behind views is discussed in Sect. 3.6.3 and a subset of the standard library views are presented in Sect. 3.6.5.

---

[24] https://github.com/ericniebler/range-v3.

[25] The differences are not relevant here.

[26] There are slight differences in the definitions, but they are not relevant here.

Since ranges are at the core of what defines SeqAn3, this entire section contains a higher degree of technical detail than others in this chapter. I will reference many of the concrete definitions in Chap. 5 and it is important to understand how the different concepts relate to each other.

3.6.2   Range Traits and Concepts

The C++20 standard library provides numerous range concepts (ISO/IEC 14882:draft, 24.4); a visual representation can be seen in Fig. 3.1. The root of this hierarchy is `std::ranges::range` which I already introduced above. Most of the other range concepts depend on concepts modelled by the range's iterator type. So analogous to `std::input_iterator` there is `std::ranges::input_range` where the type of the object returned by `std::ranges::begin()` must model `std::input_iterator`. The same is true for `std::forward_iterator` and `std::ranges::forward_range` and the remaining input iterator concepts respectively. The input range concepts also refine each other in the same way the input iterator concepts refine each other, so polymorphic range interfaces select the best given algorithm for every input.

Snippet 3.27 shows this mechanism in action. Before looking at the remaining range concepts it is important to note that some assumptions users have of ranges based on their experience with *containers* are not true:

Fig. 3.1 The range concepts hierarchy. Arrows imply refinement, all concepts are defined in the namespace `std::ranges::`

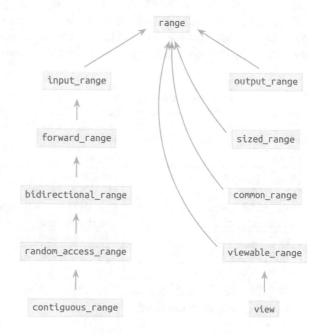

```
   void print_5th_element(std::ranges::input_range auto && rng)          // generic range interface
2  {
       auto it = std::ranges::begin(rng);
4      ++it; ++it; ++it; ++it; ++it;          // linear time
       std::cout << *it;
6  }

8  void print_5th_element(std::ranges::random_access_range auto && rng)  // refinement for random access
   {
10     auto it = std::ranges::begin(rng);
       std::cout << it[5];                    // constant time
12 }

14 print_5th_element(std::list{1, 2, 3, 4, 5, 6, 7});      // will pick first overload
   print_5th_element(std::vector{1, 2, 3, 4, 5, 6, 7});    // will pick second overload
```

Code snippet 3.27: Polymorphism based on range concepts

- The `end()` of a range need not have the same type as the `begin()`. The type returned by `begin()` is the `std::ranges::iterator_t` of a range while the type returned by `end()` is the `std::ranges::sentinel_t` (see Table 3.4). In general the requirements for the sentinel are weaker than for the iterator, but they always need to be (in-)equality comparable with each other. (To check if the end of the range has been reached.)

- There are `std::random_access_range`s that are not also a `std::ranges::sized_range`. Performing random access to arbitrary positions to the right of the current iterator may not be safe on them.

Output ranges are ranges where the individual elements can be written to, e.g. `std::vector<int>` is both an input range and an output range. `std::vector<int> const` on the other hand is only an input range. There are also output ranges that are not input ranges, although often pure output iterators are used instead (e.g. `std::ostream_iterator` that writes to a stream).

Table 3.4 The most important range and iterator traits given by their C++ 20 free-standing shortcuts. `R` is assumed to be a range type and `It` an iterator type. `r` and `it` are objects of the respective types. `decltype()` yields the type of an expression. Range traits are defined in `std::ranges::` and iterator traits are defined in `std::`.

Trait	Definition
`iter_reference_t<It>`	`decltype(*it)`
`iter_value_t<It>`	`It::value_type`[†]
`iter_difference_t<It>`	`It::difference_type`[†]
`iterator_t<R>`	`decltype(std::ranges::begin(r))`
`sentinel_t<R>`	`decltype(std::ranges::end(r))`
`range_reference_t<R>`	`iter_reference_t<iterator_t<R>>`
`range_value_t<R>`	`iter_value_t<iterator_t<R>>`
`range_difference_t<R>`	`iter_difference_t<iterator_t<R>>`

[†] These are typical, exceptions exist

Sized ranges are ranges whose size can be computed in constant time via a call
to `std::ranges::size()`. The latter looks for a `.size()` member function or
subtracts `begin()` from `end()`. `std::list` is a type where the latter would not
work, but the former does. `std::forward_list` provides neither and does not
model `std::ranges::sized_range`.

Common ranges are ranges whose iterator and sentinel type (see Table 3.4) are
the same. This is true for all ranges in the standard library prior to C++20 and
the algorithms in `std::` expect two objects of iterator type. Since C++20 most
ranges do not model this concept and the algorithms in `std::ranges::` do require
common ranges.

Viewable ranges and views will be discussed in Sect. 3.6.3.

Table 3.5 contains actual results for some described traits and concepts with
three example types. Views will be discussed below, at this point please interpret
`iota_view<int>` as simply an infinite range of the numbers 0 until ∞.

The iterator and sentinel types for vector and its `const` version are each identical,
`end()` simply returns an iterator pointing one element behind the last. In practice
the iterator types could be just `int *` and `int const *` respectively. For the
view this is already different, because, since it is infinite, an iterator may never
compare equal to the end. It should also not be possible to subtract an iterator
from the end (while it should be possible to subtract different iterators from
each other); thus iterator and sentinel have different types here. This implies that
`std::ranges::iota_view<int>` is not a `std::ranges::common_range`, while
`std::vector<int>` and `std::vector<int>` `const` are.

The value type of all three ranges is `int`, however, the type returned
when dereferencing an iterator (or calling `[]` on the range) is the *reference
type* (returned by `std::ranges::range_reference_t`). This is named so,
because historically it is a reference to the value type. It enables syntaxes like
`vec[i] = 3;` and prevents unintended copies when the element type is large. In
essence, this makes `std::vector<int>` an output range. The reference type of
`std::vector<int>` `const` is `int const &`, because it should not allow the user
to change the elements in the vector (this prevents it from being an output range).
Perhaps surprising to the reader, the reference type of `iota_view<int>` is not
actually a reference; it is the same as the value type![27] This already hints at the fact
that the view does not *hold* the elements and can expose no reference to them—
instead it generates the value on-demand and must return an rvalue, instead of an
lvalue reference. Thus, this view can also not be an output range.

The vector types both model `std::ranges::contiguous_range`, because
their elements are guaranteed to be located in memory consecutively.[28]
`iota_view<int>` on the other hand does not store any elements, so it "only"

---

[27] Prior to C++20, forward iterators were required to have an actual reference as the reference type.
This requirement has been lifted.

[28] Contiguous ranges have advantages in some algorithms, e.g. `std::copy()` may use a more
efficient implementation based on `std::memmove()` or `std::memcpy()`.

Table 3.5 The discussed traits and concepts applied to three different range types for illustrative purposes. `T` refers to the respective type. All traits, concepts and the `iota_view` template are defined in `std::ranges::`, `vector` is defined in `std::`

	`vector<int>`	`vector<int> const`	`iota_view<int>`
Trait			
`iterator_t`	`T::iterator`	`T::const_iterator`	`T::iterator`
`sentinel_t`	`T::iterator`	`T::const_iterator`	`T::sentinel`
`range_reference_t`	`int &`	`int const &`	`int`
`range_value_t`	`int`	`int`	`int`
`range_difference_t`	`ptrdiff_t`	`ptrdiff_t`	`ptrdiff_t`[†]
Concept			
`input_range` refine.	`contiguous_range`	`contiguous_range`	`random_acces_range`
`output_range`	*Yes*	*No*	*No*
`sized_range`	*Yes*	*Yes*	*No*[††]
`common_range`	*Yes*	*Yes*	*No*
`view`	*No*	*No*	*Yes*

[†] The difference type of `iota_view` may actually be larger than `ptrdiff_t`

[††] Bounded `iota_view`s can be created that are sized

models `std::ranges::random_access_range`. It is also an example of a random access range that is not sized, because, as mentioned above, the iterator cannot be subtracted from the sentinel and the range itself provides no `.size()` member. (The default iota view is infinite.)

3.6.3 The View Concept

The ranges most established in the pre-C++ 20 world are *containers* (ISO/IEC 14882:2017, 26), e.g. `std::vector` and `std::list`. There are named requirements that detail the properties of containers, but these have not been turned into concepts (yet). One important property associated with containers is *element ownership*, i.e. all elements are copied when the container is copied, and all elements are deleted when the container is deleted.

In this regard, *views* are the opposite of containers, they are required to *not* own their elements.[29] This is formalised as the requirement that copy, move and assignment of a view be in $O(1)$, i.e. independent of the number of elements in the view (ISO/IEC 14882:draft, 24.4.4). Since this is a semantic requirement not verifiable by the compiler, only types for whom the `std::ranges::enable_view` trait returns `true` are considered views.[30]

Different kinds of ranges qualify as being views, among them (ISO/IEC 14882:draft, 24.4.4):

1. A range that generates elements on-demand.
2. A range that presents the elements of another range (modified or unmodified).
3. A range that holds elements but shares them implicitly with all of its copies (e.g. using `std::shared_ptr`).

An example of the first category is `std::ranges::iota_view` that I mentioned above already. It generates a series of increasing values based on an initial value and optionally a bound. An example of the second category is `std::span`, a class template that stores only a pointer `p` to the underlying range and an offset `n`. It presents the memory region at `*p` as a contiguous range of `n` elements and is usable as a light-weight value type in place of a pointer or reference to a contiguous range (e.g. `std::vector`, `std::array`). It can also represent sub-ranges of these types. A more interesting example of the second category is `std::ranges::transform_view` which applies a transformation on every

---

[29] The range concepts make no statements about ownership and there are ranges that are neither containers nor views.

[30] There are different ways to have this trait evaluate to true, but developers can also just specialise it for their view types.

element of the underlying range. The third category is found seldom in practice due to an intrinsic performance overhead of the implementation.[31]

```
   // view of 0 to infinity                          // function takes span argument (by value!)
2  std::ranges::iota_view v1{0};                   2 void foo(std::span<int> s)
   // prints '0' and '5'                             { /* ... */ }
4  std::cout << v1[0] << ' ' << v1[5] << '\n';     4
                                                     // can be called with vector or array
6  // a function object that squares its argument  6 std::vector vec{1,2,3,4,5};   foo(vec);
   auto sq = [] (int i) { return i*i; };             std::array  arr{1,2,3,4,5};   foo(arr);
8  // view that applies lambda on v1               8
   std::ranges::transform_view v2{v1, sq};          // "sub-span" of [2,3,4]
10 // prints '0' and '25'                          10 std::span s{vec.begin() + 1, vec.end() - 1};
   std::cout << v2[0] << ' ' << v2[5] << '\n';       foo(s);
```

Code snippet 3.28: First view examples: composing and lazy evaluation are illustrated on the left; cheap construction and copy is shown on the right

The example on the left in Snippet 3.28 illustrates how two views can be combined: the iota view generates all natural numbers and the transform view returns the elements of the underlying range (the iota view) squared. The sequence generated of the combination would therefore be [0, 1, 4, 9, 16, 25, ...]—however, this sequence is not actually generated by the code in the example. It is important to remember that the views do not store their elements, they merely hold the instructions on attaining them, i.e. only when calling `std::cout << v2[5];` does the iota view generate its fifth element (5) and the transform view apply the lambda resulting in the value 25. This is called *lazy evaluation*, a core paradigm in functional programming languages (Henderson & Morris, 1976). Views might appear like a container, but in reality they represent an "algorithm" more than they represent "data".

On the right side of Snippet 3.28 one can observe a different property of views: that they are cheap to copy and (almost) always cheap to create. In fact, there is no performance overhead for using a `std::span<int>` instead of a `std::vector<int>` & in a function interface. Other than one might expect, `std::span` is not specialised over the type of the underlying range (like `std::ranges::transform_view` which is specialised over `std::ranges::iota_view`). It is only specialised over the value type. The effect can be seen in the function signature in Snippet 3.28: `foo()` is not a function template, it is a plain function. But one can still pass it different contiguous range types, because `std::span` can be constructed from all sized contiguous ranges of the same value type; the original type is *erased*. On the one hand, this reduces the amount of needed templates (and thus code generation); on the other hand it promotes safe C++-style interfaces over C-style (pointer, length)-interfaces.

---

[31] An example is `seqan3::views::persist`, introduced in Table 7.4 on p. 195. It is the only view that can hold temporary containers.

Those views that adapt other ranges (like `std::ranges::transform_view`) typically store the underlying range as a data member. This has an important implication: since the transform view promises to be copyable in constant time, it can only contain data members that are also copyable in constant time. It might appear as though views can only adapt other views, however, ranges that are not views are also acceptable as long as the view does not take ownership of the range. These requirements are denoted by the `std::ranges::viewable_range` concept which is defined as "view or lvalue reference to a non-view range". In practice, views always require that the underlying range be a view and store it by value/copy; if they are constructed from a reference to a different kind of range, they wrap this range in `std::ranges::ref_view` and store that.

```
     std::vector vec{1, 2, 3, 4, 5};
2
     // this behaves like a reference to vec but models the view concept
4    std::ranges::ref_view v1{vec};

6    auto sq = /*...*/;
     // pass a vector to the constructor by reference
8    std::ranges::transform_view v2{vec, sq}; // vec is wrapped in a std::ranges::ref_view

10   // pass a vector to the constructor as value/temporary
     // std::ranges::transform_view v2{std::vector{1, 2, 4}, sq}; // not legal
```

Code snippet 3.29: The `std::ranges::viewable_range` concept states that only views and references to existing ranges can be passed to other views. Temporary containers are not generally supported, because storing them inside the view would break the view's semantics

An example is shown in Snippet 3.29. `std::ranges::ref_view` is a special kind of view that exists exactly for this purpose. It stores a pointer to the given range, but otherwise exposes the referenced range's begin and end as its own. It does not have the type erasing effects of `std::span` but is more generic. (Any range can be wrapped, not just contiguous ranges.)[32]

3.6.4 Range Adaptor Objects

In the last subsection I showed how the transform view can adapt the iota view. Thanks to the deduction of the template arguments (Sect. 3.1.2), this definition was less complex than it would have been in C++ 98, however, it is still not very compact. Especially when many views are combined, it will lead to the definition of local variables that are only needed to create the final view object (top example in

---

[32] For the example in Snippet 3.29, one could have also manually wrapped the vector in a `std::span` and passed that to the transform view. In that case the span would have been copied into the transform view (because it is a view) and no ref view would have been created.

```
   std::ranges::iota_view v1{0};                // view of 0 to infinity
2
   auto sq = [] (int i) { return i * i; };       // a lambda that squares its argument
4  std::ranges::transform_view v2{v1, sq};       // view that applies lambda on elements of v1

6  auto divi = [] (int i) { return i % 3 != 0; }; // a lambda that checks divisibility by 3
   std::ranges::filter_view v3{v2, divi};        // view that filters out elements divisible by 3
8
   std::ranges::take_view v4{v3, 5};             // view that takes the first 5 elements
```

```
   auto v = std::ranges::take_view{
2              std::ranges::filter_view{
                 std::ranges::transform_view{
4                   std::ranges::iota_view{0},
                    [] (int i) { return i * i; }},
6                  [] (int i) { return i % 3 != 0; }},
               5};
```

```
   auto v = std::views::iota(0)                          // iota factory
2      | std::views::transform([] (int i) { return i * i; })  // transform adaptor
       | std::views::filter([] (int i) { return i % 3 != 0; }) // filter adaptor
4      | std::views::take(5);                            // take adaptor
```

Code snippet 3.30: Views and adaptor objects. The first snippet shows the "gradual" construction of a combined view. The second snippet shows how one can construct v4 from the first snippet without intermediate local variables. The third snippet shows how one can achieve the same thing more elegantly using adaptor objects

Snippet 3.30). Of course, one could also define the final object all in one statement, but this quickly becomes messy (centre example in Snippet 3.30; note how the "order" of views is reversed). The solution to this problem is *range adaptor objects* (bottom example in Snippet 3.30).

Most of the views defined in `std::ranges::` each have an adaptor object in `std::views::` [33] that can be given the same arguments as the constructor of the corresponding view. It simply forwards these arguments to the constructor and returns a new object of the view type (ISO/IEC 14882:draft, 24.7.1). However, the adaptor objects also support taking the first argument (the underlying range) via the pipe operator instead of the `()`-operator, e.g. the following three expressions are equivalent (`r` is assumed to be some other range and `l` a function object):

1. `std::ranges::transform_view{r, l}`: The constructor of the view is invoked.
2. `std::views::transform(r, l)`: The `operator()` is invoked on the adaptor object, it forwards to 1.
3. `r | std::views::transform(l)`: First `operator()` is invoked on the adaptor object, but without a range it does not invoke the view constructor, instead it returns another "intermediate" adaptor that holds `l`. Then `operator|` is

---

[33] The namespace is actually `std::ranges::views::`, but there is a namespace alias to `std::views::`.

invoked with `r` and the intermediate adaptor as arguments; it extracts `l` from the intermediate adaptor and finally invokes the view's constructor with `r` and `l` as arguments.

Since every "pipe" expression requires a range as left-hand side argument and each expression also returns a range, it is possible to create a pipeline of these expressions as can be seen in Snippet 3.30. (They are only written in separate lines for readability.)

If a *range adaptor object* requires no arguments other than a range, it is called a *range adaptor* closure *object* (abbreviated in this section to "closure object"). `std::views::transform` is not a closure object, but by providing a single argument (the function object) to it, a closure object is returned. This is the "intermediate adaptor" mentioned above (3.). Other range adaptor objects, like `std::views::reverse`, never take arguments and therefore are already closure objects.

```
    // create a custom range adaptor object by "fixing" the parameters to an existing one
2   auto my_transform = std::views::transform([] (char c) { return std::tolower(c); });

4   std::string const s = "*FOObar*";

6   auto v1 = s | my_transform;              // v1 == "*foobar*"

8   // create a combined range adaptor from two other ones
    auto my_transform2 = my_transform
10                        | std::views::transform([] (char c) { return std::isalnum(c) ? c : '_'; });

12  auto v2 = s | my_transform2;             // v2 == "_foobar_"

14  /* the following are equivalent */
    auto v3 =  s |  my_transform  | std::views::reverse ; // default evaluation order
16  auto v4 = (s |  my_transform) | std::views::reverse ; // default evaluation order (explicit)
    auto v5 =  s | (my_transform  | std::views::reverse); // combined adaptor created, then applied
```

Code snippet 3.31: User-defined range adaptor objects. `my_transform` transforms string to lower-case. `my_transform2` transforms to lower-case but also replaces non-alpha-numerical characters with `'_'`. The last example illustrates that `()` in view pipelines do not influence the outcome

This mechanism might appear complicated (and as Chap. 5 shows the implementations are indeed non-trivial), but it provides users with some very convenient features. Those adaptor objects that are not closure objects can be stored with user-provided arguments and used as regular view adaptors later on. In many cases where a developer wants to write "their own view", it is sufficient to take one of the existing adaptors and provide it e.g. with a custom lambda expression and use the resulting closure object.

Snippet 3.31 shows an example of this: a "view" that transforms strings to lower-case can be defined through a custom transform view in a single line. A feature not yet discussed is that two closure objects can be combined to form a new closure object. This is also facilitated via `operator|` (but without a leading range). An example of how this is useful can also be seen in Snippet 3.31. A side effect of this feature is that it does not matter if parentheses are placed in view pipelines, because combining the adaptors first (and applying the combined adaptor to the input range) has the same effect as passing the range to the first adaptor, then passing the return value to the next, etc.

The reader is reminded that neither the construction/combination of the adaptors, nor the application of an adaptor to a range is associated with a noticeable runtime cost and views generally do not allocate any dynamic memory. The (combined) type of the view is deduced at compile-time, the actual values are generated lazily during access.

Finally, it should be noted that not all objects in `std::views::` are range adaptor objects; some, like `std::views::iota` are *factory objects*. As explained at the beginning of Sect. 3.6.3, some views adapt other views, and some are "self-sufficient", e.g. by generating values. For the former there are range adaptor objects, for some of the latter there are factory objects that essentially just forward their arguments to the view constructor. In a view pipeline, they can only appear in the beginning.

3.6.5   Standard Library Views

The standard library of C++20 provides many useful views, a non-exhaustive overview is given in Table 3.6. For those views that do not have an adaptor or factory, the type is given. For the rest the respective adaptor or factory object is given, because this is the preferred form of interacting with them.

As can be seen in the table, there are various views that can be used to adapt a non-view (or possibly a subrange of the non-view) into a view. This variety is partly due to historic reasons,[34] but there are some noteworthy differences:

- `std::basic_string_view` can only adapt C-Strings and `std::basic_string`; it behaves differently from the other views in the following regards:

  - It is never an output range, i.e. one cannot change the values of an underlying string through the view.
  - A string can be constructed from a string view. (Usually containers cannot be constructed from views.)[35]
  - It is equality comparable with itself and with strings (usually views are not equality comparable); it follows container semantics in this regard (the elements are compared individually, not whether the views have the same underlying string object).

---

[34] `std::basic_string_view` is already part of C++17, `std::span` is part of C++20 but predates the ranges library.

[35] At least not without the help of auxiliary facilities. This is to prevent unintended dynamic memory allocation.

Table 3.6 An overview of the most useful views in the C++ 20 standard library

	Description
View types	
`std::basic_string_view`	View/subrange of `std::basic_string`; read-only; printable
`std::span`	View/subrange of contiguous range; stores pointer + size
`std::ranges::subrange`	View/subrange of any range; stores iterator + sentinel
`std::ranges::ref_view`	View of any range; stores pointer/reference to original
Factory objects	
`std::views::iota`	Generate increasing values; possibly infinite
`std::views::single`	View of size 1 with a single value
Adaptor objects	
`std::views::filter`	Filter out elements that do not pass the given predicate
`std::views::transform`	Applies an invocable on every element
`std::views::take`	The first n elements
`std::views::take_while`	The first elements that all pass a given predicate
`std::views::drop`	All, but the first n elements
`std::views::drop_while`	All, but the first elements that all pass a given predicate
`std::views::join`	Flattens an n-dimensional range into a (n-1)-dimensional range
`std::views::split`	Split a range into a range-of-subranges on the given delimeter
`std::views::reverse`	Reverse the elements

- It can be constructed from rvalues, i.e. it does not require `std::ranges::viewable_range`. This is safe for string literals which have static lifetime but might create dangling pointers for `std::basic_string` (dangerous!).

- `std::span` can only adapt ranges that are contiguous and sized. The value type is a template parameter, but otherwise the original type is fully erased.

- `std::ranges::subrange` adapts any pair of iterator and sentinel. This allows for iterator pairs that were not the original begin and end (hence the name). It can optionally also store the size if this is not computable from iterator and sentinel. It only type erases two types if their iterators and sentinels (each) have the same type.[36]

- `std::ranges::ref_view` is similar to `std::ranges::subrange` in that it can adapt any range. It does so by holding a pointer to the original. The drawback is that it needs to be specialised over the original type which eliminates any chance of type erasure. It also always represents the full underlying range. An advantage

---

[36] This would e.g. be possible for `std::vector` and `std::array`, however, it is not required.

is that the original range can be accessed via the `.base()` member function. Since its member functions call respective functions on the underlying range and do not cache iterators or size, it is one of the few views that is not invalidated by changes to the underlying range.

I already established in Sect. 3.6.3 that (almost) all views that adapt other ranges require that those ranges model `std::ranges::viewable_range`. In the bullet points above, I also stated that certain views have more requirements and indeed these can vary greatly from view to view. `std::views::reverse`, for example, requires that the underlying range is at least a bidirectional. (Otherwise it would not be able to iterate through it in reverse order.) `std::views::join` requires that the underlying range is a range of ranges, i.e. the reference type of the underlying range must itself also model the range concept. (Otherwise there is nothing to "flatten".)

But not only the requirements on the underlying range vary between views, also the concepts that a view itself satisfies depend a lot on the view—and the underlying range. `std::views::transform` preserves most of the concepts modelled by the underlying range, i.e. the transformed view will support random access if and only if the underlying range also supports random access. However, since virtually all transformations imply the computation of a new value, a transformed view is typically not an output range and not a contiguous range. The view returned by `std::views::filter` never models `std::ranges::random_access_range`, because the n-th element that satisfies the predicate cannot be computed in constant time. (All elements leading to it need to be evaluated.) On the other hand, the original reference type is fully preserved so it can be an output range.

Some of these dependencies are intuitive once one has familiarised oneself with the involved concepts, but others might still be surprising. It is important to document these properties well and design the adaptor objects in a way that they produce readable error messages when users combine incompatible adaptors.

3.7   Customisation Points

3.7.1   Excursus: Calling Conventions

How to call a function depends greatly on how it is defined (free function or member function), in which namespace it resides and from which namespace the function call is invoked. Since the specifics of this influence customisation, I want to briefly provide examples for the different cases. I am assuming one wants to call "`swap()`" with the two arguments "`lhs`" and "`rhs`". When a function call is performed, the compiler builds an "overload set" and then picks the best match out of that set. If there is ambiguity within this set, the build fails. This restriction is important, because it means that adding an overload can break existing functionality. (This is especially true for unconstrained templates, see Sect. 2.4.2 on p. 21.) Depending on how the function is called, different function definitions will be added to the overload set; these are the options:

1. Member function invocation: `lhs.swap(rhs)`

   - Member functions or a member function object of `lhs` are added. Namespaces are irrelevant.

2. Qualified (free function) lookup: `foo::swap(lhs, rhs)`

   - A match for the namespace `foo::` is searched: first within the current namespace, then in parent namespaces until the global namespace `::` is reached.[37]
   - Free functions and/or a global function object in that namespace are added.

3. Unqualified (free function) lookup: `swap(lhs, rhs)`. Leads to two kinds of lookups:

   (a) Argument-dependent lookup (ADL):

      - Function overloads from *associated* namespaces[38] of `lhs` and `rhs` are added.
      - Function overloads declared as `friend`s of either `lhs` or `rhs` are added.
      - Function objects are not considered for ADL! Current namespace is irrelevant.

   (b) Non-ADL unqualified lookup:

      - Function overloads and function objects from the current namespace are added. If none match, the parent namespace is searched. This can continue until the global namespace `::` is reached.

See *name lookup* (ISO/IEC 14882:2017, 6.4) and *overloading* (ISO/IEC 14882:2017, 16) for formal definitions. For the user invoking the function call, the technical details of lookup are usually not important and the rules specified above "do the right thing". Library designers, however, need to consider the complexities to keep their code extensible.

3.7.2 Introduction

Generic algorithms are meant to accept arguments of library-provided types but also user-provided types. They make implicit and/or explicit (concepts, see Sect. 3.4)

---

[37] To only search for a top-level match, `::foo::swap(lhs, rhs)` can be invoked instead.

[38] This includes the namespaces; these types are defined and in some cases more namespaces.

assumptions about these types, e.g. by calling certain functions on the arguments or evaluating certain type traits (e.g. `std::ranges::find()` internally calls `std::ranges::begin()` on its argument so this must be valid). To make a user-provided type (e.g. `my_vector`) compatible with the generic algorithm (in this example `std::ranges::find()`), the user needs to ensure that these assumptions are satisfied. In particular, they may need to overload/specialise certain functions and/or traits provided by the library (`std::ranges::begin()` in that case).

Such "hooks" for user-defined types are called *customisation points* (Niebler, 2014) and it is good style to clearly indicate which entities of a library are customisation points and which are not. This prevents misuse of the library by the user and forces the library developer to maintain customisation as part of the API. The following is stated for the standard library:

> The behavior of a C++ program is undefined if it adds declarations or definitions to namespace std or to a namespace within namespace std unless otherwise specified. A program may add a template specialization for any standard library template to namespace std only if the declaration depends on a user-defined type and the specialization meets the standard library requirements for the original template and is not explicitly prohibited (ISO/IEC 14882:2017, 20.5.4.2.1).

Other popular libraries, like Google's in-house C++-library *abseil*, have even stricter provisions: "You are not allowed to define additional names in namespace absl, nor are you allowed to specialize anything we provide." (Google, 2017).

There is no designated way to design such customisation points in the C++ language and the preferred form of doing so has changed over time and with the availability of certain language features (Niebler, 2014; O'Dwyer, 2018).

One of the oldest customisation points in the standard library is `swap()`, a free function that exchanges the values of its two arguments. The standard library implementation `std::swap()` uses move semantics and is (only) defined for types that are move-constructible and move-assignable (ISO/IEC 14882:2017, 23.2.3). User-defined types that do not meet these requirements (or that wish to provide specialised behaviour) typically provide their own overload of `swap()`.

How such an overload can be added depends very much on how the generic algorithm invokes the customisation point (see Sect. 3.7.1). If the algorithm performs a member call (Option 1 in Sect. 3.7.1), it is only possible to adapt a type by modifying its definition (i.e. adding such a member function). This is one of the reasons why member functions are bad in generic contexts. If the algorithm performs qualified lookup (Option 2 in Sect. 3.7.1), the user needs to add an overload for their type to that respective namespace, e.g. add an overload of `std::swap()` to the standard namespace or an overload of `seqan::length()` in SeqAn's namespace. This is discouraged (or even *illegal*, see *abseil* above), because adding an overload to an existing namespace may inadvertently interfere with other types' overload resolution, especially if the new overload is a template. If the algorithm performs unqualified lookup (Option 3 in Sect. 3.7.1), ADL is performed (Option 3a) which is usually desired; it allows the user to define an overload in their own namespace for their own type. But on the other hand, other non-ADL lookup also happens (3b) which is usually not desired and increases the chance of conflicting declarations or even recursive lookups. If the user wishes to adapt the type from a namespace they do not own, e.g. a third library, they may not even be able to satisfy ADL, because they would need to open that type's namespace. (Similar problems apply as for Option 2.)

Furthermore, the ways of invoking are mutually exclusive and it is not possible to offer multiple options of customisation to the user, e.g. allow some users to define members and others to add free functions. Unqualified lookup can perform ADL, but it cannot fallback to a specific implementation e.g. `std::swap()`. (If `std::` is not an associated namespace of the arguments.) Although a workaround for this example exists,[39] it is difficult to teach and easy to get wrong. It also does not work inside concepts, a place where customisation points are prominently used in Modern C++.

3.7.3 "Niebloids"

Eric Niebler introduced a design that solves most of the mentioned problems (Niebler, 2014). The core aspects of the design are:

1. Create a *function object* as the customisation point.

   - Generic code always invokes this function object via qualified lookup, e.g. `library::swap` or even `::library::swap`. No other overloads are considered initially.
   - Function objects are not subject to ADL, this prevents the function object from invoking itself or being invoked inadvertently.

2. The developer specifically crafts the desired overload set by defining multiple `operator()` in the function object with the desired behaviour. These delegate to the actual implementations. They can include:

   - Lookup of an argument's member functions [Option 1 in Sect. 3.7.1].
   - Qualified lookup in a specific namespace (multiple of these can be specified) [Option 2 in Sect. 3.7.1].
   - Unqualified lookup (to perform ADL) [Option 3 in Sect. 3.7.1].

This design gives users the possibility to adapt their types via different means but keeps the invocation simple. It is even possible to prevent non-ADL forms of

---

[39] It is possible to do the following: `using std::swap; swap(a, b);`. This imports the definitions from the standard namespace and adds them to unqualified lookup. In fact, the formal definition of the Swappable *named requirement* is to perform this (ISO/IEC 14882:2017, 20.5.3.2).

unqualified lookup (Option 3b in Sect. 3.7.1) via a "trick" that poisons the lookup with a deleted overload.[40]

```
   namespace library::detail
2  {
   template <typename T>
4  void swap(T, T) = delete;               // prevent non-ADL forms of unqualified lookup

6  struct swap_t                           // type of the function object
   {
8      template <typename T>               // overload that performs unqualified lookup
           requires (requires (T & lhs, T & rhs) { {      swap(lhs, rhs) }; })
10     void operator()(T & lhs, T & rhs) const    {      swap(lhs, rhs);   }

12     template <typename T>               // overload that performs qualified lookup in std::
           requires !(requires (T & lhs, T & rhs) { {      swap(lhs, rhs) }; }) &&
14             (requires (T & lhs, T & rhs) { { std::swap(lhs, rhs) }; })
       void operator()(T & lhs, T & rhs) const    { std::swap(lhs, rhs);   }

16
       /* further overloads for calling e.g. lhs.swap(rhs) */
18 };
   } // namespace library::detail
20
   namespace library
22 {
   inline constexpr detail::swap_t swap{}; // the function object
24 }
```

Code snippet 3.32: Customisation point objects/"Niebloids"

Snippet 3.32 shows an example of this technique. One defines the function object's type in a detail or implementation namespace; it does not have state, it only provides the operators for being called in function notation. The operators are constrained in such a way that they are only defined if the function they are invoking is defined and that an order between them is introduced.[41] The namespace also includes the deleted overload to restrict unqualified lookup to performing ADL.

The actual function object (a constant of the previously defined type) is then declared in the main namespace of the library.

As Snippet 3.33 shows, it is very easy to use such customisation point objects: the user always does a qualified call and does not need to add `using` declarations. The user can easily provide a specialisation by adding an appropriately called friend or free function. Had I added a check for member functions to the CPO in Snippet 3.32, all possible implementations would be "found". For types that do not have a specialisation, the fallback to `std::swap()` will be chosen by the CPO. (If the requirements for `std::swap()` are met.)

---

[40] It may surprise the reader that non-member functions can be explicitly deleted, but this is by design. The deleted overload is initially accepted as valid by the compiler so no further namespaces are searched, but is then discarded, because it is deleted. Thus, no functions from the surrounding namespaces are searched.

[41] To make ordering of multiple overloads easier it is advisable to use priority-tags (O'Dwyer, 2018), but they are omitted here for simplicity.

```
   namespace user
2  {
   struct T
4  {
      /* ... */
6     friend void swap(T &, T &) { /* ... */ }
   };
8  }

10 int main()
   {
12    int i1 = 5;
      int i2 = 4;
14    library::swap(i1, i2); // resolves to std::swap via qualified call (fallback)

16    user::T t1;
      user::T t2;
18    library::swap(t1, t2); // resolves to friend defined above via ADL
   }
```

Code snippet 3.33: Using and specialising customisation point objects

   Note that free functions outside the user-defined type's namespace would not be found and that the user cannot add an overload to the library's namespace, because defining a function and a function object with the same name in the same namespace is not legal. This greatly reduces the possibility for error and enforces a clean separation of code (user code is in the user's namespace; the library's namespace is not opened). Since non-ADL forms of unqualified lookup are prevented, the overload set is smaller which can speed-up compile-times. Another benefit is that one can make the CPO enforce more constraints on the parameters and/or return types of potential overloads—this further increases code-correctness.

   A use-case that was not yet covered are third party types, i.e. there is library code in the namespace `library::`, user code in the namespace `user::`, but the user wants to adapt a type from some other library in namespace `third::`. The user cannot add an overload to `user::`, because this will not be found by ADL, and they are not able/supposed to add to namespaces `library::` and `third::` so as not to break the above-mentioned separation. The solutions to this problem are discussed in length by O'Dwyer (2018). In essence, a dedicated class template for customisation is added to `library::` (either one per CPO or a single one for some/all CPOs). This is explicitly marked as being "customisable by users of the library" and users can specialise the template with the type from `third::` and provide functions as (static) members of that type. An extra overload inside the CPO looks for potential overloads in respective specialisations of this template.

   C++ 20 formally defines *customisation point objects* (ISO/IEC 14882:draft, 16.4.2.2.6) and introduces many as part of the ranges library. This includes `std::ranges::begin` and `std::ranges::end` which behave similar to `std::begin()` and `std::end()` but follow the above design and protect against certain misuses like being called on temporary containers.

   In combination with `constexpr` functions (Sect. 3.3.3), variable templates (Sect. 3.3.1) and alias templates (Sect. 3.3.1) these customisation point objects can also be used to define transformation/type traits of any kind.

We will see later how customisation points are used in SeqAn3.

3.7.4 Future Standardisation

While the technique introduced in the previous subsection is easy to use for users of
a library and provides much better code separation and enforcement of correctness,
it is significantly more difficult to implement and maintain. An alternative to the
current library-based "workarounds" would be a dedicated C++ language feature.

There currently is a proposal to WG21 that would add such *customisation
point functions* (Calabrese, 2018). According to this proposal the code shown in
Snippet 3.34 would be sufficient to define and specialise a customisation point.

```
   namespace std
2  {

4  // A customization point named "swap", including a default definition.
   template<typename T>
6  virtual void swap(T & lhs, T & rhs) { /*...*/ }

8  } // namespace std

10 // A user's override (may appear in their own namespace)
   void swap(Foo & lhs, Foo & rhs) override : std::swap { /*...*/ }
```

Code snippet 3.34: Customisation point functions according to P1292

The paper describes a customisation point function as. . .

> [. . .] a "virtual", but statically-dispatched, non-member function that declares a logical entry
> point for user-customization. Users may explicitly override such a function for their type
> from their own namespace, in which case qualified calls to the customization point will
> dispatch to the user's customized implementation whenever appropriate (Calabrese, 2018).

In my opinion the paper elegantly solves the problem of customisation points and
could greatly simplify some parts of the SeqAn library. Future developers of SeqAn
should keep track of this proposal and—if adopted—consider it for when they are at
a point where they can perform breaking changes and raise the revision of the C++
standard required for SeqAn.

3.8   Concurrency & Parallelism

Prior to C++ 11, C++ -programs implemented parallelism by directly accessing platform specific facilities (POSIX threads, Windows threads) or non-standard language
extensions (e.g. OpenMP; Dagum & Menon, 1998). This was often perceived as an
obstacle to bringing widespread parallelisation to applications and libraries.

```
   // define worker function (object)                      // define "loop" body
2  auto worker = [] ()                              2      auto sq = [] (auto & val)
   {                                                        {
4      /*... do one thing ...*/                     4          val = val * val;
   };                                                        };
6                                                   6
   // start background thread or defer                      // define data
8  auto f = std::async(worker);                     8      std::vector vec{1, 7, /*...*/};
10 /* do something else in main thread */           10     // invoke parallel "for-loop"
                                                            std::for_each(std::par,
12 // wait for worker to finish                      12                 vec.begin(), vec.end(),
   f.wait();                                                            sq);
```

Code snippet 3.35: Concurrency and parallelism in Modern C++. Simple concurrency is shown on the left (C++ 11): The worker function may be executed in parallel to the main thread. A "parallel for-loop" is shown on the right (C++ 17): elements in vec are squared utilising all available threads

C++ 11 first introduced support for threads, locks, futures and promises. A simple example of concurrent programming is displayed on the left side of Snippet 3.35. C++ 17 then introduced execution policies and parallelised versions of the standard library algorithms. std::for_each() can thus be used with the std::par policy to create a "parallel for-loop", similar to what OpenMP is often used for (shown on the right side of Snippet 3.35).

C++ 20 will deliver minor additions and fixes in this area and very large changes are again expected for C++ 23 with the unified executors proposal (Hoberock et al., 2020). The details of the threading support in C++ 17 are not very novel or exciting by today's standards, however, it is very important that they are available, because this enables SeqAn3 to perform threading inside the library without depending on third party solutions.

3.9 C++ Modules

C++ Modules are a huge addition to C++ 20 (Smith, 2019). They fundamentally change how programs are built and will likely have a significant impact on the entire C++ ecosystem. Since they are not supported by the compilers initially targeted by SeqAn3, there is currently no dedicated Modules support in SeqAn3. But through the use of certain compatibility macros, it could be added as an optional feature to a later SeqAn3 release without breaking compatibility with older compilers.

The core aspect of modules is that, instead of including headers (which just "copy'n'pastes" the source code), a structure of *exported* symbols and names is created. This can be *imported* which is cleaner, because only those symbols that are actually exported are visible to the user, and all implementation detail can be hidden. As a result the amount of symbols that need to be parsed shrinks and, importantly, the compiler can cache the entire module structure in an intermediate format—even code that cannot be pre-compiled ordinarily like templates. This promises a very significant reduction in compile-times and thereby counters the main criticism of header-only libraries.

An important side effect of C++ Modules is that future compilers will be able to cache the result of parsing regular headers, too. This means that even if not actively supported by SeqAn3, users may still automatically benefit from the feature in the future.

3.10   Utility Types

Many auxiliary types defined in SeqAn1/2 are now provided by the standard library. This includes the templates `std::pair<T1, T2>` (since C++ 98; ISO/IEC 14882:2017, 23.4) and `std::tuple<T1, T2, ... >` (since C++ 11; ISO/IEC 14882:2017, 23.5). Elements in a pair or tuple `t` can be accessed via `std::get<I>(t)` (where `I` is the index of the element) or `std::get<T>(t)` (where `T` is the type of the element if that is unique). There are several more auxiliary functions for concatenating tuples, calling functions on all of a tuple's elements and for using tuples in metaprogramming.

C++ 17 introduces several new utility types:

`std::optional<T>`   Can hold an object of type `T` or be in a "null-state" (ISO/IEC 14882:2017, 23.6). In contrast to a pointer that can also have a null-state, the object is always allocated within the optional (potentially on the stack).

`std::variant<T1, T2, ... >`   A type-safe replacement for the C `union`, i.e. an object that can alternatively hold values of different types (ISO/IEC 14882:2017, 23.7). The object is guaranteed to be stored inside the variant without additional dynamic memory.

`std::any`   A type that can store the value of any type (ISO/IEC 14882:2017, 23.8). In contrast to `void*`, that is also used for this purpose, memory is managed by the object. If possible, dynamic memory allocation is avoided.

All of these types have safe interfaces that perform checks and `throw` on an error, but they also offer zero-overhead (unsafe) access functions. A big advantage is that these types communicate programmer intent more clearly than the other solutions mentioned above.

`std::variant` is of particular interest to SeqAn3, because it has a powerful mechanism of reifying the stored type via the *visitor pattern*. An example of this is shown in Snippet 3.36 where a variant is defined that can either store an `int` or a `std::string`. This variant has a distinct static type independent of its value. It cannot be printed with `std::cout`, because the compiler would not know which overload of the stream operator to pick. However, the variant can be "visited" with a function object that accepts all possible variant types. All template-paths are instantiated and the correct one is selected at runtime (typically via a table lookup similar to a vtable). Within the function object's `operator()` the type is then either

```
   auto visitor = [] (auto & val)
2  {
       using val_type = std::remove_cvref_t<decltype(val)>;
4      if constexpr (std::integral<val_type>)
           std::cout << val << " is a number\n";
6      else if (std::same_as<val_type, std::string>)
           std::cout << val << " is a string\n";
8  };

10 std::variant<int, std::string> v = 3;          // declare variant and initialise to int-state
   std::visit(visitor, v);                        // prints "3 is a number"
12
   v = "foobar";                                  // destruct current value and store string
14 std::visit(visitor, v);                        // prints "foobar is a string"
```

Code snippet 3.36: Variants and the visitor-paradigm. Note that a simple
`std::cout << val;` inside the `visitor` would also be valid

`int` or `std::string`. This mechanism is a form of dynamic polymorphism and
an elegant alternative to the tag-dispatching used in SeqAn2 that loops over tags
recursively to select the correct overload.

3.11   Discussion

In this chapter I demonstrated that the C++ of today is in many ways an entirely
different language than C++ 98. Many of the changes were novel when introduced,
some were borrowed from other programming languages and again others have been
in the making for as long as C++ exists—only to become part of the language now.

Stroustrup (2017) wrote that "[w]e need to simplify generic programming in
C++. The way we write generic code today is simply too different from the way
we write other code." With C++ Concepts and type deduction, simplified generic
programming has become reality. Working with templates in general has become
much simpler—for the developer writing (and maintaining) a generic codebase but
also for the application developer that uses generic interfaces. Generic interfaces
with concepts are more expressive and still look more similar to object-oriented
interfaces. Parts of the API previously expressed in documentation are now part of
the language leading to better diagnostics and cleaner code. Static polymorphism is
now possible without language hacks or exotic techniques and at the same time is
much better at adapting third party types. This means that generic programming is
not only easier to do now (and prettier to look at!), but the level of *genericity* that is
attainable by the average programmer is higher than before.

C++ Ranges provide a new foundation for dealing with data in any kind of
collection. This is particularly important for sequence analysis as *sequences* are
at the heart of this domain of bioinformatics and all algorithms deal with them in
one way or another. Views present the possibility to express many algorithms on
ranges as the combination of small single-purpose adaptors. Defining those may
not be trivial and the syntax for using them is novel for C++ developers, but they

enable the developer to express problems in a much shorter, more concise style known from functional programming languages: "By operating declaratively and functionally instead of imperatively, we reduce the need for overt state manipulation and branches and loops. This brings down the number of states your program can be in, which brings down your bug counts." (Niebler, 2019). Unnamed functions in the form of lambda expressions and parameter packs that can be folded (introduced later) complete the arsenal of functional programming aspects in Modern C++.

Since `constexpr` functions have been added to C++, computing single constants and even large tables at compile-time is possible without using the template preprocessor. This is a paradigm shift that removes the "meta" from a lot of metaprogramming and makes it accessible to more developers. Where (template-) metaprogramming is still needed, it can resort to a considerable quantity of standard library traits.

Performance is at the core of C++ and move semantics play an important role in avoiding unnecessary copy operations (and thereby memory allocations). They allow optimisations on a fundamental language level that were not possible before. And they also permit a more natural style of programming where return values are actually returned from functions and not taken as "out-parameters". While the technical details of move semantics are arguably not trivial to understand, the amount of exposure to these details is rather limited for the average programmer.

I strongly believe that the features summarised here allow for a radically different, much improved library design that I will introduce in the next chapters. This does not mean that current C++ is perfect. I have highlighted which aspects I find lacking and which proposals I think might address these problems in the future. This will hopefully aid future developers in updating the designs presented here. I have also detailed my involvement in the C++ standardisation process and while it may not always be feasible to wait for an updated version of the standard, I do think that being involved provides great long-term value for C++-focused projects.

Part II
SeqAn3

The second part of this book describes the design and implementation of SeqAn3. It begins with the fourth chapter which integrates the results of Chaps. 2 and 3 into a new library design as well as covering questions relating to SeqAn as a project and its interactions with other libraries and applications. Subsequent chapters then discuss the implementation of the library with schematic overviews, code examples and benchmarks. Each of these chapters concludes by discussing the respective module's contribution to achieving the formulated design goals.

Chapter 4
The Design of SeqAn3

4.1 Design Goals

As I elaborated in Sect. 2.4, the design goals set previously were the right ones at the time, even if they were not all achieved to the same degree. I would still argue that in general the primacy of *performance* as a design goal over the others is valid for SeqAn as a C++ library. Acknowledging that previous SeqAn's biggest weaknesses were in *simplicity* and *integration* means that any new version will have to address these issues head-on. Seeing how related the goals of *generality*, *refineability* and *extensibility* are, I want to from now on subsume them under the term *adaptability*. Finally, I want to introduce *compactness* as its own goal to forestall some developments that previously burdened the project.

I anticipate that at times the design goals will conflict with each other and I have stated that *performance* is given a certain precedence over other goals. This is, however, not a golden rule: depending on the context, a minor performance overhead might be acceptable if alternatives would severely violate other goals. Eventually, the goals are weighed against each other to deliver a good and consistent experience to users of the library.

Regarding the general *direction* of SeqAn, I want to shift to a stronger professional focus and move away from being an "academic testbed". This does not mean that algorithms research does not happen in the context of SeqAn,[1] but the focus of *the software that is shipped* is to represent the results of this research and not necessarily the entire spectrum of academic questions surrounding it. In practice this means that if there are two algorithms or data structures, and rigorous research and testing have shown one to always be inferior to the other, it will not be included in the library just for demonstration purposes. For example, uncompressed suffix

---

[1] On the contrary—some entirely novel algorithmic approaches have been developed in recent times (Pockrandt et al., 2017).

H. Hauswedell, *Sequence Analysis and Modern C++*, Computational Biology 33, https://doi.org/10.1007/978-3-030-90990-1_4

arrays are no longer part of SeqAn, because compressed suffix arrays/FM-indexes have replaced them in all use-cases. This decision furthers the design goals of *compactness* and *simplicity*.

Some design goals are referenced strongly in this chapter; others will become clearer when the actual implementation is discussed in the following chapters.

4.1.1   Performance

I explained in Sect. 2.4.1 that SeqAn's performance has been excellent in the past. Thus, maintaining performance as a main goal in SeqAn3 primarily means achieving a similar performance to SeqAn2 with the new code. In regard to the various aspects of concurrency and parallelism (including vectorisation), I want to pursue a deep integration into the library that on the one hand exploits modern computers' potential optimally but on the other hand hides the involved complexity from the user. This also means relying on the now standardised threading capabilities of C++ and not using technologies like OpenMP (Dagum & Menon, 1998) that are visually distinct from regular C++ code and difficult to configure by traditional means. The specifics of parallelisation should be configurable at runtime and not depend on compile-time flags or even macros or environment variables. For certain interfaces a "parallel-by-default" approach will be chosen. And if the algorithm design enforces a preference for either the serial case or the parallel case, the latter should be preferred.

The intent of all these measures is to make a high performance accessible to as many of SeqAn's users as possible. Parallelism is not considered an advanced feature but an integral part of the library design and is promoted by the default configuration choices. Of course, library-side parallelisation should still be optional for users that choose to implement parallelism on an application level and want to prevent conflicts/over-subscription.

4.1.2   Simplicity

C++ is widely considered one of the more complex programming languages and, as I described in Sect. 2.4, SeqAn1/2 did many things on a technical and organisational level that increased the inherent complexity notably. The greatest challenge of SeqAn3 is to overcome this complexity and deliver a library that is *simpler*—without compromising the other design goals.

There are different dimensions of simplicity: simplicity from the perspective of the user is the most obvious, but simplicity for the contributors and especially the maintainers is also important for the long-term viability of the library. In many cases these goals are dependent on each other, but this may not always be true. Where they conflict, providing a simpler user experience should be the priority.

I have described aspects of Modern C++ in Chap. 3 that help write code that is simpler to use and maintain. Some techniques used will appear more "traditional" and should thus pose less of a problem than SeqAn2's code. But other programming techniques used will be new to C++ experts and SeqAn veterans, so I do not expect everyone to have a trivial start into SeqAn3. However, once the core principles have been learned, new parts of the library should open up quickly to the user. This is part of *conceptual integrity* (Brooks, 1995), i.e. delivering a design that is "of one piece" and in itself consistent. The library design intentionally does not cater to previous users of SeqAn and instead focuses on being as accessible as possible to new users. This includes users coming from other programming languages with experience in other biological software libraries or frameworks such as BioPython (Cock et al., 2009).

Beyond the actual programming techniques, simplification also needs to happen on a structural level, i.e. the organisation of the codebase itself needs to become simpler so that users (and developers) quickly find what they are looking for. This will be an important difference to SeqAn1/2 where it was notoriously difficult to find the interface that would solve a given problem. Documentation is central to guiding through the codebase and SeqAn's documentation has some room for improvement in this area. This includes the API documentation and accompanying tutorials and HowTos that gently introduce the design principles used by SeqAn3.

Finally, I will introduce many simplifications on a "project-level", i.e. changes in the organisation and processes of the project and the management of the code. These should help to get, install and use SeqAn but also to contribute to and maintain SeqAn.

4.1.3 Integration

In Sect. 2.4.4 I discuss two aspects of integration: "source-code level integration" and "project-level integration". Regarding the former, I concluded that SeqAn1/2 facilitates integration of single third party types well but failed at properly handling third party libraries. In particular, its interoperability with the standard library was far from optimal. SeqAn3 will use many of the Modern C++ techniques introduced in Chap. 3 to allow for a better integration on source-code level and truly deliver on the promise given for SeqAn1: "The library is able to work with other libraries and built-in types" (Gogol-Döring, 2009).

I have already noted on the progress made in regard to project-level integration between SeqAn1 and SeqAn2. SeqAn3 must build on this and improve in those areas where I identified deficits. This includes clearly communicating which parts of SeqAn constitute the API and developing explicit guidelines on the stability of the API and other project properties. Furthermore, the repository structure needs to be cleaned up and the process of including SeqAn (with and without build system) be improved. SeqAn3 should be able to coexist with SeqAn2 on one computer and ideally also be used from within the same application—both of which were not possible with SeqAn1 and SeqAn2.

4.1.4 Adaptability

SeqAn1/2 offered a near unlimited degree of adaptability with its lack of access control and its free (meta-)functions that allowed overriding the behaviour of every entity in the library. However, as I discussed in Sect. 2.4.4, this is indeed not the best design as it opens up the possibility of undesired customisation and severely complicates the process of adaptation for the user, because it is not clear which interfaces are best adapted and in which manner.

For SeqAn3 I envision a fundamentally different system where customisation points are clearly marked as such and there is comprehensive documentation of how these can be used. Allowing for multiple methods of customisation provides for a high degree of flexibility and accommodates for the different requirements a user may have. On the other hand, typical instruments of the C++ language shall be employed to prevent users from mistakenly relying on or changing implementation detail. Extension through type traits and traits classes will be simpler than overriding metafunctions, because this is much more established as a mechanism.

Giving up on *template subclassing* will allow for even better refineability and generality/genericity.

4.1.5 Compactness

SeqAn was first released at a point in time when much of what is considered essential C++ today was not yet available, so it contained a lot of functionality that later became part of the standard (under similar or completely different names). Furthermore, the use of *template subclassing* as a method for polymorphism (see Sect. 2.3.2) suggested the definition of custom types for many use-cases where there actually were comparable data structures in the standard library (e.g. `std::vector<alph_t>` vs `seqan::String<alph_t, Alloc<>>`). Additionally, the (over-)consistent use of *generic programming* and global function interfaces (in contrast to object-oriented designs) made it difficult to rely on usual C++ tooling like documentation and testing systems.

Together these factors fostered a project culture of *not-invented-here* (Piezunka & Dahlander, 2015) where it was normal to create in-house solutions for all problems, be they of algorithmic or organisational nature. As discussed in Sect. 2.4, SeqAn not only had its own data structures for things like vectors or pairs, but it also had a custom documentation system, a custom testing framework and needed multiple different scripting languages to build.

This led to a dramatic growth of the library and the project as a whole. In itself, this is already detrimental to the quality of software as the number of defects is highly correlated with software size. Research has repeatedly shown that even simple complexity measures like lines of code (LOCs) can be used as indicator for the quality of software and that this is independent of programming language (McConnell, 2004; Zhang, 2009).

In this context *compactness* describes the opposite trend; it means focusing on the core task: providing data structures and algorithms for sequence analysis, relying on other libraries where possible— most importantly the standard library—and using well-established state-of-the-art tooling. It also means that any addition to the library needs to have strong motivation and that the cost of an increased codebase and API is considered in this process.

4.2 Programming Techniques

This section defines the technical basis for achieving the aforementioned goals. That includes all those design decisions that relate to the "style of C++" used in writing SeqAn3.

4.2.1 Modern C++

The programming techniques available to the developer depend strongly on the version of C++. I elaborated on the development cycle of C++ in Chap. 3 and I also showed that new versions of the C++ standard typically come with significant new features and/or simplifications that can help meet the design goals discussed above. It is therefore not easy to decide which version to base a project on and one might be tempted to always wait for the next standard and delay a project indefinitely in the pursuit of the newest features.

In practice one also cannot target "the standard" but only implementations of the standard, i.e. specific compilers. As can be seen in Table 4.1, adherence to the standard has improved notably in the last decade with major compilers shipping full support for a standard in the year of publication or not much later. This is a significant difference to the pre-Modern-C++ era where support for C++ 98 came very late and compiler-specific language extensions were common. However, it is important to take into account that not all the target audience may be able to update to a new compiler when released and that operating systems with long-term support may be particularly outdated in regard to the available software.

It was clear to me that a fundamental re-design of SeqAn would need to be based on C++ concepts and the existing solutions for emulating concepts without language support were so complex on the one hand and so limited on the other hand that only C++ 20 or the Concepts TS would suffice. Waiting for C++ 20 was not an option as it

Table 4.1 History of C++ versions and compiler support. The years given represent the time when the GNU Compiler Collection (GCC), the LLVM Clang compiler and Microsoft Visual Studio (MSVC) released versions of their compiler that claimed support of the respective C++ standard (complete language level support and at least significant support of the standard library)

ISO standard		GCC	Clang	MSVC
1998	C++ 98	2004[†]	2010	Never[††]
2011	C++ 11	2013	2013	2015
2014	C++ 14	2015	2014	2017
2015	Concepts TS	2016	Never	Never
2017	C++ 17	2017	2018	2018

[†] Release of GCC-3.4 is first to claim to be "much closer to ISO/ANSI C++ standard"
[††] Important C++98 aspects like two-phase name lookup only arrived in 2017

would mean delaying the first release until 2020 and releasing in a situation where almost no one would be able to try out the library much less deploy applications built with it. Moreover, it would have meant having to develop the library without means of testing it—a rather difficult task.

Development of SeqAn3 started in late 2016 and the GNU project had just shipped support for the Concepts TS in GCC6—with GCC7 (to be released in April 2017) promising to support most of C++ 17. **Thus, it was decided to establish C++ 17 and the Concepts TS as the basic C++ requirements for SeqAn3.** This would allow SeqAn3 to rely on a wealth of new features while still having a usable compiler for the development. Since GCC is the most important compiler on Linux-based operating systems there would also be a reasonable adoption rate with the largest part of the target audience when publishing the first release 2–3 years later.

Ideally, other major compilers would have picked up support for the Concepts TS, as well, but this did not happen. Furthermore, there was no guarantee that Concepts would indeed become part of C++ 20 and whether the final implementation would be compatible with the Concepts TS or not. This was one of the reasons I became involved in the standardisation process: it was important to keep track of changes to the standard and gauge sentiment in regard to future decisions. This allowed me to adapt the design of the library while it was evolving so that it would reflect not only the capabilities of GCC7 but would also be compatible with C++ 20 and respective compilers, once released.

4.2.2   Programming Paradigms

SeqAn3 follows the paradigm of generic programming but in a less dogmatic fashion than SeqAn1/2. C++ is a multi-paradigm language which means it combines aspects of different programming paradigms and one is free to choose the techniques that best solve a problem—where "best" depends on the situation and design goals. Some techniques described below are typically associated with object-oriented programming and others are aspects of functional programming, but generic programming is still at the core of SeqAn3. All algorithms and most types are strongly parameterised via template parameters even if these are not always as visible as before. I would argue that in many ways the code is *more generic* than that of SeqAn1/2 due to the use of concepts and other techniques described in Chap. 3.

4.2.3 Polymorphism and Customisation

Static polymorphism is essential to reaching the design goal of a high *performance* and SeqAn3 uses concepts-based polymorphism as a form of static polymorphism. As discussed in Sect. 3.4.4, concepts-based polymorphism is more generic than template subclassing (the technique used in SeqAn1/2), it therefore furthers the goals of *integration* and *adaptability*. Using customisation points (Sect. 3.7) in combination with concepts goes even further towards both of these goals. Concepts-based polymorphism is easier to write and maintain but also easier for the users of polymorphic interfaces. This is an important part of making SeqAn *simpler*. It has virtually no drawbacks over other approaches.

In certain situations runtime polymorphism is necessary, because the decision which implementation to use must happen at runtime, e.g. picking the right function in file I/O after the format, has been detected from a user-provided file. SeqAn1/2 used recursive *tag-dispatching* which implies that in the worst-case all possible tags are compared before the right dispatch is chosen (linear complexity). If possible, SeqAn3 uses virtual functions for runtime polymorphism since the overhead of a virtual function call is constant.[2] If the interfaces that are being dispatched to are themselves templates, this is not possible, because function templates cannot be marked `virtual`. In this case a `std::variant` is used to select the correct overload via the visitor pattern (see Sect. 3.10). This dispatching happens in constant time.

Virtual functions are a widely used and well-understood mechanism, `std::variant` and `std::visit()` are new in C++ 17 but as part of the standard library comparatively well-known. Both approaches are *simpler* than the tag-dispatching approach, and they *perform better*.

---

[2] This is an example of using the right paradigm in the right situation: when the runtime overhead can be avoided, it is better to use static polymorphism, but if the choice is a runtime decision by definition, there is no reason not to use virtual functions. In fact, they are faster than the solution used by SeqAn1/2.

4.2.4 Aspects of Object-Orientation

As discussed in Sects. 2.3.3 and 2.4.2 the biggest obstacle to *simplicity* in SeqAn1/2 is the excessive use of free functions, the lack of information hiding and the non-locality/distributed-ness of implementation. While most concept definitions in SeqAn3 use customisation points and/or free functions as *requirements* (for improved *adaptability*), there is no reason that types should not have member functions. In fact, SeqAn3 uses member functions and access specifiers (`public`, `private`) for the definition of class types. If possible, all behaviour that constitutes the type is implemented as members or friends of the type.[3] Any members that are not part of the public interface are marked `private` or `protected`. If code is shared with a base type, this is done through inheritance (see below). All headers in SeqAn3 are self-contained, i.e. they include all other headers that they require, as a result all headers can be included individually.

Inheritance is the natural way of reusing code in C++ and since SeqAn3 gives up on template subclassing and decouples code reuse from polymorphism (see Sect. 3.5), the full arsenal of inheritance is available to SeqAn3. The form of inheritance most widely used is CRTP (Sect. 3.5.1), but when only data members are inherited or member functions are known not to depend on each other, simple public inheritance is preferred. The few places of the codebase that are guaranteed not to influence runtime (e.g. the argument parser which is ever only evaluated once at program start) may also use virtual functions instead of CRTP. This follows a pragmatic approach where the least complicated solution to a problem is chosen that still aligns with the design goals.

These design decisions likely present one of the most visible changes from SeqAn1/2. The result is code that is much *simpler*, because all important aspects of a type are defined in the same place (or at least in very few well-defined places). In case inheritance is involved in the definition of the type, the header that defines the base type can be easily found from the header that defines the derived type.

While CRTP is not as simple as I would like, all techniques described here are *simpler* in the sense that they are less surprising for the majority of the developers, because the majority of developers are familiar with object-oriented programming and idioms like encapsulation and information hiding.

It should be noted that the only reason this more "traditional" approach does not jeopardise *integration* and *adaptability* is that concepts are used to constrain algorithms.

---

[3] A positive side effect of this is a reduction in compile-times, because building a large overload set of free functions and picking the correct one can be avoided.

4.2.5 Ranges and Views

Biological sequences, e.g. DNA sequences and protein sequences, are at the heart of sequence analysis. In C++ 20, sequences, containers, collections, etc. are abstracted as *ranges* (Sect. 3.6). SeqAn3 uses the containers of the standard library and only adds new containers when absolutely necessary. This is part of pursuing *compactness*, but it is also a good indicator for *integration* and it improves *simplicity* since users do not need to learn about new data structures when widely available and widely known ones suffice.

As discussed in Sect. 3.6.3, algorithms on ranges can be modelled as ranges themselves through the definition of views. These are useful in common contexts (e.g. returning the first five elements of a range) but also in specific biological applications (e.g. generating the reverse complement of a DNA sequence). SeqAn3 relies on views from the standard library but also provides many views of its own. Using views and their corresponding adaptor objects will be new to most developers, however, once learned, the mechanics are much *simpler* and less error-prone than other solutions to the same problems. Since views are novel, their impact on *performance* will have to be measured.

Views represent an algorithmic adaptation of other ranges, they do not contain data proportional to their size. But range adaptors that annotate another range with (substantial) data are also possible and useful. An example would be an adaptor that annotates a DNA sequence with gap characters to represent (a part of) an alignment. They are called *decorators* in SeqAn3. They are neither containers, nor views, but they can model the remaining range concepts like any other range.

These ranges are all biological sequences or adaptations thereof, but the notion of ranges is much more general. In SeqAn3, files are modelled as ranges over file/format specific records, and algorithms like sequence alignment return ranges of alignment results. This allows applying the declarative/functional programming style of chaining views (e.g. transformations, filters) on the records of files or the results of algorithms. Examples of this will be shown later. Treating files as ranges of records is not novel outside C++, e.g. BioPython uses a very similar design.[4] I conclude that it is at least as *simple* as the imperative/procedural design of SeqAn1/2—if not more so.

At the beginning of this section, I established C++ 17 + the Concepts TS as the baseline for SeqAn3, but the entire ranges machinery has only become part of C++ 20 and is not included in previous versions of GCC. However, since it is entirely made of library code and requires no extra language features, it can be emulated by a stand-alone library. SeqAn3 prefers an official ranges library but automatically falls back to the stand-alone library if necessary, see Sect. 4.4.1.

---

[4] http://biopython.org/DIST/docs/tutorial/Tutorial.html#htoc49.

4.2.6   "Natural" Function Interfaces

SeqAn3 follows certain policies for the definition of functions and function templates to ensure that interfaces are *simple* to understand and maintain. A general rule is that one should try to minimise the number of function parameters and the number of overloads. This reduces the potential for user error and enables contributors reading the code to quickly assess which implementation is picked for which combination of arguments.

One way of reducing the number of parameters and an important *simplification* in its own right is the policy that functions should return their output by value instead of writing to an "out-parameter". This incurs no performance penalty in Modern C++ (see Sect. 3.2.3) and is more "natural" for programmers coming from other programming languages. Multiple return values can be wrapped in a custom `struct` or `std::tuple`.

Strongly coupled parameters shall also be wrapped in a custom `struct` or `std::tuple` to indicate this coupling. Resulting interfaces are more expressive by clearly communicating the relation of the parameters to each other. This reduces the possibility for user error (wrong "mixing" of arguments) and also reduces the total number of parameters.

Function templates can make use of *conditional instantiation* to perform refined or type-specific behaviour (see Sect. 3.3.4). This helps avoid the number of overloads needed.

There is no strict order for function parameters in SeqAn3.[5] Developers should follow the "natural" order implied by the function name, e.g. for the function `assign_char_to(a, b)`, a should be an in-parameter and b the (in)-out-parameter; or by importance of the parameters, e.g. an algorithm would take first the "data" parameters and later configuration options.

Functions that would take multiple parameters of the same type instead usually take *strong types* in SeqAn3 (introduced in Sect. 9.2.1). This also prevents silent user error and makes invocations of the interface much easier to read.

In general these policies will make the SeqAn3 API much more *compact* and working with it a lot *simpler*. Most of the policies are widely accepted best practices, some are taken directly from the CppCoreGuidelines.[6]

4.2.7   `constexpr` if Possible

When they were introduced in C++ 11, functions evaluated in constant expressions were very limited. Every standard released since has relaxed these restrictions and

---

[5] SeqAn1/2 had the strict (and unintuitive) order of "out" → "in-out" → "in".

[6] http://isocpp.github.io/CppCoreGuidelines/CppCoreGuidelines#S-functions.

it is likely that most functions can be declared `constexpr` in the future (see Sect. 3.3.3). SeqAn3 follows the policy to declare any free function `constexpr` that possibly can be and to declare the set of all member functions of a class type `constexpr` if possible (this makes objects of that type usable in constant expressions).

This might seem excessive at first, but it has proved useful in a surprising number of contexts. One example are biological sequence alphabets that are not likely to be processed by applications at compile-time, because sequence content is provided at runtime (e.g. via files or user input). However, since they can be created and manipulated at compile-time, new types can be generated from them, including composite alphabets and optimised scoring schemes. This is a form of metaprogramming that relies solely on the regular interface of the type.

Constant expressions are an integral part of Modern C++ and help deliver a better *performance* and *simpler* metaprogramming.

4.3   Administrative Aspects

This section discusses the design decisions that impact using SeqAn on an administrative level (all tasks surrounding the inclusion of SeqAn3 in an application or other software library). Often overlooked, these are crucial to achieving the goals introduced at the beginning of this chapter.

4.3.1   Header-Only Library

SeqAn3 is a header-only library like SeqAn1/2, i.e. the library is always distributed in source form and cannot be compiled into an object file for dynamic or static linking. This design is mostly due to the nature of the code—templates must be defined in headers, so only very few entities could possibly be pre-built. But distributing a library as header-only has some other notable advantages:

1. The library only has to provide API stability (compatible interfaces), not ABI stability (compatible memory layouts); see also Sect. 4.3.4.
2. Whenever an application is built using the library, it can generate machine-optimised instructions also for the library code; this is important for making use of e.g. vectorisation.
3. Inclusion of header-only libraries can happen from inside C++ source code whereas separately compiled libraries need to be linked which typically involves a build system.
4. Since C++ 17 it is also possible to *detect* the presence of header-only libraries from inside C++ source code and conditionally include the library; separately compiled libraries require a build system and/or package manager to handle

detection (because the presence of a header file does not guarantee presence of the shared object).

The first point makes developing SeqAn3 *simpler* and the second point ensures a good *performance*. While using a build system is recommended (see Sect. 4.4.1), they are often a source of frustration for the user. Thus, providing the best-possible experience without depending on a build system (points 3. and 4.) makes using SeqAn3 *simpler*.

The notable disadvantage of header-only libraries is an increased compile-time. This may be mitigated to a large degree by future compilers and C++ Modules (see Sect. 3.9).

4.3.2   Licence

SeqAn3 is licensed under the terms of the 3-clause BSD license, the same as SeqAn2. This allows SeqAn to be used without any legal requirements other than attribution. In practice, it means that closed source software can use SeqAn but also any Free/Open Source Software, because the 3-clause BSD license is "compatible" with all other Open Source licences. For software libraries, this is particularly important, because they are always combined with other code which may be distributed under different terms.

4.3.3   Platform Support

Platform support influences (and is influenced by) the design goals in several ways. First and foremost, SeqAn3 should work on all platforms that its (potential) users use, this is part of *simplicity* (from the perspective of the user) and ensures that SeqAn-based applications can be used in existing workflows and setups (better *integration*). On the other hand a diverse set of platforms might result in a large set of platform-specific codepaths (optimisations, workarounds, etc.). This decreases *simplicity* for the maintainer and reduces *compactness*. Finally, producing the most efficient possible machine code on every CPU is crucial to delivering the best *performance*.

Compiler

SeqAn3 as a C++ header-only library primarily depends on the capabilities of the C++ compiler and standard library involved. Conceptually, it targets C++ 20 compilers and has fallback solutions (Concepts TS, *STD* module/range-v3) for GCC7, GCC8 and GCC9. Since C++ 20 is not final, yet, no compiler officially supports C++ 20 and GCC7–9 are the only officially supported compilers at the time of writing.

However, Microsoft Visual Studio 2019 16.3 (MSVC) has just acquired support for C++ 20 Concepts[7] and there is an experimental branch of LLVM Clang that supports these, too.[8] The development branch of GCC has recently also gained support for C++ 20 Concepts (instead of just the Concepts TS). Support for these compilers is currently being worked on and is likely to land before the first stable release of SeqAn3.

During the development of SeqAn3 the library was tested only with GCC, but I put an emphasis on compiler-agnostic code early on. This means SeqAn3 was developed without GNU language extensions (-std=c++17, not -std=gnu++17) and with very high warning levels and language correctness (Wall Wextra -pedantic). The only part that currently relies on non-standard C++ are SIMD specific intrinsics and these are supported by most C++ compilers including MSVC and Clang. There should therefore be no fundamental barrier to supporting the common C++ compilers as soon as they advertise C++ 20 support.

However, experience in working with GCC has shown that a certain amount of bugs is to be expected from compilers in regard to new language features. This might delay full support of SeqAn3 on that specific compiler or make compiler-specific workarounds necessary.[9]

Operating System

Since Modern C++ encompasses many APIs that were previously specific to the operating system (e.g. threading and filesystem support), SeqAn3 is currently independent of the operating system and does not contain codepaths specific to POSIX (Austin Common Standards Revision Group, 2014) or Microsoft Windows. But developer experience shows that operating system specific behaviour sporadically does appear, so rigorous testing cannot be avoided. SeqAn3 is currently tested on various flavours of GNU/Linux, macOS and FreeBSD. As soon as MSVC is supported as a compiler, SeqAn3 will routinely be tested on Microsoft Windows, too. There are patches for supporting the use of GCC on Windows (MinGW)[10] that will likely be integrated soon, as well.

---

[7] https://devblogs.microsoft.com/cppblog/c20-concepts-are-here-in-visual-studio-2019-version-16-3/.

[8] https://github.com/saarraz/clang-concepts-monorepo.

[9] In general, I discourage compiler-specific workarounds and would suggest to wait for bug-free releases of new compilers before adding official support. However, in cases like regressions, workarounds may not always be avoidable.

[10] http://www.mingw.org/.

An important POSIX-specific feature that was available in SeqAn1/2 and is currently not found in SeqAn3 is memory-mapping.[11] Memory-mapped I/O allows for faster reading/writing of data under certain circumstances. It is likely that this feature will be added to SeqAn3 in a later release again.

Machine Architecture

As mentioned in the beginning of this section, almost all of SeqAn3 is standards-conforming C++. The only code that is not standards-conforming and specific to the machine architecture is the aforementioned SIMD code. Since the 64bit x86 architecture ("amd64" or "Intel64") has virtually replaced all other architectures in desktop, workstation (IHS Markit, 2010), server and high-performance domains (Meuer, 2008; Fig. 4.1), it is clear that the focus of SeqAn3 is optimising for this architecture.

Within this architecture, different CPU generations support different features, especially in regard to SIMD. Ideally, SeqAn3 will make use of any features available, but in how far this is possible generically is still part of ongoing research. Right now it is capable of generating optimised instructions for SSE4, AVX2 and AVX512; this covers the entire range of current and upcoming x86 processors.

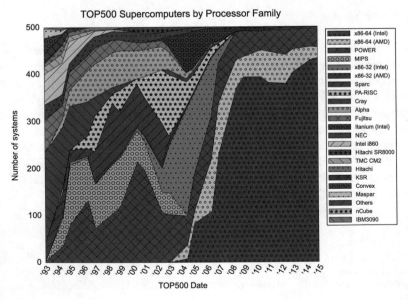

Fig. 4.1 Prevalence of different CPU architectures in the Top500 supercomputers. ©2015 by Dan Lenski, licensed under ⊜①◎

---

[11] http://man7.org/linux/man-pages/man2/mmap.2.html.

In the past, the PowerPC architecture played a more important role in desktops, workstations (it was used in Apple hardware) and especially in high-performance computing (HPC) where it still plays smaller but notable role today (Meuer, 2000, 2008; Fig. 4.1). With the OpenPOWER initiative (Gschwind, 2014), this architecture may or may not become more important again in the future. For this reason SeqAn3 is tested on PowerPC, although PowerPC does not yet receive the same amount of optimisation.

Historically PowerPC is a *big-endian* architecture, i.e. it has a different *byte-order* than x86, but recent generations of PowerPC can be configured to be *little-endian*, as well. Testing on a big-endian platform would certainly be useful as it might reveal architecture dependent code in SeqAn (usually in the context of computing with bit-masks or reading/writing of binary formats).

Offloading to secondary processing units (GPUs, compute cards or FPGAs) is an important part of HPC nowadays, however past efforts to make SeqAn use these—while maintaining its generic nature—have proven to be difficult (see Sect. 2.4.1). Future research will have to show whether generic approaches are possible and feasible for a software library. Note should also be taken on how the C++ standardisation efforts in this domain play out.

But even in the realm of very standards-conforming C++ there are several pitfalls to architecture-independent C++ code. Two notable examples identified by Koenig (1988) for C (and inherited by C++) are:

- "7.2. How Big is an Integer?"—The width of standard integral types is not fixed.
- "7.3. Are Characters Signed or Unsigned?"—It is not mandated whether the `char` type is signed or unsigned.

To solve the first problem C99 (and thereby C++11) introduces fixed width integers (e.g. `int32_t` is always 32bit whereas `int` can vary in width). SeqAn3 has a very strict policy of only using fixed width integers in interfaces and as member variables. The only exceptions are when performing arbitrary counts (`size_t` is used) and pointer arithmetic (`ptrdiff_t` is used).[12] To solve the second problem, it is forbidden to use `char` for arithmetic purposes.

4.3.4 Stability

Whenever somebody uses a software library in their project, they have implicit assumptions about that library. In commercial software, these assumptions can be made explicit via a contract or licence agreement. With most academic and open source software the opposite is usually the case and the licensing terms state that "THE IMPLIED WARRANTIES OF MERCHANTABILITY AND FITNESS FOR A PARTICULAR PURPOSE ARE DISCLAIMED" (or similar).

---

[12] These types are inherently connected to the machine word size.

SeqAn3 strives to be better than most academic and open source software and voluntarily engages in a user contract that attempts to clarify what the developers think they can promise and what users can and cannot expect. This is especially important with regard to the *stability* of the software library, i.e. whether and which parts of the library are subject to (breaking) change. In turn, this greatly impacts *integration* with other libraries and applications and is a requirement for *simplicity* in dependency management.

These self-imposed rules are part of the SeqAn3 documentation[13] and only summarised here. Some rules are adapted from the "compatibility guidelines" of Google's *abseil* library.[14] See also the paragraph on semantic versioning in Sect. 2.4.4.

API

The application programming interface (API) of a library is the sum of the interfaces (functions, classes, templates, etc.) exposed by the library for use in applications. For SeqAn3, the API is defined as all entities that are part of the user documentation—unless explicitly marked as not being part of the API. In particular, this includes names in the namespace `seqan3::` and its sub-namespaces, unless the sub-namespace is called `detail::` and except for members declared `private` and some members declared `protected`.

These interfaces are promised to be *stable* within the entire SeqAn3 series unless they are marked as experimental (this is reserved for new interfaces in the first release they are present in). *Stable* in this context means that applications developed with one release of SeqAn3 can be compiled against any later release of SeqAn3 without errors.

This only implies that interfaces can be *used as specified*, but not necessarily that they do not *change in a compatible way*, e.g. new overloads can be added to free functions, parameters can be defaulted, members can be added to a type, a.s.o. This restricts certain (more or less obscure) use-cases that rely on the exact given signature via metaprogramming tricks or rely on the uniqueness of the address of a free function.

As a special rule, the first minor release of SeqAn3 and its patch level releases (3.0.x) are all marked as experimental. This gives the SeqAn team time to evaluate the usability of interfaces before setting them in stone.

Once an API is declared stable, it should under no circumstances be changed; this is vital to a good user experience and to establishing SeqAn3 as a reliable component in professional software ecosystems. In the unlikely, unforeseeable, but unavoidable event that an API is broken, the SeqAn team promises to provide a clean upgrade path, i.e. a new minor release where the deprecated and the new API can be used in

---

[13] http://docs.seqan.de/seqan/3-master-user/about_api.html.

[14] https://abseil.io/about/compatibility.

parallel (before the deprecated API is removed in a later minor release). If possible, tooling shall be provided that transforms deprecated user code to conforming user code. Considering SeqAn's history, I can only stress that this should be only the absolute last resort (e.g. in a case were such a change is forced upon SeqAn by a dependency).

ABI

The application binary interface (ABI) of a library encompasses the representation of library entities in machine code. Since SeqAn is a header-only library it is not distributed as machine code and it makes **no promises regarding its binary representation**. In particular, this means that the memory layout and size of all class types is subject to unannounced and undocumented change. This gives a reasonable amount of freedom to the SeqAn developers, but it also means other parties cannot create compiled representations of SeqAn (or other libraries using SeqAn) and expect them to be stable.

Platform

Another dimension of stability is the long-term viability of a specific platform for development and/or deployment of SeqAn-based applications. As described above, SeqAn3 primarily depends on the compiler and not on operating system specifics. Currently, GCC7, GCC8 and GCC9 are considered *stable compilers*. To guarantee that SeqAn-based applications developed and/or deployed on a particular operating system can be upgraded, the SeqAn team promises to not discontinue support for a stable compiler on a given operating system until a newer supported compiler is easily attainable on that operating system.[15] The list of operating systems considered relevant can be seen in Table 4.2.

Table 4.2 Operating systems and releases supported by SeqAn3

Operating system	Supported releases
RedHat Enterprise Linux	The latest release
CentOS Linux	The latest release
SUSE Linux Enterprise Server	The latest release
Debian GNU/Linux	"Stable" and "old-stable"
Ubuntu Linux	The two latest LTS releases
macOS	The two latest releases
FreeBSD	The latest stable release

---

[15] A trustworthy third party package repository would be acceptable—having to build a compiler from source would not be.

New operating systems and compilers will likely be supported during the release cycle of SeqAn3, however, not all new compilers will be considered *stable compilers* and not all new operating systems will be *officially* supported. SeqAn intentionally promises to only require that a single combination of stable compiler and supported operating system is maintained, e.g. while GCC7 is likely to be supported on GNU/Linux distributions for a long time, it may be discontinued on macOS or FreeBSD earlier, because they have shorter release cycles and quickly provide access to new compilers.[16]

4.3.5   Availability

SeqAn3 is primarily distributed via its page on GitHub.[17] It has a distinct repository from SeqAn2 to underline that version three of the library is independent and can be installed in parallel. GitHub was chosen to host SeqAn3, because it proved to be very helpful in the development and promotion of SeqAn2. Should it become necessary at any point to choose a new platform for hosting, the usage of Git makes transferring a project easy.

The repository contains the library, documentation and unit tests since it makes sense to version these together. It does not contain any application code and the software for creating the documentation and running the unit tests is also separate (see Sects. 4.4.2 and 4.4.3). This makes the repository *compact* in contrast to SeqAn2's which included a majority of code irrelevant to library users (application code, tooling) and improves *integration*. On the other hand the repository includes *git-submodules*[18] of the *sdsl*, *range-v3*, *cereal* and *lemon* (see Sect. 4.4.1). This means users who checkout the repository receive the dependencies automatically. As a result, the dependencies do not make installing SeqAn3 less *simple* than SeqAn2.

I have previously explained that I think distribution via package managers is also important to reach SeqAn's target audience (p. 30). The first release of SeqAn3 is already available for Debian GNU/Linux and Ubuntu, BioConda (Grüning et al., 2018) and Easybuild (Hoste et al., 2012). I expect wide availability in common package managers (similar to SeqAn2) once the first stable version (3.1) is released and more applications make use of it.

---

[16] As previously mentioned, there *should not* be behaviour specific to the operating system that would suggest such a step, but experience has shown that platform-specific quirks accumulate over time and this rule prevents the proliferation of such workarounds.

[17] https://github.com/seqan/seqan3.

[18] These can be thought of as "links" to other repositories.

4.3.6 Combining SeqAn2 and SeqAn3

SeqAn3 was designed in a way that allows installing and using it in parallel with SeqAn2. This may be useful when a specific feature is not yet available in SeqAn3 or a codebase is too large to port in a single step. To this end, SeqAn3 uses a different folder for its headers (`include/seqan3/` instead of `include/seqan/`) and namespace for its code (`seqan3::` instead of `seqan::`). The CMake modules are also distinct and treat the two as entirely different packages.

Since SeqAn2 and SeqAn3 follow different naming conventions (and a different style in general) there should only be few naming conflicts. However, I still recommend not importing any namespaces via `using namespace FOO;` (which is good advice in any case). In how far entities of one version can practically be combined with the other (e.g. using one version's types with another version's functions) is explored in Part III.

In any case the mere possibility of using them on the same computer and even in the same codebase is a significant improvement over the situation with SeqAn1 and SeqAn2 where this was impossible and led to considerable conflicts for package maintainers. A minor release of SeqAn2 that is scheduled after SeqAn-3.1 is planned to contain auxiliary code that will help with any such integration attempts.

4.4 Dependencies and Tooling

This section covers libraries and applications used either by SeqAn3 directly or related processes like documentation building and testing; Fig. 4.2 shows a visualisation.

The dependencies of the SeqAn3 library are not contained in the SeqAn3 repository source code, but are referenced via *git-submodules* which allows for easy inclusion (see above). Documentation and testing code is versioned together

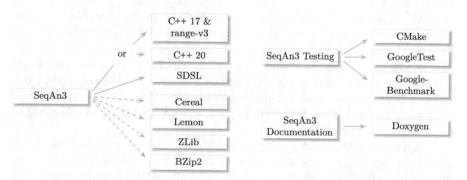

Fig. 4.2 The dependency graph of SeqAn3. Dashed lines imply optional dependencies. "C++ 17" and "C++ 20" refer to the respective standard libraries

with the library in one repository, but the related dependencies and tooling are not included to keep the repository *compact*.

4.4.1 Library Dependencies

An important difference of SeqAn3 compared with previous versions is its focus on *compactness*. To reach this goal, it relies on other libraries whenever this is feasible. Since adding dependencies increases the complexity of the build process (which reduces *simplicity* for developers and users of the library), each case needs to be deliberated carefully.

The following criteria are used when evaluating a third party library for inclusion:

1. It must be useful as a whole, not just single aspects of it (larger libraries entail greater complexity).
2. It must be header-only.
3. It may not impose stricter licensing terms then SeqAn.
4. It must either be actively maintained or have a very clear-cut purpose and a history of stability.

The first point ensures that the value a library presents (reduction of own code/maintenance) is proportionally larger than its cost in added complexity. Points 2. and 3. guarantee that the advantageous properties introduced in the previous two subsections are not offset by dependencies. The second point enables SeqAn itself to perform detection of its dependencies from within the source code and disable certain codepaths for *optional* dependencies when they are missing—or give usable diagnostics in the case of missing *required* dependencies. The last point expresses the need for being able to report/fix bugs in dependencies.

In cases were only a part of a library is required (e.g. a single header file) or the library itself is very small and no changes are expected to the library, the source code of the library may be integrated into SeqAn3 as a sub-folder in the *Contrib* module and within the namespace `seqan3::contrib::`. This furthers the goal of *compactness* by relying on existing, well-established solutions but does not impose the overhead of full dependency management (improved *simplicity*). A clear separation inside the source code (folder and namespace) preserves the clean structure of the code and highlights that certain files follow different conventions and possibly different (but compatible) licence terms. It should be noted that this is only a solution for very small external pieces of code. Whenever upstream changes are expected (and indeed desired) it is preferable to use regular mechanisms of dependency management.

To keep usage of SeqAn as *simple* as possible and to ease *integration*, dependencies are typically not exposed to users of SeqAn3, i.e. the dependency is used by SeqAn3 internally and (except for the standard library and two types of from Cereal) users are not expected to use names from a dependency's namespace or read third party API documentation.

As mentioned in Sect. 4.3.1, SeqAn3 can be used without a build system and required dependencies are also detected from within the source code. Nevertheless, a CMake module is provided since CMake is arguably the most important build system for C++ applications (Wojtczyk & Knoll, 2008). This module enables easy *integration* of SeqAn3 into CMake-based applications and it is *simpler* to use than manually invoking the compiler, because it takes care of setting certain required flags (e.g. C++-standard and threading-support). It also automatically detects and links (de-)compression libraries (see below). If there is demand, configuration files for other build systems (e.g. the increasingly popular Meson[19]) will be added.

The C++ 20 Standard Library

The C++ standard library provides many useful algorithms and data structures that have usually gone through very rigorous testing and optimisation. The first step in avoiding duplication of effort and possibly inferior custom solutions should therefore be relying on the standard library. As I have noted previously, SeqAn3 relies strongly on aspects of the standard library that are only introduced with C++ 20, most importantly C++ Ranges. These are not available on GCC7–GCC9, so they need to be provided by other means.

SeqAn3 has a module called *STD* that provides these missing parts of the C++ 20 standard library and some parts of the C++ 17 standard library that are not available in certain versions of GCC. Most of the functionality is provided as aliases to entities in the range-v3 library, the reference implementation for C++ Ranges. The range-v3 library promises to conform fully to implementations in the standard. Some required standard library entities not provided by range-v3 are provided as full definitions in the *STD* module (either as copies from existing C++ 20 standard libraries or as custom implementations).

The header files in seqan3/std/ are named just as their counterparts in the standard library, e.g. <seqan3/std/iterator> instead of <iterator> which makes including them *simple*. I also decided to place them in namespace std:: so they can be used as "drop-in replacements". This is a minor violation of the standard,[20] but it allows using the C++ 20 standard library if shipped by the compiler and only falling back to SeqAn's aliases/definitions when this is not the case. I believe that the alternative (providing all these names in SeqAn's namespace or even a third one) would have been highly confusing for users. With this solution, users of SeqAn3 can e.g. use std::views::filter independent of whether it is provided by the standard library or by SeqAn3 or indirectly by the range-v3 library. This provides the highest possible degree of *integration* with the standard library and is as *simple* as possible for users of SeqAn. It results in a SeqAn API that is

---

[19] https://mesonbuild.com/.

[20] The standard asks users not to define names in namespace std::.

much more *compact* than it would otherwise be, and users can be directed to the standard library documentation for said names.

As such, the range-v3 library is a hard requirement, but only for pre-C++ 20 compilers. The mechanisms for detecting this and enabling/disabling respective codepaths are cross platform (no hard-coded compiler-specific behaviour) and built into the headers—they do not have to be performed by the build system. Should SeqAn3 at some point in time drop support for pre-C++ 20 compilers, one can simply delete the *STD* module folder and remove the redundant include-statements from the library code.

The Succinct Data Structure Library

The Succinct Data Structure Library (SDSL) is a C++ library with the focus of providing space-efficient data structures to bioinformatics, information retrieval and related areas (Gog et al., 2014). Among other data structures, it provides compressed bitvectors and compressed suffix arrays/FM-indexes. In the latter regard it could be seen as competing with SeqAn1/2 which also provided efficient FM-index implementations.

Due to the new focus on *compactness* in SeqAn3, it was decided to form an academic and technical cooperation with SDSL and its authors. As a result, the SDSL has become a header-only library, changed its licence and accepted contributions from SeqAn team members. On the other hand, SeqAn3 now requires the SDSL as a dependency and no longer provides own implementations of FM-indexes.

The changes performed in the SDSL include adding EPR-dictionaries (Pockrandt et al., 2017), serialisation support through cereal (see below) and various modernisations of the codebase and smaller fixes. More contributions are planned in the future; current and former SeqAn developers are now the most active contributors to the SDSL.

Cereal (Optional Dependency)

Serialisation is the process turning an object (including its members) into a (text or binary) representation that can later be used to recreate (deserialise) the object in its original state (Standard C++ Foundation, 2019). An example in bioinformatics would be to create a full-text index (a computationally expensive task) and store a serialised version on-disk so that later invocations of the program can deserialise the index without needing to recompute it. Many programming languages including Python and Java have built-in support for serialisation, but for C++ this is not the case.

SeqAn1/2 had its own very rudimentary serialisation implementation. It only worked for some SeqAn types and did not support standard library types. It also did not operate recursively and it serialised members into individual files, e.g. an FM-index would be serialised to more than ten files. This was confusing for users of SeqAn-based applications and made for poor *adaptability* and *integration*.

In the pursuit of more *simplicity* and *compactness*, SeqAn3 relies on a well-established solution to this problem. The decision fell on the Cereal library[21] for the following reasons:

- Satisfies aforementioned criteria (header-only, BSD licence, etc.).
- Based on the well-known design of the Boost serialisation library (Schäling, 2011; Cogswell, 2015), but does not require Boost.
- Uses Modern C++ which promises better *integration* with SeqAn3.
- Choice of different serialisation formats:

  - JSON or XML (text formats)
  - binary formats (also compatible between little-endian and big-endian architectures).

- Better *performance* than most other libraries.[22]

With Cereal, it is possible to easily create archives that store all of an application's state in a single file. Since regular C++ I/O streams are used, compression can also be applied to serialised archives (see below). The dependency on Cereal is optional and not required if serialisation is not used by the application.

Lemon (Optional Dependency)

SeqAn has had its own graph module since the first release although only few applications used it. Out of the applications that did use the graph module, graph-based (re-)alignment was the most relevant feature. Implementations of different graph algorithms (Dijkstra, Ford-Fulkerson, etc.) seemed to have followed a more theoretical interest with comments as such found in multiple places:

```
// WARNING: Functionality not carefully tested!
```

Moreover, many applications that relied on SeqAn for core functionality still preferred other libraries to realise graphs—an indication that the graph module was not as well-received as other SeqAn features. Consequently, it was decided to not implement a custom graph module in SeqAn3. Instead, SeqAn3 makes use of the Lemon library (Dezso et al., 2011).

The Lemon library meets the required criteria and was found to be of high code quality. It receives updates only infrequently but has responsive maintainers. Since only multiple-sequence alignment (MSA) is planned to rely on graphs, the

---

[21] http://uscilab.github.io/cereal/.

[22] Comparison of different libraries: https://github.com/thekvs/cpp-serializers.

dependency on Lemon is also optional. This means that applications that do not perform MSA do not require the Lemon library to be present at build-time.

(De-)Compression Libraries (Optional Dependencies)

An important aspect of SeqAn has always been input/output, i.e. the support for reading and writing typical biological file formats, like FASTA (Lipman & Pearson, 1985) and SAM (Li et al., 2009). Often text-based formats are stored compressed on-disk to save space and some formats like the BAM format (Li et al., 2009) are compressed as part of their specification. Thus, it is very important for SeqAn to handle (de-)compression of files.

SeqAn1/2 provided C++ input/output streams that interface with the GZip/ZLib (Deutsch, 1996) and BZip2[23] libraries. These stream interfaces are among the very few pieces of SeqAn2 code adopted in SeqAn3.

Neither of the two libraries are header-only, however, they are present on most modern UNIX-like operating systems in the default install. SeqAn3's CMake module automatically detects their presence and adds required linker flags. The libraries are optional; if they are not available at build-time, the respective codepaths are deactivated in the code (this can be detected by a developer using macros).

It should be noted that many new compression algorithms have been published in recent years and SeqAn3 would most certainly profit from broader support in this area. Most notably, ZStandard compression (Collet & Kucherawy, 2018) has become very popular, because it outperforms the de facto standard GZip in all metrics, i.e. compression ratio, compression speed and decompression speed are all better. Other previously popular algorithms like LZMA[24] and LZ4[25] yield better results in one or two areas but perform worse in others (Table 4.3). However, LZMA is e.g. suggested by the now popular CRAM alignment format (Fritz et al., 2011), so it may be important to attain support for this algorithm independent of it how it compares to others.

Table 4.3 Comparison of modern compression algorithms. Results produced by *lzbench*, quoted from https://engineering.fb.com/core-data/smaller-and-faster-data-compression-with-zstandard/

	Ratio	Comp. speed	Decomp. speed
lz4	2.10	444.69 MB/s	2165.93 MB/s
zstd (ZStandard)	3.14	136.18 MB/s	536.36 MB/s
zlib (GZip)	3.11	23.21 MB/s	281.52 MB/s
xz (lzma)	4.31	2.37 MB/s	62.97 MB/s

---

[23] https://sourceware.org/bzip2/.

[24] https://7-zip.de/sdk.html.

[25] https://lz4.github.io/lz4/.

4.4.2 Documentation

As I discussed in Sect. 2.4.2 (p. 22), documentation is an integral part of a software project, especially a library. To avoid the shortcomings of SeqAn1/2's documentation efforts, it was decided to use the well-established Doxygen documentation generator (van Heesch, 2008) for SeqAn3. Not having to maintain a custom documentation generator reduces work for the SeqAn team and makes the project more *compact*. Using a well-established system also reduces the learning curve for new contributors (Fig. 4.3).

Doxygen's versatility enables the SeqAn team to have API documentation and extensive Tutorials/HowTos in the same place—they were separate in SeqAn1/2 and required distinct steps to be generated (each using a different assortment of Python scripts and packages). On the other hand, the support for GitHub-flavoured Markdown[26] in Doxygen means that the different flavours of documentation and all project communication (GitHub issues, wikis, pull-requests, etc.) happen in the same markup language. This is a significant *simplification* compared to SeqAn1/2 where API documentation (dddoc/dox/HTML), Tutorials (ReStructured Text) and reviews/issues (Wiki markup in Trac) were all written in distinct languages.

One of the main technical problems identified with the documentation generators of SeqAn1/2 was their lack of any source code parsing, i.e. the independence of

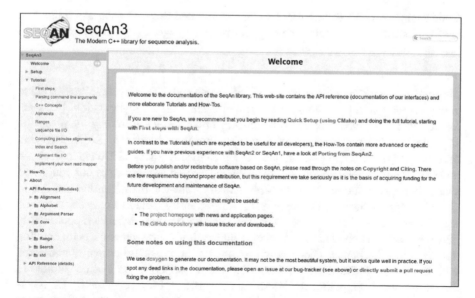

Fig. 4.3 Screenshot of the API documentation of SeqAn3 (built with Doxygen). Screenshot taken by myself; content is part of SeqAn documentation, see Sect. A.3.1

---

[26] https://github.github.com/gfm/.

```
   /*!\brief Adds an option to the seqan3::argument_parser.
2   *
   * \tparam option_type Must have a formatted input function (stream >> value).
4   *                    If option_type is a container, its value type must have the
   *                    formatted input function (exception: std::string is not
6   *                    regarded as a container).
   *                    See [FormattedInputFunction](LINK_OMITTED).
8   * \tparam validator_type The type of validator to be applied to the option
   *                    value. Must satisfy seqan3::validator.
10  *
   * \param[in,out] value    The variable in which to store the given command line argument.
12  * \param[in]    short_id  The short identifier for the option (e.g. 'a').
   * \param[in]    long_id   The long identifier for the option (e.g. "age").
14  * \param[in]    desc      The description of the option to be shown in the help page.
   * \param[in]    spec      Advanced option specification, see seqan3::option_spec.
16  * \param[in]    val       The validator applied to the value after parsing (callable).
   *
18  * \throws seqan3::parser_design_error
   */
20  template <typename option_type, validator validator_type = detail::default_validator<option_type>>
   //!\cond
22      requires (argument_parser_compatible_option<option_type> ||
                 argument_parser_compatible_option<std::ranges::range_value_t<option_type>>) &&
24               std::invocable<validator_type, option_type>
   //!\endcond
26  void add_option(option_type & value,
                   char const short_id,
28                 std::string const & long_id,
                   std::string const & desc,
30                 option_spec const & spec = option_spec::DEFAULT,
                   validator_type val = validator_type{}) // copy to bind rvalues
32  {
       /* ... */
34  }
```

Code snippet 4.1: An example of Doxygen-annotated SeqAn3 code. An inline link has been omitted so that this code snippet fits on the page. Code taken directly from SeqAn3 source code, see Sect. A.3.1

the source-code comments from the actual source code. This made it impossible to verify the way an interface was documented against the implementation. Doxygen has a C++ parser which is not very modern but worked surprisingly well on the strongly templatised codebase of SeqAn3. This may be due to SeqAn3 relying on more orthodox coding techniques or due to the advances of Doxygen itself, in any case the problems anticipated by previous SeqAn authors were not confirmed. The only major issue of Doxygen with SeqAn3's code is the definition and use of C++ Concepts. However, Doxygen has a similar entity called *interfaces* (normally used for annotating Java source code), and SeqAn's concept *definitions* could easily be expressed as such. This led to a nice integration in the general documentation and even in inheritance/specialisation graphs (if the types are correctly annotated with the concepts they model). Regarding the *use* of concepts and constraints in the definition of templates, the intermediate syntax is fully accepted by Doxygen, but the verbose syntax needs to be escaped so that Doxygen ignores it. As a result, the requirements on template parameters are always stated as part of the textual description inside the API documentation. This is a compromise, but it could not reasonably be expected that any documentation generator supports a C++ feature that was not yet officially part of C++.

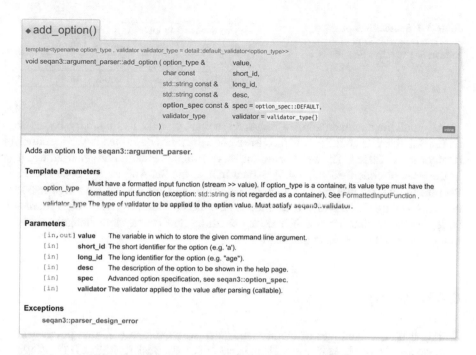

Fig. 4.4 Screenshot of the HTML rendering of Snippet 4.1. Note that the function signature is parsed from the code and is not part of the annotations. Screenshot taken by myself; content is part of SeqAn documentation, see Sect. A.3.1

Snippet 4.1 shows an example of Doxygen-annotated source code and Fig. 4.4 shows the html rendering. Parameters and template parameters are individually annotated and parameters indicate whether being *in*, *out* or *in+out*. One constraint is given in the intermediate syntax (`validator validator_type`), the other constraints are given in the verbose syntax (a *requires* clause) and are hidden from Doxygen (`//!\cond` and `//!\endcond`) but expressed in text.

The most important aspect of a parser integrated into the documentation generator is the ability to verify the documentation; had a parameter in Snippet 4.1 been misspelled or forgotten, Doxygen would have generated a warning. Furthermore, Doxygen can be configured to warn for every entity in the codebase that is not documented. SeqAn3 follows this approach and requires that absolutely every entity must be documented—even those that are not visible to users. This greatly improves the quality of the documentation and allows for two different documentation targets to be generated (similar to different make targets):

User documentation   This documentation defines the API; it is used by application developers.

Developer documentation   This documentation additionally includes all names declared `private` or in namespace `seqan3::detail::`; it is used by library developers/contributors.

Table 4.4 Source-lines of code and comment-lines of code in SeqAn1/2 vs Seqan3

SeqAn release	SLOC	CLOC	CLOC in %
SeqAn-1.0	88,332	36,578	29.28
SeqAn-2.0	168,488	94,635	35.97
SeqAn-3.0	32,209	32,049	49.88

All-in-all the documentation of SeqAn3 is a main pillar of its *simplicity*, both, for its users and for its contributors. It follows good practices described in the literature (Geiger et al., 2018). The strict requirements for documentation are also reflected in the lines of code (see Table 4.4). About half of the SeqAn3 library code is API documentation and this does not include Tutorials and HowTos. This is a significant boost over SeqAn2 for which the numbers also include comparatively more source-code comments that are not API documentation, and (to my knowledge) this is unique in the realm of academic software.

4.4.3   Testing

Testing is an essential part of quality control in software and common practice in professional software businesses (Runeson, 2006; Garousi & Zhi, 2013). While some companies have dedicated software testers as part of their teams (Garousi & Zhi, 2013), SeqAn, as an academic project, needs to manage software testing as part of the development process. Software testing is an area often neglected (similar to documentation) if there is no strict process enforcing it. I would argue that some barriers to software testing in academia are comparable to those in industry, e.g. time and cost constraints (Garousi & Zhi, 2013). Other problems are specific to academia (see Sect. 4.5).

The aim of software testing is to prove the ability of the software to perform as advertised and to detect/fix bugs as part of the development process before they are encountered by a user. There are different aspects of testing and different testing methodologies. SeqAn3 comes with the following test suites:

Unit   The unit tests call the public interfaces in a specified automated manner and compare the results to a predefined set of expected answers. The goal is to reach complete or at least high *coverage*, i.e. every line of code in the library should be covered by at least one unit test.

Snippet   The snippet tests build all code snippets from the API documentation and the tutorials to verify that all examples from the documentation actually compile. This prevents typos and regressions in the instruction materials.

Performance   The performance tests contain benchmarks for performance-critical (and some low-level) interfaces. This allows detecting and investigating performance regressions down to a per-commit level.

Header   The header tests primarily test that every header file in the library can be included multiple times safely.[27] They are also used to enforce certain guidelines, e.g. that every header should include `<seqan3/core/platform.hpp>`, and to detect certain bugs.

Documentation   The documentation test builds user and developer documentation (see Sect. 4.4.2) and verifies that Doxygen produces no warnings or errors (e.g. undocumented entities).

Test Metrics

One metric shared by all test suites is that respective tests should build without errors or warnings; this already tests many core aspects of the library. It is especially important for a header-only library since it has to be rebuilt with every application and cannot be shipped pre-compiled. Thus, it is also important to take into account different setups the application developer might have (more on this below). This metric also contributes to detecting API breakage, i.e. having to change a test to accommodate for changes in the library is an obvious indicator for an API break.

In addition to "compileability", the *unit tests* also perform checking of values at runtime, e.g. function return values. This verifies the runtime aspects of the API.

The third and very central metric for unit tests is a high *code coverage*. This proves that indeed a test covers all codepaths in a specific module or header file. There are certain limitations to code coverage analysis in combination with C++, because code coverage analysis is a runtime analysis. Consequently, templates that are never instantiated are not considered for analysis at all and code only executed during compile-time will appear as not being covered. These restrictions notwithstanding, code coverage is a very useful metric for detecting unexecuted statements or unevaluated branches. See Fig. 4.5 for an example.

SeqAn3 currently boasts over 97% code coverage and new code is required to have a coverage of 100% with only very few exceptions allowed. This is a big difference to SeqAn1/2 where different (custom) mechanisms for tracking coverage where tried, but ultimately no process with a clear coverage goal was implemented.

The *performance tests* are the only tests that do not have a clear metric as the time it takes to execute an operation depends highly on the operation tested and also the environment. While benchmarks (also against SeqAn2) are helpful during development, the focus of the performance tests is to aid in regression analysis, i.e. to find the responsible commit(s) after a performance regression has been detected. They can also be used to verify before a new release of the library that no such regressions have occurred. This plays an important role in delivering and maintaining a good *performance*.

---

[27] This implies correct *linkage* of all entities and using "header guards" or `pragma once` to avoid violations of the One-Definition-Rule.

```
68        // Insertion
69   ○    if (error_left.insertion > 0)
70        {
71   ○        search_param error_left2{error_left};
72   ○        error_left2.insertion--;
73   ○        error_left2.total--;
74
75            // always perform a recursive call. Abort recursion if and only if recursive call found a hit and
76            // abort_on_hit is set to true.
77   ○        if (search_trivial<abort_on_hit>(cur, query, query_pos + 1, error_left2, delegate) && abort_on_hit)
78                return true;
79        }
80
81        // Do not allow deletions at the beginning of the query sequence
82   ○    if (((query_pos > 0 && error_left.deletion > 0) || error_left.substitution > 0) && cur.extend_right())
83        {
84   ○        do
```

Fig. 4.5 Example of code coverage tests detecting an untested statement in a pull-request against SeqAn3. Screenshot taken by myself; content is SeqAn3 code annotated by *gcov* and https://codecov.io, see also Sect. A.3.1

Implementation

The infrastructure for building and running the tests is implemented with CMake (and its testing component CTest), similar to SeqAn2.

All tests rely on *GoogleTest*[28] to provide infrastructure (mostly in the form of macros) for defining test cases, verifying expected values and generating human readable test reports. Different other test frameworks were evaluated, including *doctest*[29] and *Catch2*[30] which promise a more modern look and feel (Modern C++, less macros, header-only). But both were found to be inadequate for testing SeqAn3, because they lacked one or more important features, like type/value parameterisation of tests (important for testing templates) or exception tests (evaluate that a certain exception is thrown). Boost.Test[31] would have also met the requirements, but there seemed to be no clear advantages and cooperating projects like the SDSL use GoogleTest so this was chosen. The important decision was to use a framework at all—and not maintain a custom solution like SeqAn1/2.

Calculating the coverage for the unit tests happens with the help of *gcov*, the coverage analyser of GCC[32] and the more user-friendly front-end *lcov*.[33]

The performance tests rely on *GoogleBenchmark*,[34] a benchmark framework closely tied to GoogleTest. GoogleBenchmark does not directly measure the individual runtime of a single invocation, instead it tests how often it can perform a

---

[28] https://github.com/google/googletest.

[29] https://github.com/onqtam/doctest.

[30] https://github.com/catchorg/Catch2.

[31] https://github.com/boostorg/test.

[32] https://gcc.gnu.org/onlinedocs/gcc/Gcov.html.

[33] https://github.com/linux-test-project/lcov.

[34] https://github.com/google/benchmark.

```
Running core/char_operations/char_predicate_benchmark
Run on (12 X 3600.16 MHz CPU s)
Load Average: 1.34, 0.95, 0.65
-----------------------------------------------------------------
Benchmark                 Time              CPU   Iterations
-----------------------------------------------------------------
simple<true>            7.77 ns         7.75 ns     90427593
simple<false>           1.13 ns         1.13 ns    621046374
combined<true>          14.6 ns         14.6 ns     48032388
combined<false>         1.13 ns         1.13 ns    621184154
```

Fig. 4.6 Output of a SeqAn3 performance test run with GoogleBenchmark. Screenshot taken by myself; text is generated by GoogleBenchmark

call or iteration in a fixed amount of time. This is more robust to disturbances and can better visualise the relative differences of very short runtimes (Fig. 4.6).

It should be noted that while all tests require GoogleTest and the performance tests require GoogleBenchmark; these are not listed as dependencies of the library and are also not given as git-submodules. This means application developers do not needlessly receive them when cloning the library. They are, however, automatically cloned by CMake when needed, so contributing to SeqAn3 stays as *simple* as possible.

Execution

As mentioned above, compile-testing is a central part of testing a header-only library. It is therefore important to build the unit tests with all supported compilers, on all supported operating systems and in all possible configurations. This includes Release vs Debug mode, with and without optional dependencies. Beside exceptionally high standards conformance (pedantic compiler flags, no GNU extensions) and warning levels (-Wall -Wextra) there are also tests with *sanitisers* enabled.

Sanitisers insert instrumentation into the program during build-time that usually incurs notable performance penalties but that allows to detect certain bugs at run-time that would otherwise go unnoticed. Most importantly the memory sanitiser[35] detects memory leaks and the use-after-free and out-of-bounds errors that frequently occur in C and C++ programs. The undefined behaviour sanitiser, on the other hand, detects many cases of undefined behaviour, e.g. signed integer overflow, floating point divide-by-zero and dereferencing of null pointers. Building unit tests with sanitisers has uncovered and helped prevent several severe issues. While SeqAn2 briefly included tests with *valgrind*,[36] these were never relied upon in production, because of a high false-positive rate in connection with OpenMP that was also used in SeqAn2.

---

[35] https://github.com/google/sanitizers/wiki/AddressSanitizer.

[36] A stand-alone memory debugger and profiler that can detect problems similar to the address sanitiser. http://valgrind.org/.

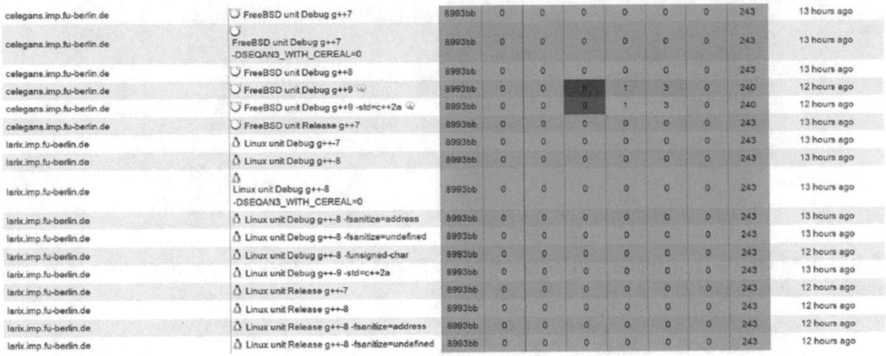

celegans.imp.fu-berlin.de	FreeBSD unit Debug g++7	8993bb	0	0	0	0	0	0	243	13 hours ago
celegans.imp.fu-berlin.de	FreeBSD unit Debug g++7 -DSEQAN3_WITH_CEREAL=0	8993bb	0	0	0	0	0	0	243	13 hours ago
celegans.imp.fu-berlin.de	FreeBSD unit Debug g++8	8993bb	0	0	0	0	0	0	243	13 hours ago
celegans.imp.fu-berlin.de	FreeBSD unit Debug g++9	8993bb	0	0	8	1	3	0	240	12 hours ago
celegans.imp.fu-berlin.de	FreeBSD unit Debug g++9 -std=c++2a	8993bb	0	0	8	1	3	0	240	12 hours ago
celegans.imp.fu-berlin.de	FreeBSD unit Release g++7	8993bb	0	0	0	0	0	0	243	13 hours ago
larix.imp.fu-berlin.de	Linux unit Debug g++-7	8993bb	0	0	0	0	0	0	243	13 hours ago
larix.imp.fu-berlin.de	Linux unit Debug g++-8	8993bb	0	0	0	0	0	0	243	13 hours ago
larix.imp.fu-berlin.de	Linux unit Debug g++-8 -DSEQAN3_WITH_CEREAL=0	8993bb	0	0	0	0	0	0	243	13 hours ago
larix.imp.fu-berlin.de	Linux unit Debug g++-8 -fsanitize=address	8993bb	0	0	0	0	0	0	243	13 hours ago
larix.imp.fu-berlin.de	Linux unit Debug g++-8 -fsanitize=undefined	8993bb	0	0	0	0	0	0	243	13 hours ago
larix.imp.fu-berlin.de	Linux unit Debug g++-8 -funsigned-char	8993bb	0	0	0	0	0	0	243	12 hours ago
larix.imp.fu-berlin.de	Linux unit Debug g++-9 -std=c++2a	8993bb	0	0	0	0	0	0	243	13 hours ago
larix.imp.fu-berlin.de	Linux unit Release g++-7	8993bb	0	0	0	0	0	0	243	12 hours ago
larix.imp.fu-berlin.de	Linux unit Release g++-8	8993bb	0	0	0	0	0	0	243	12 hours ago
larix.imp.fu-berlin.de	Linux unit Release g++-8 -fsanitize=address	8993bb	0	0	0	0	0	0	243	12 hours ago
larix.imp.fu-berlin.de	Linux unit Release g++-8 -fsanitize=undefined	8993bb	0	0	0	0	0	0	243	12 hours ago

Fig. 4.7 Excerpt of nightly build results displayed in CDash Screenshot taken by myself; content is created by the CDash application

The possible combinations of compilers, flags, operating systems and dependencies yields a large matrix of tests that are run every night to detect regressions in the codebase (Fig. 4.7). Beyond these *nightlies* there is also continuous integration (CI), i.e. whenever a contributor opens or updates a pull-request against the SeqAn repository, a subset of the aforementioned tests is run. This helps to detect problems early on and prevents changes from being merged that would break tests. But since it is not feasible to run all the tests during CI, the nightly builds are still an important factor in ensuring the continued quality of SeqAn3.

Future

While the current mechanisms for testing are extensive and considerably more thorough than SeqAn2's (except for a smaller list of currently supported compilers), there are some areas where I think testing could still be improved.

Fuzzing

The unit tests described above are a form of *glassbox testing*, because the tests are based on the knowledge of the implementation, i.e. "because I know how the algorithm works, I know that it shall return 42 for the input 23". A different form of testing is *fuzzing* which is a form *random testing* (Duran & Ntafos, 1981). Fuzzing can be defined as:

> A highly automated testing technique that covers numerous boundary cases using invalid data (from files, network protocols, API calls, and other targets) as application input to better ensure the absence of exploitable vulnerabilities. The name comes from modem applications' tendency to fail due to random input caused by line noise on "fuzzy" telephone line (Oehlert, 2005).

Fuzzing is especially important for software security, but is also a part of general software quality assurance, because it helps detect problems that would otherwise go unnoticed (Takanen et al., 2018). One can directly fuzz the API by providing random or semi-random input[37] to e.g. function calls, but for a project like SeqAn, a particular area that would benefit from fuzzing is the input/output module. This is because there is a high variation in file formats, partly due to a lack of standardisation in bioinformatics, but also due to random deviations in files resulting from storage or transmission errors.

Applications that crash upon opening such a file are a common and frustrating user experience in bioinformatics and finding/fixing these problems is not trivial for developers if the user does not or cannot provide the responsible file to the developer, e.g. because it contains sensible scientific or medical data. While it may not always be feasible to enforce all aspects of a file format for performance reasons, the library should never crash the application. Fuzzing can help prevent such errors and I strongly suggest to future SeqAn maintainers that they setup infrastructure for this. Since fuzzing tests are computationally expensive, it may only be feasible to perform them once per week instead of *nightly* and/or only on a subset of platforms/configurations But I am sure they will uncover hidden issues in the code.

Automatic Performance Analysis

The current performance tests are most useful during development and after a performance regression has been detected (to find its origin). If the nightly performance tests were run in a controlled environment without external influences, one could setup a system that also automatically detects any performance regressions and reports them immediately. This would reduce debug times and ensure that regressions do not go unnoticed.

Code Coverage and Templates

I have mentioned the limitations of coverage analysis in combination with strongly templatised code above. There are some analysers that suffer less from this than *gcov* and there are tools that mitigate the deficiencies of existing coverage analysers by annotating the original source code and performing certain post-processing steps on their output. One such tool is *force-cover*.[38] Since code coverage is important, I recommend that further options be explored to achieve an even higher actual coverage of SeqAn3's templates.

---

[37] Not all fuzzing is completely random as it is often necessary to pass certain early checks to reach further parts of the program. This is required to reach a high code coverage through fuzzing.

[38] https://github.com/emilydolson/force-cover.

4.5   Project Management and Social Aspects

Project management was not part of the author's job description and, this book being a work of science and engineering, I do not want to spend too much time on the social aspects of the SeqAn project. A few things I do, however, want to mention to help the reader understand the history of my own involvement and the uncertainties that projects like SeqAn face in general.

Research software and the developers of research software, the so-called research software engineers (RSEs),[39] face some unique challenges in academia (Fig. 4.8). First and foremost this is a lacking recognition of the value of software for science in general and/or a lacking recognition of the means for producing such software (Goble, 2014). The latter includes time and resources required for quality assurance, reproducibility, maintenance and sustainability. This in turn makes it harder for RSEs to earn reputation within the academic system and results in a significant amount of research software being developed in unsustainable ways, e.g. a "bus factor"[40] of 1 (Philippe et al., 2016). In turn this leads to many software projects being abandoned after a short period of time—as I described in Sect. 2.4 this is now the case for many of SeqAn's former "competitors".

SeqAn is being developed in an academic group that is primarily concerned with the development of computational methods so it suffers less from a lack of visibility (within the group) than research software developed in groups where all software is just a "by-product". However, members of the group all have their individual focus which is usually a concrete tool or application. After Andreas Gogol-Döring left the project, there was no one fully dedicated to the improvement and maintenance of the library. David Weese, Manuel Holtgrewe and Enrico Siragusa did end up spending significant amounts of time on the library, but they worked under the constraint of

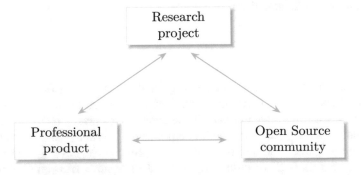

Fig. 4.8 The different aspects of SeqAn as a project

---

[39] https://researchsoftware.org/.

[40] The bus factor describes the number of people required to fail ("be hit by a bus") for the entire project to fail.

achieving a degree that had a different or narrower scope. This highlights that typical problems of research software apply: even though the group focuses on software, the library itself is more of a by-product or means to an end.

All main developers over the last ten years were doctoral students, so the turnover in the team has been significant. This made it difficult to build a functioning team that fully understands all aspects of a project the size of SeqAn. And it led to a fragmentation of styles and paradigms inside the library and conceptual regressions over the original design (see Chap. 2).

Knut Reinert provides excellent leadership in an overarching sense. He gives important insight and direction in matters of current research and guides the team members in their own research topics; he has been very successful in acquiring funding for the continued development of SeqAn and related tools; and he finds new motivated students to pursue work on SeqAn every year. But a professional software product also needs leadership on a more technical level that can only be provided by someone involved in the day-to-day development of the software.

Since the end of 2016 this has been my role, although René Rahn shared many responsibilities and I relied strongly on his advice. We later formed the "SeqAn core team" to involve more developers in the decision-making process. This increases the "bus factor" (decreasing the dependence on a single developer), institutionalises a form of knowledge transfer and formalises handover procedures. According to Philippe et al. (2016) these are the most important metrics for sustainability.

Beyond the core team, there are many regular team members that join library development on a case-by-case basis but primarily work on applications. New team members receive introductory training and pair-programming to integrate them into the team and teach them basic requirements, e.g. the Modern C++ techniques used.

The team meets at least weekly and the core team has an additional meeting to discuss questions in-depth; the results of meetings are recorded in writing. In the spirit of Open Science and Open Source, all of this is public. We experimented with different kinds of project management and agile software development methodologies; finding out what works best is still an ongoing process.

To guarantee a high quality of the source code, a double-review process was established, i.e. all changes made to the library have to be made in the form of pull-requests that need the approval of two different team members before being merged. This can imply multiple review steps by each reviewer and not only ensures quality but also improves the skills of the submitter and the reviewer's knowledge of different parts of the codebase.

Being an Open Source project means also working well with users and contributors not part of the team. To this end I have written a contributor guide[41] and various templates for issues and pull-requests[42] that are automatically presented to new contributors. I have also introduced a code of conduct[43] to make everyone feel

---

[41] https://github.com/seqan/seqan3/blob/master/CONTRIBUTING.md.

[42] https://github.com/seqan/seqan3/tree/master/.github/ISSUE_TEMPLATE.

[43] https://github.com/seqan/seqan3/blob/master/CODE_OF_CONDUCT.md.

welcome and safe. SeqAn relies on receiving bug reports from external contributors and having well-documented processes reduces the time needed for all involved parties.

I would conclude that the process of developing and maintaining SeqAn has improved significantly over previous versions. I have strong hopes that the team can uphold the level of activity as well as add many more features that are useful to computational biologists. However, I also think that a project like SeqAn would profit strongly from having staff that is invested in SeqAn for longer than the duration of a PhD and has no obligations to publish. This is because the academic system does not fully recognise the value of software engineering and it is difficult for doctoral students to justify spending the necessary amount of time on these tasks.

Chapter 5
Library Structure and Small Modules

Based on the analysis of SeqAn2 in Chap. 2 and the ambitious plans developed for SeqAn3 in Chap. 4, the first thing to establish is that SeqAn3 will have to be a new library, not a mere improvement on SeqAn2. While many of its designs will be inspired by SeqAn2, the fundamental shifts in the employed programming techniques mandate starting with a clean slate.

This chapter introduces the new module structure, the namespace hierarchy and library-wide naming conventions. It also covers the small, foundational modules of the library. The following chapters will then each introduce one of the larger, important SeqAn3 modules. Each module will be presented with a short module overview and individual sections for important submodules. Finally, a discussion section will summarise the module and analyse in how far the design goals described in Sect. 4.1 were reached. This section will then also compare against SeqAn2 where it will be especially important to see if the good performance of SeqAn2 can be maintained with the new designs and programming techniques. Where this makes sense, I decided to also compare against Python/BioPython since these are very popular with Bioinformaticians. If the new library can shed the complexity of SeqAn2 and attain some of (Bio-)Python's simplicity, this will likely have the most profound impact on user satisfaction and possibly also user numbers.

The state presented here is the current state of the library as found in its GitHub repository at the beginning of 2020. There are some exceptions to this where design changes (like moving a certain class from one module to another) have been decided upon but not yet executed or where implemented change is waiting to be merged. This is true, in particular, of many benchmark results presented here during whose production I performed optimisations and fixes that have not yet been merged into the master branch. As discussed in Sect. 4.3.4, SeqAn3 is not yet stable, so anything presented in this book might still change before the 3.1-release—although I hope, of course, that any such changes will be in the spirit of this work.

It should be noted that, although comprehensive, this book does not cover every class and function of SeqAn3 and cannot replace the API documentation. I cover

H. Hauswedell, *Sequence Analysis and Modern C++*, Computational Biology 33, https://doi.org/10.1007/978-3-030-90990-1_5

all modules but to different degrees, and I select examples that I feel demonstrate the new designs and programming techniques well. Preference is also given to those parts of the library where my own influence has been strongest. In a few places code snippets have been taken directly from the API documentation, see also Sect. A.3.1.

5.1   Library Structure

5.1.1   Files and Directories

There are multiple ways to organise source code and/or to express structure. The most common one is the hierarchy of files and directories, i.e. the physical organisation of the source code. Being a header library, SeqAn3 consists only of header files that are all contained within `include/seqan3/`. A versioned include path allows coexistence with SeqAn2 on the same filesystem.[1] The headers have the extension `.hpp` and not `.h` (as in SeqAn2) to underline that they contain C++ code and not C code. This is particularly common among header-only libraries (e.g. Boost) where there is no accompanying `.cpp`-files that could make this distinction.

The source code is subdivided into directories and sub-directories (but not sub-sub-directories). Only directories called `detail` may be added to sub-directories and thus introduce a third level in the directory-hierarchy. No headers other than `version.hpp` may appear at the root of the include-directory, i.e. every other header must be in a directory or sub-directory. Every directory or sub-directory (other than those called `detail`) provides an `all.hpp`-file that includes all other headers in that directory. These are not used within the library itself but allow convenient access to large parts of the library from small snippets. On the other hand, every header is also self-contained (i.e. includes all other headers it requires), so it can be included individually which may reduce compile-times.

5.1.2   Modules and Submodules

SeqAn3 is *logically* subdivided into *modules* (see Table 5.1) and *submodules*. These correspond to the directory structure almost 1-to-1. Having such a similar logical structure (also represented in the API documentation) improves the usability of the library, because users immediately know which headers to include and where to find source code should they want to look at the implementation. This was not given for SeqAn2 which mainly organised the API documentation around several "use-cases". SeqAn2 also did not have any "submodules", instead relying on a very large number of modules (49) with 2-90 header files each.

---

[1] SeqAn1 and SeqAn2 cannot coexist, because they both install into `include/seqan/`.

Table 5.1 Overview of the modules of SeqAn3, submodules are not shown

Module	Description	Discussed in
Alignment	Pairwise and multiple sequence alignment.	Chapter 10
Alphabet	Biological and related alphabets.	Chapter 6
Contrib	Imported third party code.	This chapter
Core	Library code shared by multiple modules.	This chapter
I/O	Input/output related tools and file formats.	Chapter 8
Range	General-purpose and alphabet-specific containers & views.	Chapter 7
Search	Data structures and algorithms for indexed search.	Chapter 9
STD	C++ 20 standard library emulation layer.	This chapter
Utility	Stand-alone utility code used by multiple modules.	This chapter

The only directories that do not represent modules/submodules are those called `detail`. They may be found in any module/submodule and the contained headers and names are considered part of that "parent" module/submodule. Header files that only provide names in the `seqan3::detail::` namespace (see below) are required to be placed in a detail folder or have "detail" in their name.

While the directory names are always lowercase/"snake_case", module/submodule names are usually given in "Sentence case" in the API documentation. When a module-, directory- or filename can be either singular or plural (e.g. "Alphabet" or "Alphabets"), SeqAn3 prefers singular. The only exception is strongly established terms like "type_traits" and the *Views* submodule, because the latter is named in accordance with the `seqan3::views::` namespace which in turn is modelled after the `std::views::` namespace. These measures help reduce surprises as are currently found in the range-v3 library where folder and namespace names often diverge in numerus.

Note that "SeqAn3 modules" do not (yet) represent C++ 20 Modules (Sect. 3.9). However, if support for the latter is added in a future release, I expect these to map exactly onto the current modules.

The module structure also allows defining rules for the inter-dependence of the modules:

Tier 0 modules   Core, Utility, Contrib and STD may each not depend on any module but themselves.

Tier 1 modules   Alphabet may depend on Tier 0 modules; Range may depend on Alphabet and Tier 0 modules.

Tier 2 modules   Alignment, I/O and Search may depend on any Tier 0 or Tier 1 modules but not on each other.

Enforcing these rules simplifies the include-graph and reduces compile-time. It significantly improves maintainability of the library, because "small" issues appear as such and do not cause failures in the entire test suite.

5.1.3   Names and Namespaces

Another form of code structure is given by C++ namespaces. Using namespaces is important to avoid name clashes between libraries and it helps differentiate names provided by libraries from those defined locally by the application. Namespaces also assist tooling (e.g. associating function calls with the correct definitions) and make it easier to refactor code at a later point in time. SeqAn3 uses the versioned top-level namespace `seqan3::` which means that it can be used in the same application with SeqAn2.[2]

While certain C++ projects create a hierarchy of namespaces that corresponds to the module/submodule structure of the project, it was decided to use the more common C++ approach for SeqAn3 where most names defined by the library are in the main namespace. Additional namespaces are only created for one of the following reasons:

1. To hide names from the user that are not meant to be part of the API. This is primarily achieved through the namespace `seqan3::detail::` and its sub-namespaces (not shown in Table 5.2), but the namespace `seqan3::contrib::` is also hidden from the API.
2. To prevent naming conflicts/collisions within the main namespace. This is true for most other namespaces, e.g. `seqan3::views::` helps distinguish view adaptors from algorithms or other function objects of the same name.[3]

Table 5.2 Namespace overview

Namespace	API	Description
`seqan3::`	✓	Main SeqAn3 namespace; most things are here.
`seqan3::align_cfg::`	✓	For config elements in alignment, see Sect. 10.4.
`seqan3::contrib::`		All entities defined in the *Contrib* module.
`seqan3::custom::`	✓	"Upload space" searched by customisation point objects.
`seqan3::detail::`		Implementation detail, auxiliary functions, etc.
`seqan3::list_traits::`	✓	Traits on `seqan3::type_list`.
`seqan3::literals::`	✓	Custom literal definitions of the library [inline].
`seqan3::pack_traits::`	✓	Traits on parameter packs.
`seqan3::search_cfg::`	✓	For config elements in the search, see Sect. 9.5.
`seqan3::views::`	✓	View adaptor objects are defined here.

---

[2] Again, this was not possible for SeqAn1 and SeqAn2 which both used `seqan::`.

[3] An example: `seqan3::complement` is a function object that returns the complement of a single nucleotide object, and `seqan3::views::complement` is an adaptor that applies this operation to a range of nucleotide elements.

One exception to this rule is `seqan3::custom::` which specifically exists for customisation purposes (see Sect. 3.7). Users may not explicitly specialise templates in any other (sub-)namespace of SeqAn3.

Another exception is `seqan3::literals::` which contains all of SeqAn3's user-defined literals (introduced in Sect. 6.1.1 on p. 148), e.g. `'A'_dna4` (user-defined char literal) or `''AGATTA''_dna4` (user-defined string literal). It is an `inline` namespace which means that names defined inside appear as being in the top-level namespace `seqan3::` for all regular purposes. But, importantly, users can import the literals-namespace without importing the rest of SeqAn3. So while I strongly discourage to do `using namespace seqan3;`, because it forfeits all of a namespace's advantages described above, doing `using namespace seqan3::literals;` is much less problematic and indeed required to use the literal operators easily. This is in line with the CppCoreGuideline rule SF.7.[4]

The *names* inside SeqAn3's namespaces (i.e. names of functions, class types, templates, etc.) follow `snake_case` naming convention and not `camelCase` like in SeqAn2. This is a big (visual) change, but it brings SeqAn3 closer to the standard library, the Boost libraries, the SDSL and many other important projects.

5.2 "Small" Modules

5.2.1 Argument Parser

SeqAn2 shipped an argument parser that was quite versatile and offered the following features beyond argument parsing:

- Auto-generation of program help-page.
- Auto-generation of manual pages.
- Option to export program interface description in CTD format.
- Update-checks and telemetry.

For SeqAn3 it was long unclear whether an argument parser should be provided with the library. On the one hand, argument parsing is necessary in almost all applications and the features listed above were highly regarded by SeqAn2-users. On the other hand, it is a very common task and—except the CTD support—not particularly "bioinformatical". Hence, it was considered a violation of the *compactness* goal to include such a general-purpose module in SeqAn3. A completely modernised argument parser was initially shipped with SeqAn-3.0 but will be split into a stand-alone library before the 3.1-release.

---

[4] https://github.com/isocpp/CppCoreGuidelines/blob/master/CppCoreGuidelines.md#sf7-dont-write-using-namespace-at-global-scope-in-a-header-file.

```
   int main(int argc, char const ** argv)
2  {
       // Setup ArgumentParser.
4      seqan::ArgumentParser parser("modify_string");

6      addArgument(parser, seqan::ArgParseArgument(seqan::ArgParseArgument::STRING, "TEXT"));

8      addOption(parser, seqan::ArgParseOption("i", "period", "Period to use for the index.",
                                               seqan::ArgParseArgument::INTEGER, "INT"));
10     addOption(parser, seqan::ArgParseOption("U", "uppercase", "Select to-uppercase as operation."));

12     // Parse command line.
       seqan::ArgumentParser::ParseResult res = seqan::parse(parser, argc, argv);
14
       // If parsing was not successful then exit with code 1 if there were errors.
16     // Otherwise, exit with code 0 (e.g. help was printed).
       if (res != seqan::ArgumentParser::PARSE_OK)
18         return res == seqan::ArgumentParser::PARSE_ERROR;

20     // Extract option values and print them.
       unsigned period = 0;
22     getOptionValue(period, parser, "period");
       bool toUppercase = isSet(parser, "uppercase");
24     seqan::CharString text;
       getArgumentValue(text, parser, 0);
26
       std::cout << "period   \t" << period << '\n'
28               << "uppercase\t" << toUppercase << '\n'
                 << "text     \t" << text << '\n';
30     return 0;
   }
```

Code snippet 5.1: Argument parsing in SeqAn2

```
   int main(int argc, char const ** argv)
2  {
       /* Setup options */
4      uint32_t period = 0;
       bool toUppercase = false;
6      std::string text;

8      /* Setup argument parser */
       seqan3::argument_parser parser{"modify_string", argc, argv};
10
       parser.add_option(period, 'i', "period", "Period to use for the index.");
12     parser.add_flag(toUppercase, 'U', "uppercase", "Select to-uppercase as operation.");
       parser.add_positional_option(text, "The text to be transformed.");
14
       /* Run parser */
16     parser.parse();

18     /* Rest of program */
       std::cout << "period   \t" << period << '\n'
20               << "uppercase\t" << toUppercase << '\n'
                 << "text     \t" << text << '\n';
22     return 0;
   }
```

Code snippet 5.2: Argument parsing in SeqAn3

As such, I will only give a very brief demonstration of its features here. Snippet 5.1 shows the introductory example of the argument parser in SeqAn2 and Snippet 5.2 is the same code transformed to SeqAn3's argument parser. The most important structural change is that there is no separate "extraction phase" after parsing, variables are passed to the argument parser immediately when options/flags are added. This allows the argument parser to also deduce the type of these options which was previously done via extra `enum` arguments. It further enables setting a default value based on the value of that variable (in contrast to handling default values via extra arguments). Both were previous sources of error, because users would set conflicting types or define default values in the wrong place. It is also no longer possible to forget extracting a value, because the target variable needs to be given on definition of the option/flag.

Errors (e.g. a provided value that is out-of-range) are now expressed as C++ exceptions instead of return values. Parser-formats that lead to intended early termination do so via `std::exit()`.[5] This makes using the parser much less verbose.

Policy-wise, the argument parser now adheres to POSIX conventions[6] and implements popular GNU extensions.[7] This includes the following features not available in SeqAn2:

- Flags can be written en-bloc: `-xzf` is equivalent to `-x -z -f`.
- Options have short (single dash + single-character) and long (double-dash + string) specifiers.[8] These are equivalent: `-s TEST`, `-sTEST`, `-s=TEST`, `--str TEST` and `--str=TEST`.

Releasing the argument parser as a separate library will make it available to users not interested in SeqAn and will on the other hand ensure that SeqAn3 only picks up features that are necessary in the context of bioinformatics. This is important as software increasingly depends on multiple libraries and users may already employ a different mechanism for argument parsing in their project.

5.2.2 The Core Module

The *Core* module provides utilities that are strongly coupled to SeqAn3 and required by more than one module. See Table 5.3 for an overview. Most are not relevant to average users, but advanced users may benefit from having access to e.g. the facilities provided by the *Parallel* and *SIMD* submodules. Currently, these submodules are only used internally within SeqAn3 algorithms (alignment and search), but they will become part of the API in a future release, because more sophisticated setups (e.g. distributed computing and device offloading) require exposing them to the user.

---

[5] Parser-formats like the help-page printer or manual creator should not result in regular program execution, they quit the program after being run.

[6] http://booksonline.nl/tutorial/essential/attributes/_posix.html.

[7] https://www.gnu.org/software/libc/manual/html_node/Argument-Syntax.html.

[8] In SeqAn2, both could be strings.

Table 5.3 Core module overview

Core module	
Submodules	Algorithm cfg, Debug stream, Parallel, SIMD
Important headers	`platform.hpp`

The *Algorithm cfg* submodule provides base classes and operators for the algorithm configuration system that is also used by alignment and search. It is introduced in detail together with the search in Sect. 9.3.

```
   // The alphabet normally needs to be converted to char explicitly:
2  std::cout << seqan3::to_char('C'_dna5);              // prints 'C'

4  // The debug_stream, on the other hand, does this automatically:
   seqan3::debug_stream << 'C'_dna5;                    // prints 'C'
6
   // The debug_stream can also print vectors which std::cout cannot:
8  std::vector vec = "ACGT"_dna5;
   seqan3::debug_stream << vec;                         // prints "ACGT"
```

Code snippet 5.3: The Debug stream is primarily used for debugging (hence the name) but may be used for general console output, as well. The underlying stream object can be made to point to `std::cout`, `std::cerr` or a custom output stream

The *Debug stream* submodule is the only part of Core with larger user exposure. It provides an output stream similar to `std::cout` that is able to print many types typically not printable by `std::cout` or `std::cerr`. This includes many standard library types like `std::vector` (in fact, any input range), but also SeqAn3 types like alignments can be printed descriptively. The submodule provides the stream object for printing and various overloads for generic concepts and standard library types. Overloads for specific SeqAn3 types are usually provided together with those.

The *Core* module provides many more resources that are shared between SeqAn3 modules but that are not relevant to end-users. They are thus located in a detail directory and placed in namespace `seqan3::detail::`.

A noteworthy header file in the *Core* module is `platform.hpp`. It checks whether the compiler supports all necessary C++ features, whether all required dependencies are found (in the correct versions), and it defines certain platform specific macros. The header is required to be included by every other header in the library (directly or indirectly).[9] This ensures that readable diagnostics are printed when a misconfigured system tries to use SeqAn3.

---

[9] This is enforced by the library test suite.

Table 5.4 Utility module
overview

Utility module	
Submodules	Char operation, Tuple, Type list, Type traits

5.2.3 The Utility Module

The *Utility* module is similar to Core but contains general-purpose utilities with no direct connection to bioinformatics.[10] See Table 5.4 for an overview. They could well be defined by an external library or a future version of the standard library, but they are provided in this module since multiple other SeqAn3 modules require them.

The *Char operation* and *Type list* submodules are briefly introduced below. The *Tuple* submodule provides two custom tuple types and defines a concept that encompasses all tuple types (`std::pair`, `std::tuple` and multiple SeqAn3 types). It also provides utilities for working with tuples, e.g. a function to split a tuple into two. The *Type traits* submodule contains various type traits that are used in metaprogramming. Most of these are not public but some are, e.g. `seqan3::is_constexpr_default_constructible` which behaves like `std::is_default_constructible` but additionally checks that the default constructor is `constexpr`-qualified.

Operations on Characters

The *Char operation* submodule reimplements certain standard library features specific to the type `char`, but with better performance:

1. Transformations: `seqan3::to_lower(c)` and `seqan3::to_upper(c)`
2. Predicates that check if a character falls into certain sub-ranges of the ASCII code (Table 5.5).

When standard library utilities do not perform optimally, it is a difficult decision whether one should reimplement them or not, because obviously one needs to violate either the design goal of *performance* or *compactness*. In this case, I decided to reimplement them, because the given transformations and predicates are used heavily in input/output and the performance difference is very notable (see Table 5.6).

SeqAn3's functors achieve a superior performance by being implemented as tables, i.e. every functor internally holds a `static constexpr std::array<bool, 256>` that indicates which elements of the ASCII code are in the respective set. These tables are not hard-coded, instead they are generated by a constant expression

---

[10] Splitting the *Core* module into Core and Utility was not yet performed in the master branch at the time of writing.

Table 5.5 Char predicates in the standard library, in SeqAn3 and in SeqAn2. Please see the standard library documentation for definitions of which characters exactly are satisfied by which predicate

std::	seqan3::	seqan::
iscntrl	is_cntrl	
isprint	is_print	
isspace	is_space	IsWhitespace
isblank	is_blank	IsBlank
isgraph	is_graph	IsGraph
ispunct	is_punct	
isalnum	is_alnum	IsAlphaNum
isalpha	is_alpha	IsAlpha
isupper	is_upper	
islower	is_lower	
isdigit	is_digit	IsDigit
isxdigit	is_xdigit	
	is_char<'X'>	EqualsChar<'X'>
	is_in_interval<'A', 'Z'>	IsInRange<'A', 'Z'>
	is_eof	

at compile-time (if the respective codepaths are used by the user). The `operator()` only looks up the value in this table—an operation that has minimal overhead.

Furthermore, the functors in SeqAn3 can be combined via `||` or negated via `!` which will create a new functor with a new table **at compile-time**. Evaluating such a new functor is also a single table lookup, i.e. in SeqAn3 n chained adaptors can be evaluated in $O(1)$ whereas SeqAn2 and the standard library require $O(n)$.

To reduce the added complexity of a custom implementation, care has been given to use exactly the same names and definitions for SeqAn3 functors as the standard library uses for its functions.[11] The only exception is that SeqAn3 names contain an underscore to satisfy the naming requirements (and improve readability).

Excursus: Variadic Templates, Parameter Packs and Folds

Before type lists are covered, I want to briefly introduce variadic templates (ISO/IEC 14882:2017, 17.5.3) and fold expressions (ISO/IEC 14882:2017, 8.1.6). The former is available since C++ 11, the latter were added in C++ 17.

Template parameter packs are template parameters that accept zero or more template arguments, and *function parameter packs* are function parameters that accept zero or more function arguments (ISO/IEC 14882:2017, 17.5.3). On the left side of Snippet 5.4 the definition of a simple tuple-type is shown. The base

---

[11] The definitions can be found here: https://en.cppreference.com/w/cpp/string/byte.

```
     template <typename ...types>                  template <typename ...types>
 2   struct tuple                              2   auto sum_all(types && ...values)
     {};                                           {
 4                                             4       return (0 + ... + values);
     template <typename type0>                     }
 6   struct tuple<type0>    // recursion anchor 6
     {                                             auto a0 = sum_all(42, 23);     // == 65 (int)
 8       type0 _head;                          8   auto a1 = sum_all(1.3, 3, -4); // == 0.3(double)
     };
10                                            10   template <typename ...types>
     template <typename type0, typename ...types>  auto sum_squares(types && ...values)
12   struct tuple<type0, types...>            12   {
     {                                                 return sum_all(values * values...);
14       type0          _head;               14   }
         tuple<types...> _tail;
16   };                                       16   auto s1 = sum_squares(1, 2, 3); // ==  14 (int)
                                                   auto s2 = sum_squares(0.5, 1);  // == 1.25(double)
```

Code snippet 5.4: Variadic templates and folding. The left shows the definition of a custom tuple-type. The right shows folding based on `operator+` (top) and pack expansion with element-wise multiplication (bottom)

template takes a template parameter pack denoted by `...` and the actual definition happens recursively via two specialisations. l. 15 shows an *expansion* of the template parameter pack into individual arguments that are passed to the definition of the `_tail` member.

On the right side of Snippet 5.4 two function templates are shown that both take template parameter pack and a function parameter pack. The types in the template parameter pack are automatically deduced from the types of the arguments in the function parameter pack. l. 13 shows an expansion of the pack that also performs the squaring of the elements. In general, element-wise operations can be performed during pack expansions (e.g. arithmetic operations, function calls), but the number of elements remains the same. Similar in appearance is the *fold* displayed in l. 4. This is not an element-wise operation; it sums up all elements of the pack (an initial value of `0` is only provided so that the function call is also valid for empty parameter packs).

These features are very powerful and allow for functional-style programming that is much simpler to read and write. Notably, it is also easier to process for the compiler and reduces compile-time compared to more sophisticated template metaprogramming used previously.

Type Lists

Working with lists of types is one of the areas notably absent from the standard library and although I do not expect most of SeqAn's users to need this feature, it is used internally frequently and the interfaces are simple and stable enough to expose them. SeqAn3 originally used the external *meta*-library[12] for this purpose but later switched to custom solutions.

This decision was again a trade-off as *compactness* would have suggested relying on third party library, but because *meta* relied on C++ 11 it was significantly less *simple* to use than what was achievable with C++ 17 and concepts. Moreover, SeqAn3 required only a subset of meta's features, so it was decided to implement this code "in-house".

```
namespace seqan3::detail
2 {
template <ptrdiff_t idx, typename head_t, typename ...tail_t>
4 auto at()
{
6     if constexpr (idx == 0)
          return std::type_identity<head_t>{};
8     else
          return at<idx - 1, tail_t...>();
10 }
}

12
namespace seqan3::pack_traits
14 {
template <ptrdiff_t idx, typename ...pack_t>
16     requires (idx >= 0 && idx < sizeof...(pack_t))
using at = typename decltype(detail::at<idx, pack_t...>())::type;
18 }
```

```
  //    first argument is the index | rest is the pack to be searched
2 using t = seqan3::pack_traits::at<2, bool, double, int, long>;
  // t is int
```

Code snippet 5.5: Traits of type lists. This example shows the implementation of the `at<>` trait that works like `[]` on a type list or pack. First is a function definition that performs recursion on the parameter pack; then comes the actual definition of the trait as an alias template. The smaller snippet at the end shows how this trait can be used

Working with C++ 17 allowed me to use a metaprogramming style popularised by Hana Dusíková, author of the compile-time regular expression library (CTRE)[13] which is also proposed for standardisation in C++ 23 (Dusíková, 2019). It relies on `constexpr` functions for compile-time computations (not novel) and *function signatures* for transformation traits (rather novel, although possible since C++ 11).

An example of this style is shown in Snippet 5.5. Perhaps surprisingly, the function template is never *evaluated*, not even at compile-time (it is not marked `constexpr`!). The template is only *instantiated* during compile-time to determine its signature, i.e. the compiler's rules for the deduction of the return type are what results in the desired semantics. Wrapping the returned type in `std::type_identity<>` is required to prevent possible `const`-qualification from

---

[12] https://github.com/ericniebler/meta.

[13] https://github.com/hanickadot/compile-time-regular-expressions.

being lost[14] and to ensure that default construction is always valid. The trait, which is defined as an alias template, unwraps the type identity and presents the actual type. Implementing the trait as an alias template means that it can be used without the infamous `typename foo<T>::type`.

The *Type list* submodule provides an empty class template that encodes a type list through its template parameters:

```
template <typename ...element_ts> struct type_list {};
```

It then defines many traits on type lists (in namespace `seqan3::list_traits::`) and also on template parameter packs (in namespace `seqan3::pack_traits::`). The latter allows using many traits directly without additionally instantiating the list type beforehand. Traits defined include the following (each in namespace `seqan3::list_traits::`, for packs in the equivalent namespace):

Traits that return a value: `size<list_t>`, `count<q_t, list_t>`, `find<q_t, list_t>`,...

Traits that return a type: `at<i, list_t>`, `front<list_t>`, `back<list_t>`

Traits that return a type list: `concat<list1_t, list2_t, ... >`, `drop<i, list_t>`, `take<i, list_t>`,...

All of these directly return values/types; they do not need `_t` or `_v` shortcuts to access `::type` or `::value` members. The very common names should also highlight why extra namespaces are required.

5.2.4 The STD Module

As explained in Sect. 4.4.1, SeqAn3 relies on many parts of C++ 20's standard library that are not available with GCC7, GCC8 and GCC9, the only currently supported compilers. Since these components are not just used internally by SeqAn3 but are crucial to using SeqAn3's algorithms and data structures, it is necessary to make them available to SeqAn's users in some way. The most obvious solutions are:

1. Reimplement or import them from an existing codebase (e.g. LLVM standard library or stand-alone range-v3)[15] and expose them as part of SeqAn3's namespace.
2. Rely on a third party library like range-v3 and teach SeqAn3 users to use that library's facilities and consult that library's API documentation.

The first approach is *simpler* for SeqAn3 users but makes the library significantly less *compact*. For the second approach it is the other way around, *simplicity* is

---

[14] `const` on non-reference return values is always ignored by the language.

[15] Both libraries' licence is compatible with SeqAn3's.

considerably reduced due to the overhead of learning an additional library and, in particular, due to the poor state of range-v3's documentation.

Both solutions seemed far from optimal considering that the facilities in question were exactly those that are novel to SeqAn3's users and likely the most difficult to teach. Moreover, I expected that most compilers released in 2020 and later would ship these components, so most developers adopting SeqAn3 after its first stable release would have access to "official implementations" and any workarounds for GCC≤ 9 would only be required in deployment. It would thus be a shame to burden the library for years to come with significant amounts of duplicated code or a complicated third party dependency that many users would not even need.

The solution to this problem is the *STD* module which contains several headers named exactly like standard library headers, e.g. `<seqan3/std/type_traits>` and `<seqan3/std/ranges>`. These headers include the respective standard library headers and evaluate *feature test macros*[16] to detect whether the required C++ 20 features are provided by the header; if not, they include relevant headers from the range-v3 library and alias the respective functions/classes/objects into namespace `std::`/`std::ranges::`/`std::views::`. Since *range-v3* provides dedicated and stable namespaces with C++ 20 entities, these namespaces can even be imported as a whole.

In effect, users and developers of SeqAn3 can always use the C++ 20 entities as if they were provided by the standard library; in the case where they are not, they resolve to fallback implementations from range-v3 or custom implementations (e.g. copied from the LLVM standard library).

This is a slight violation of the C++ -standard which forbids adding names to namespace `std::` (to avoid collisions and undefined behaviour), but under the circumstances it is the best solution with respect to the design goals. The feature test macros prevent any collisions and as additional protection against naming conflicts, any aliases defined by the *STD* module are not placed directly into `std::`, `std::ranges::` or `std::views::` but into *anonymous* sub-namespaces[17] within these namespaces. This has the effect that in the unlikely case that a name is imported into `std::` that is previously defined there by the standard library (the feature test macro returned a wrong value), the name already in `std::` is given precedence over the alias, and there is no ambiguity.

5.2.5   The Contrib Module

The *Contrib* module has also been introduced briefly in Sect. 4.4.1. It contains small portions of code directly imported from other projects. This happens only when it is not desirable to depend on the third party project as a whole but when it is also not

---

[16] https://isocpp.org/std/standing-documents/sd-6-sg10-feature-test-recommendations.

[17] https://en.cppreference.com/w/cpp/language/namespace#Unnamed_namespaces.

feasible to reimplement the respective code. The module is rarely used as a staging area when it is planned to later reimplement a feature inside the library.

Code within this module must satisfy SeqAn3's licence and header-only requirements but need not adhere to style-guides and is not part of SeqAn3's API (and also does not appear in the documentation). It is imported into the namespace `seqan3::contrib::` to prevent conflicts with system-wide instances of the same code.

Currently, this module only contains SeqAn2 code for stream compression and decompression (GZip, BZip2, BGZF), see also Sect. 4.4.1. It is likely that this code will at some point be reformatted and included into SeqAn3's *I/O* module or be replaced by an external library.

5.3   Discussion

This chapter laid the practical foundation for the implementation of SeqAn3 and it introduced several small modules that are part of this foundation.

5.3.1   *Performance*

Since this chapter deals primarily with utilities and auxiliary data structures and functions, most of it is not relevant for performance. One exception are the character predicates that are reimplemented in SeqAn. Table 5.6 shows benchmark results between the standard library, SeqAn3 and SeqAn2. And it highlights why it is right in this case to perform this reimplementation and "micro-optimisation".

For a single invocation of a predicate, SeqAn3 is already faster than the standard library by factor of 7x. For a combined invocation of three predicates, SeqAn3 is 13x faster if properly invoked, because it is not more expensive to invoke multiple combined predicates in SeqAn3. Even if the three predicates are invoked individually in SeqAn3, the speed-up is almost 9x.

Table 5.6 Char predicate benchmarks. See Snippet 5.7 for implementations of the measured calls. "Single" means a single predicate check per iteration; "Three individual" means three single checks that are ||-ed. "Three combined" means that the three predicates are combined in some way prior to performing the check

	STL	SeqAn3	SeqAn2
Single	7.75ns	1.13ns	1.76ns
Three individual	15.10ns	1.68ns	2.12ns
Three combined	–	1.13ns	2.12ns

SeqAn2 already provided a measurable speed-up over the standard library. A possible explanation is that the standard library calls are not properly `inlined` by the compiler. However, due to performing equality checks and not table-lookups, SeqAn2 is still slower than SeqAn3. Furthermore, SeqAn2's runtime increases with the number of combined predicates even if a new type is created statically that encompasses multiple predicates. For three combined predicates the speed-up of SeqAn3 over SeqAn2 is almost 2x.

5.3.2 Simplicity

Many of the structural simplifications of SeqAn3 have been introduced in this chapter. These include naming conventions that are well-defined and more strongly enforced resulting in fewer surprises. By switching to `snake_case` they have also become more similar to the standard library and other popular C++ libraries. Furthermore, the module structure has come to include submodules which enables the library to keep related content in closer proximity. This is one aspect of reducing the total number of (user-visible) top-level modules from 49 to 8. The way the "small modules" are divided clearly states their role and simplifies maintaining them, however, most users of SeqAn3 will not use these modules often.

Where SeqAn2 had only namespace `seqan::`, SeqAn3 has distinct namespaces for certain purposes. Most importantly it provides `seqan::detail::` which hides many names from the user. This differentiation makes it simpler to understand the library and where (not) to look for solutions. It makes maintaining the library easier, because it is obvious which parts of the library must remain stable and which may change.

```
  #include <seqan3/std/algorithm>
2 #include <seqan3/std/ranges>

4 void foo(std::ranges::random_access_range auto && range)
  {
6     std::ranges::sort(range);
  }
```

Code snippet 5.6: Using the *STD* module. The definition of the concept and algorithm come from the standard library if possible, otherwise from *range-v3*. Users that only target C++20 compilers can simply replace `<seqan3/std/algorithm>` with `<algorithm>` (and ranges respectively)

The way in which the *STD* module is defined is simpler than the alternatives, because it means users can use names and documentation from the standard library (whether the implementation originates there or not). This reduces the burden for developers knowledgeable of the C++20 standard library and enables inexperienced users to rely on standard tutorials and other widely available material.

```
   /* Standard library, std:: assumed*/
2  bool b0 = isalpha(c);
   bool b1 = isalpha(c) || isblank(c) || isdigit(c);        // individual
```

```
   /* SeqAn3, seqan3:: assumed */
2  bool b0 = is_alpha(c);
   bool b1 = is_alpha(c) || is_blank(c) || is_digit(c);     // individual
4  bool b2 = (is_alpha || is_blank || is_digit)(c);         // combined
```

```
   /* SeqAn2, seqan:: assumed */
2  bool b0 = IsAlpha{}(c);
   bool b1 = IsAlpha{}(c) || IsBlank{}(c) || IsDigit{}(c);  // individual
4  bool b2 = OrFunctor<OrFunctor<IsAlpha, IsBlank>, IsDigit>{}(c);  // combined
```

Code snippet 5.7: Usability of character predicates. `char c;` assumed before the snippets. SeqAn3's predicates are used and named almost identically to the standard library's. Combining SeqAn3's functors into a new (faster) one is very simple. SeqAn2 defines functor *types* that need be constructed via `{}` before usable. Combining SeqAn2's functor types involves manual metaprogramming and there is no speed advantage

Earlier in this chapter I explained that character predicates were reimplemented for performance reasons and provided benchmark results in the last subsection. Snippet 5.7 shows that using SeqAn3's character predicates is also much *simpler* than using SeqAn2's. Combining multiple predicates into one (which happens at compile-time and enables $O(1)$ evaluation of multiple predicates) is achieved by simply putting parentheses around the functors and supplying the argument afterwards (line 4, second snippet in Snippet 5.7).[18]

```
   /* Meta library */
2  struct to_value_t
   {
4      template <typename T>
       using invoke = std::ranges::range_value_l<T>;
6  };

8  using input_list_t = meta::list<std::vector<int>, std::list<double>>;
   using output_list_t = meta::transform<input_list_t, to_value_t>;
10 // == meta::list<int, double>
```

```
   /* SeqAn3*/
2  using input_list_t = seqan3::type_list<std::vector<int>, std::list<double>>;
   using output_list_t = seqan3::list_traits::transform<std::ranges::range_value_t, input_list_t>;
4  // == seqan3::type_list<int, double>
```

Code snippet 5.8: Usability of SeqAn3 type lists. The type lists and respective traits in SeqAn3 are designed to directly work with alias templates and do not require defining extra "invocable" types like for the *meta*-library

---

[18] The semantics of the functor are encoded in its type and a custom `operator||` generates the combined type.

Another feature introduced in this chapter are type lists and traits on them. Snippet 5.8 illustrates how it is *simpler* to use these instead of the *meta*-library. This simplification may seem trivial, but in more elaborate metaprogramming contexts these kinds of simplifications have a large impact on the maintainability of the codebase.

5.3.3 Integration

Many of the structural decisions presented in this chapter allow for easy integration of SeqAn2 and SeqAn3. I consider this very important, because SeqAn3 does not (yet) cover all use-cases of SeqAn2 and at least some applications will want to depend on both.

Separating *core* and *utility* will make it easier to integrate the utilities individually with other applications or move them to stand-alone libraries if there is the demand. For the argument parser this is already being done, allowing for it to be integrated into other projects independent of SeqAn3.

The design of the *STD* module allows for the best possible integration with the standard library, because the standard library *is* being used when possible—and if not it appears as if. Using `snake_case` throughout SeqAn3 makes it appear much closer in style to the standard library which hopefully also underlines that these should be used together.

5.3.4 Adaptability

The design for character predicates in the *utilities* module is an example for improved adaptability. Because combining predicates does not increase runtime, it is now possible to easily create new predicates that do not exhibit any overhead, e.g. `auto is_fob = is_char<'F'> || is_char<'O'> || is_char<'B'>;` and then using `is_fob(c)` to check characters. Working with objects of deduced type here is also much more expressive than working with combined types like in SeqAn2.

5.3.5 Compactness

Deciding to move the *argument parser* into a separate library is a prime example of SeqAn3's focus on compactness. While two examples given in this chapter (character predicates and list traits) might seem like preference is not given to this design goal, I want to point out that the overall size of utility code in SeqAn3 is

Table 5.7 Code sizes of utility code

	Lines of code	Lines of comments
SeqAn3's small modules[†]	5319	6694
SeqAn2's *basic* module[††]	15,933	13,206
SeqAn3's type list subm.	247	568
meta library	2488	995

[†] Without *contrib* as streaming code is not utility code (also separate in SeqAn2)

[††] Without alphabet-specific code since this is separate in SeqAn3

much lower than in SeqAn2. This can also be seen in Table 5.7 which shows a 3x larger utility codebase for comparable functionality in SeqAn2.

The solution found for the *STD* module is a large factor in keeping the utility code as compact as it is.

Table 5.7 also highlights that while adding the *type list* submodule with its traits to SeqAn3 increases the code size, this is only a tenth of the code that would have been imported by adding the meta library to the *Contrib* module. Interestingly, the *type list* submodule achieves a very relevant subset of *meta*'s features with much less code.

Chapter 6
The Alphabet Module

Biological sequence data is at the heart of sequence analysis and each sequence is composed of individual *letters*. The type of these letters is called an *alphabet*. Letters can be in a finite, non-empty amount of states (*values*). A typical example is the DNA alphabet which encompasses the values A, C, G, T. Similar to the DNA alphabet, many alphabet types directly represent biological compounds, but others represent associated data like base-pair quality information or stronger abstractions like the role of a compound in a 2D or 3D structure.

While alphabets are a seemingly simple thing, their ubiquity in all sequence analysis software demands that special care be taken in their design. The simplicity of the underlying matter makes the *Alphabet* module a good blueprint for the rest of the library and helps illustrate the overall design choices well. Therefore, I will spend more time on this module than on most of the others. See Table 6.1 for an overview.

Submodules in the *Alphabet* module include:

Adaptation Provides alphabet adaptations of some standard character and unsigned integer types.

Aminoacid Provides the amino acid alphabets and functionality for translation from nucleotide.

CIGAR Provides (semi)alphabets for representing elements in CIGAR strings.

Composite Provides templates for combining existing alphabets into new alphabet types.

Gap Provides the gap alphabet and functionality to make an alphabet a gapped alphabet.

Mask Provides the mask alphabet and functionality for creating masked composites.

Nucleotide Provides the different DNA and RNA alphabet types.

Quality Provides the various quality score types.

Structure Provides types to represent single elements of RNA and protein structures.

© The Author(s), under exclusive license to Springer Nature Switzerland AG 2022
H. Hauswedell, *Sequence Analysis and Modern C++*, Computational Biology 33,
https://doi.org/10.1007/978-3-030-90990-1_6

Table 6.1 Alphabet module overview

Alphabet module	
Submodules	Adaptation, Aminoacid, CIGAR, Composite, Gap, Mask, Nucleotide, Quality, Structure
Concepts	`seqan3::semialphabet,` `seqan3::writable_semialphabet,` `seqan3::alphabet,` `seqan3::writable_alphabet`
Class types	`seqan3::alphabet_base`
Traits	`seqan3::alphabet_char_t,` `seqan3::alphabet_rank_t,` `seqan3::alphabet_size`
Function objects	`seqan3::assign_char_strictly_to,` `seqan3::assign_char_to,` `seqan3::assign_rank_to,` `seqan3::char_is_valid_for,` `seqan3::to_char,` `seqan3::to_rank`

I will first explain the general design (Sect. 6.1) and illustrate this with small user-defined alphabets and adaptations (Sect. 6.2). Subsequently, I will introduce alphabet submodules based on their importance and complexity. Since the first submodules are covered in more detail, the later, more complex submodules should be easier to understand. I begin with nucleotides (Sect. 6.3) and amino acids (Sect. 6.4) and then cover composite alphabets (Sect. 6.5) in whose context the gap and mask submodules are also discussed. Finally, I introduce the quality alphabets (Sect. 6.6). The *CIGAR* and *Structure* submodules are not discussed here since their design is similar to the other modules.

6.1   General Design

The design goals for the alphabet module are part of the general design goals discussed for the library in Chap. 4. But since alphabets are such an integral part of sequence analysis, in particular the design should be easy to use and difficult to misuse. Library developers should furthermore be able to quickly define new alphabets so that SeqAn3 offers alphabets for all typical use-cases. For those cases where the library does not have the optimal alphabet, an application developer should be able to easily add one without changing the library.

6.1.1 *Character and Rank Representation*

Since most alphabets could be represented by a simple `char`, I should clarify first why this is a poor design:

1. A `char` is always[1] one byte (it can have 256 different values), but most alphabets are smaller and can be represented by less than a byte under certain circumstances. Moreover, it is essential to know the actual size of an alphabet at compile-time in many situations (optimisation of algorithms, size of conversion tables etc.).
2. C++ is a statically typed language and *type safety* is an important aspect of this, e.g. an object that represents a nucleotide should not be usable as an amino acid because that would be a source of bugs.
3. There should be mechanisms to transform invalid values to valid values or prevent invalid values from being assigned to an alphabet. This is especially true when handling unverified user input as it would be difficult to reason about a DNA letter with the value of for example `'!'` later on.

For these reasons, alphabets are distinct types in SeqAn3 and have a fixed size at compile-time.

Independent of the specific type, it is desirable to be able to create alphabet objects from characters, because they are usually provided as characters by user input or user-provided files, e.g. DNA input would typically be given as one of `'A'`, `'C'`, `'G'` or `'T'`. This is the *character representation* of an alphabet.

From an implementer's point of view, it is, however, much more useful to consider alphabet values as numbers between 0 and $alphabet\_size - 1$, because these provide a total order and can be used to index arrays. This is the *rank representation* of an alphabet, because every value of the alphabet is represented by its numeric rank.

```
  dna d;
2 d = 'A';      // assign via char; we expect "letter A"
  d = 1;        // assign via rank; we expect "letter C", but actually assign ASCII code 1 (SOH)
4
  static_cast<int>(d); // does this return rank or character representation?
```

Code snippet 6.1: Ambiguity of characters and integers in C++. *Note that this is not valid SeqAn3 code!*

Since, unfortunately, characters in C and C++ are integral types[2] and all integral types are also implicitly convertible to each other, it is difficult to provide

---

[1] It is even allowed to be larger although it is one byte on all major platforms.

[2] `uint8_t` is indeed the same type as `unsigned char` in most implementations.

interfaces that clearly differentiate between character representation and numerical representation (see Snippet 6.1).

```
   seqan3::dna5 d{};           // seqan3::dna5 can be "letters" A, C, G, T, N
2
   d.assign_char('A');         // assign via char; d is in the state "letter A"
4  d.assign_rank(1);           // assign via rank; d is in the state "letter C"

6  d.to_char();                // convert to char; returns 'C'
   d.to_rank();                // convert to rank; returns 1
8
   // d = 'C';                 // does not work!
10 d = 'C'_dna5;               // this works, see excursus on user-defined literals
```

Code snippet 6.2: Character and rank representations of alphabets. In general, no assignment operator or conversion operators are defined for/to either representation

Alphabets in SeqAn3 thus provide two separate interfaces for character and rank representation; they are not implicitly convertible to/from either. This is visible in Snippet 6.2.

Excursus: User-Defined Literals

The C++ language defines certain literals (ISO/IEC 14882:2017, 5.13) including integer literals (3), character literals ('C') and string literals (''FOO''). C++ 11 adds the so-called *user-defined literals* (ISO/IEC 14882:2017, 5.13.8) that are introduced by a certain suffix following the literal. The standard library for example provides such a user-defined string literal denoted by the suffix s that is of type std::string. User-defined literals defined outside the standard library are required to have their suffix start with an underscore. Snippet 6.3 shows examples of standard language, standard library and SeqAn3 literals. Many literal operators are provided for alphabets in SeqAn3,[3] but this is not required by the concept as they are considered mostly a convenience feature. User-defined literal operators do not

```
   auto c0 = 'A';              // type is 'char'
2  auto s1 = "ACGT";           // type is 'char const *'

4  using namespace std::literals;
   auto s2 = "ACGT"s;          // type is 'std::string'
6
   using namespace seqan3::literals;
8  auto c1 = 'A'_dna5;         // type is seqan3::dna5
   auto s3 = "ACGT"_dna5;      // type is std::vector<seqan3::dna5>
```

Code snippet 6.3: User-defined literals. The first two literals are defined by the C++ language (not "user-defined"), and the third is a "user-defined" literal (although defined by the standard library). The last two are defined by SeqAn3

---

[3] The definition of these operators is not shown here; they simply call assign_char() .

exhibit the previously mentioned confusion between character and integer types.[4] This means that while `dna5 d = 'A';` is not valid, `dna5 d = 'A'_dna5;` is.

6.1.2   Function Objects and Traits

The type `seqan3::dna5` and its member functions `.assign_rank()`, `.assign_char()`, `.to_rank()` and `.to_char()` were introduced in Snippet 6.2. These are part of the type's interface, and if one works with a very specific given type, they are the simplest way to access and modify the type.

However, in generic contexts and especially in contexts that are meant to be used and extended with user-defined types, member functions are not the best choice (see Sect. 3.4.4). Instead, customisation point objects (CPOs) are used that delegate to the actual implementation (see Sect. 3.7).

For a given object, a customisation point typically picks one of the following implementations:

1. A member (function) of the object
2. A free function that accepts the object as argument, found via argument-dependent lookup for example in the namespace where the type of the object is defined
3. A (static) member of the class `seqan3::custom::alphabet<T>`, where `T` is the type of the object

Almost all alphabets in SeqAn3 implement their functionality as members. See Sect. 6.2 for the different ways to satisfy the customisation points and add user-defined alphabets.

The different parts of rank and character interfaces are described below. Examples for all interfaces are given in Table 6.2.

`seqan3::alphabet_size<T>`   holds the size of the alphabet `T`. It is defined as a trait in the form of a variable template, although the underlying implementation is a CPO. Calls to `alphabet_size` are always constant expressions. The type of the variable depends on the alphabet but is required to model `std::integral`.

`seqan3::to_rank(alph)`   returns the rank representation of an alphabet object. It is defined as a function object/CPO. Implementations are required not to throw exceptions and be marked `noexcept`.[5]

`seqan3::alphabet_rank_t<T>`   is a transformation trait that exposes the rank type of the alphabet `T`. It cannot be customised; it is always defined as the type

---

[4] The user-defined literal operator for `char` does not imply the user-defined literal operator for `int`. Thus, `'A'_dna5` works as expected, but `3_dna5` is not valid and there is no ambiguity.

[5] This is enforced by the CPO to make optimisations possible.

Table 6.2 Generic alphabet interfaces and CPOs. Examples are given on the left, the implementation that they resolve to is given in the centre (also valid to call in non-generic contexts), and the result of the expression is on the right. The examples assume that `dna5 d{'C'_dna5};` is defined before each line. Namespace `seqan3::` is assumed

Rank interface	Implementation	Result
`alphabet_size<dna5>`	`dna5::alphabet_size`	5
`to_rank(d)`	`d.to_rank()`	2
`alphabet_rank_t<dna5>`	`decltype(to_rank(dna5{}))`	`uint8_t`
`assign_rank_to(3, d)`	`d.assign_rank(3)`	d (now is letter G)
`assign_rank_to(42, d)`	`d.assign_rank(42)`	Undefined behaviour/assertion
Character interface		
`to_char(d)`	`d.to_char()`	`'C'`
`alphabet_char_t<dna5>`	`decltype(to_char(dna5{}))`	`char`
`char_is_valid_for<dna5>('A')`	`dna5::char_is_valid('A')`	`true`
`char_is_valid_for<dna5>('!')`	`dna5::char_is_valid('!')`	`false`
`assign_char_to('A', d)`	`d.assign_char('A')`	d (now is letter A)
`assign_char_to('!', d)`	`d.assign_char('!')`	d (now is letter N)
`assign_char_strictly_to('A', d)`	`d.assign_char_strictly('A')`	d (now is letter A)
`assign_char_strictly_to('!', d)`	`d.assign_char_strictly('!')`	Exception is thrown

of whatever `seqan3::to_rank()` returns for a parameter of type `T`. For almost all alphabets in SeqAn3, this type is `uint8_t`—the smallest unsigned integer type able to represent all values in rank representation. It is required to model `std::integral`.

`seqan3::assign_rank_to(r, alph)`   changes the value of `alph` by assigning a new rank. It is defined as a function object/CPO. The type of `r` is required to be convertible to the rank type of `alph`. If `r` is larger than the alphabet size, the behaviour is undefined; most alphabets in SeqAn3 throw an assertion in debug mode and perform no check in release mode. The reason for this design choice is that ranks are primarily written within algorithms and library detail functions. On the one hand, it is important to not convert to a valid value silently, because an invalid value is always indicative of a bug. On the other hand, it is performance critical to not perform checks in release code and to be able to annotate the function with the `noexcept` keyword to allow strong optimisations. The CPO enforces the latter on all implementations.

`seqan3::to_char(alph)`   returns the character representation of an alphabet object. It is defined as a function object/CPO. Implementations are required not to throw exceptions and be marked `noexcept`.

`seqan3::alphabet_char_t<T>`   is a transformation trait that exposes the character type of the alphabet `T`. It cannot be customised; it is always defined as the type of whatever `seqan3::to_char()` returns for a parameter of type `T`. This type is `char` for almost all alphabets in SeqAn3; it is required to be either `char`, `char8_t`, `wchar_t`, `char16_t` or `char32_t`.

`seqan3::char_is_valid_for<T>(c)`   returns whether `c` is in the set of valid characters for the alphabet `T` or not. It is defined as a function object/CPO. The type of `c` is required to be convertible to the character type of `alph`. Specifying this CPO for user-defined alphabets is optional under most circumstances. If no user-defined customisation is provided, all those characters where assignment (via `seqan3::assign_char_to()`) and retrieval (via `seqan3::to_char()`) are bijective (the same character is returned as was originally assigned) are considered *valid*.[6]

`seqan3::assign_char_to(c, alph)`   changes the value of `alph` by assigning from a character. It is defined as a function object/CPO. The type of `c` is required to be convertible to the character type of `alph`. The alphabet is expected to silently convert any values it considers invalid to a valid value and thus create a valid state for any character input. This is based on the assumption that many alphabets have an extra state for "unknown" or "any"[7] and that non-standard characters do not indicate user or programmer error. This mapping of input characters to valid characters is implemented as a single table lookup in all SeqAn3 alphabets although other implementations are possible. Since all inputs

---

[6] Similar functionality is not necessary for the rank representation where it always holds that the values between [0, *alphabet_size* − 1] are valid and the rest is not.

[7] For example, "N" for nucleotides and "X" for amino acids.

to this CPO are considered valid, the function can also be marked `noexcept` and
this is required of all implementations.

`seqan3::assign_char_strictly_to(c, alph)`   changes the value of `alph`
by assigning from a character. It is defined as a function object but **cannot** be
customised, because its semantics are entirely provided by other CPOs. The function object calls `seqan3::char_is_valid_for<decltype(alph)>(c)` and
throws an exception if the result is false; otherwise it calls `seqan3::assign_`
`char_to(c, alph)`. This function object can be used as an alternative to the
previous one if silent conversions to valid values are not desired. It is the only
basic operation whose implementation must contain a branch (`if`-check) and
that cannot be marked `noexcept`.

6.1.3 Concepts

To use alphabets in generic programming and enable polymorphism, C++ concepts
need to be defined that encompass the typical and more refined use-cases. When
first working with concepts, one is tempted to design concepts very closely on the
interface of the types that one is designing. But experience and multiple iterations
of re-designs have shown that it is better to base the concept definitions on actual
use-cases.

```
 2   // constrained function template that        Dna d;
     // accepts types that model 'alphabet'   8   Dna const cd;
     void foobar(alphabet auto && a)
 4   {                                        10   foobar(d);       // type seen by foobar is "Dna &"
         /*...*/                                   foobar(cd);      // type seen by foobar is "Dna const &"
 6   }                                        12   foobar(Dna{});   // type seen by foobar is "Dna"
```

Code snippet 6.4: The effect of `const` and `&` on concepts. `foobar()` is constrained
to only accept types that model the hypothetical `alphabet` concept. The actual type
of `a` depends highly on the input, see Snippet 3.8 on p. 42 for how `auto &&` can
become anything from `T` to `T const &`. *Note that this is not valid SeqAn3 code*

The example in Snippet 6.4 illustrates one of the lessons learned from this: it
is important to remember that `T`, `T &`, `T const &`, ... are all distinct types in
C++, and while one may satisfy a given concept, this is not necessarily true for
other "related" types. Ignoring CPOs for simplicity for now and assuming that
one defines an `alphabet` concept that requires exactly the four member functions
(`.to_rank()`, `.to_char()`, `.assign_rank()`, `.assign_char()`), the code in
Snippet 6.4 would *not compile*. This is due to the second invocation (line 11) passing

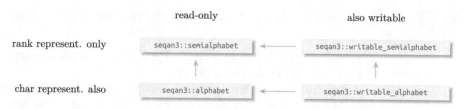

Fig. 6.1 The main alphabet concepts. Arrows imply concept refinement, so `seqan3::writable_alphabet` is the strongest concept, and it subsumes all others

a **constant** to the function on which the member functions `.assign_rank()` and `.assign_char()` cannot be called (they change the object).[8]

Now, if the function `foobar()` indeed calls either `.assign_rank()` or `.assign_char()` in the function body, the described behaviour is the intended behaviour, and we expect the compiler to reject `Dna const &` as an input type. But there are many use-cases where just the "read-only" aspects of an alphabet are important, e.g. both sequence alignment and indexed search require only to read from alphabets. It is thus advisable to split the concept into two: a read-only concept that is satisfied by variables and constants, and a second concept that refines the first one with additional requirements met only by variables and not by constants.[9]

These two concepts are named `seqan3::alphabet` and `seqan3::writable _alphabet`, respectively, as can be seen in Fig. 6.1. The rules for these are simple: use the first one by default, and use the second one only if the constrained template needs to modify objects of the given type.

The figure also introduces the so-called *semialphabets*. These require only the interfaces that are part of the rank representation (thus "one half" of what is considered a full alphabet). They are far less important for application developers/users of the library but play an important role in the definition of some alphabets (see Sect. 6.5).

As previously pointed out, the concepts also do not rely on member functions but on the aforementioned customisation points. The following paragraphs give the definitions of the concepts—very closely to how they appear in the API documentation.

---

[8] Similar to how concepts evaluate differently for `const` vs non-`const`, some also evaluate differently for `&` vs non-`&`; e.g. a type `T` might model `std::default_constructible`, but `T &` never does, because reference types need to be constructed from another object.

[9] While it is possible to work around the whole problem by removing `const`-ness from the type information before performing the concept check (via traits), this negates the simplicity of concepts (i.e. describing the operations actually available on the given type) and was a frequent cause for bugs in the first versions of SeqAn3.

`seqan3::semialphabet` is the root of the alphabet concept hierarchy. It requires the following:

- The `seqan3::alphabet_size` and `seqan3::to_rank` customisation points must find valid implementations for the given type.
- The type must not allocate dynamic memory and be efficiently copyable. This is an implicit assumption by many algorithms working with alphabets and ensures that one can pass-by-value instead of pass-by-reference and that copy is not slower than move in generic code. This requirement is semantic and cannot be checked by a concept fully, and however it is checked that the type models `std::copy_constructible` and satisfies the `std::is_nothrow_copy_constructible_v` trait. It is recommended (but not enforced) that types satisfy `std::is_trivially_copyable_v`.
- The type must model `std::totally_ordered`. This implies existence (and sane implementation) of all the comparison operators. The result of comparing two objects that model semialphabet shall be identical to the result of comparing the ranks of those two objects.
- It is recommended that types also model the `std::regular` concept (default-constructible, assignable from self) and satisfy the `std::is_standard_layout_v` trait; the combination of these two defines what is frequently called a "plain-old-datatype" (POD). However, this is not enforced as e.g. the requirement for default-constructability is irrelevant for many use-cases and would prevent any reference type from being a (semi)alphabet.

`seqan3::writable_semialphabet` is the writable version of `seqan3::semialphabet`. It requires the following:

- The type must model `seqan3::semialphabet`.
- The `seqan3::assign_rank_to` customisation point must find a valid implementation for the given type.

`seqan3::alphabet` is the "main" alphabet concept, and it adds character representation requirements to `seqan3::semialphabet`. It requires the following:

- The type must model `seqan3::semialphabet`.
- The `seqan3::to_char` customisation point must find a valid implementation for the given type.
- The `seqan3::char_is_valid_for` customisation point must find a valid implementation for the given type. Note that for most types,[10] a default implementation will be generated for this CPO.

`seqan3::writable_alphabet` is the most refined (strictest) alphabet concept. It requires the following:

---

[10] The types need to be default-constructible.

- The type must model `seqan3::alphabet`.
- The type must model `seqan3::writable_semialphabet`.
- The `seqan3::assign_char_to` customisation point must find a valid implementation for the given type.

Table 6.3 shows how the alphabet concepts evaluate on some example types with different `const` and reference qualifications. The example types will be introduced in the next sections; at this point, it is only important to know that `dna5` is a typical SeqAn3 alphabet and that `mask` has no char representation. One can then observe in Table 6.3 how `const` and `&` influence the satisfaction of the respective concepts and traits. This should highlight why different concepts are necessary for different use-cases and why some concepts and traits mentioned above are not subsumed by SeqAn3's concepts. For the full definitions of these concepts and traits and explanations why `const` and `&` influence the ability to satisfy them, I recommend reading the respective sections in the standard or online documentation.

6.2 User-Defined Alphabets and Adaptations

Before discussing the alphabets available in SeqAn3, I will show how to implement a custom alphabet and adapt an existing type as an alphabet. This should highlight why the different abstraction layers previously introduced make sense and how simple it is in practice to work with them (even if the theory behind them may not be trivial). They also show the full extent of *adaptability* in SeqAn3.

Table 6.3 Alphabet concepts and example types. The `dna5` alphabet will be discussed in Sect. 6.3, and the `mask` semialphabet will be discussed in Sect. 6.5

Alphabet concepts	dna5		dna5		mask		mask	
	dna5	const	dna5 &	const &	mask	const	mask &	const &
semialphabet	✓	✓	✓	✓	✓	✓	✓	✓
writable_semialphabet	✓		✓		✓		✓	
alphabet	✓	✓	✓	✓				
writable_alphabet	✓		✓					
Other concepts/traits								
std::is_trivially_copyable_v	✓	✓			✓	✓		
std::is_standard_layout_v	✓	✓			✓	✓		
std::regular	✓				✓			

6.2.1   User-Defined Alphabets

```
     #include <seqan3/alphabet/concept.hpp> // only needed for CPO / concept checks
 2
     namespace my_namespace
 4   {

 6   struct my_alph
     {
 8     uint8_t s; // the state of the object

10     /* Required for std::totally_ordered (required for seqan3::semialphabet) */
       constexpr friend bool operator==(my_alph lhs, my_alph rhs) noexcept { return lhs.s == rhs.s; }
12     constexpr friend bool operator!=(my_alph lhs, my_alph rhs) noexcept { return lhs.s != rhs.s; }
       constexpr friend bool operator<=(my_alph lhs, my_alph rhs) noexcept { return lhs.s <= rhs.s; }
14     constexpr friend bool operator>=(my_alph lhs, my_alph rhs) noexcept { return lhs.s >= rhs.s; }
       constexpr friend bool operator< (my_alph lhs, my_alph rhs) noexcept { return lhs.s <  rhs.s; }
16     constexpr friend bool operator> (my_alph lhs, my_alph rhs) noexcept { return lhs.s >  rhs.s; }

18     /* Required for seqan3::semialphabet */
       static constexpr uint8_t alphabet_size = 2;                                  // size of alph.
20     constexpr uint8_t to_rank() const noexcept { return s; }                     // read rank rep.

22     /* Required for seqan3::alphabet */
       constexpr char    to_char() const noexcept { return (s == 0) ? 'A' : 'B'; }  // read char rep.
24
       /* Required for seqan3::writable_semialphabet */
26     constexpr my_alph & assign_rank(uint8_t r) noexcept { s = r; return *this; }  // write by rank

28     /* Required for seqan3::writable_alphabet */
       constexpr my_alph & assign_char(char c) noexcept { s = (c == 'B'); return *this; } // write by char
30   };

32   } // namespace my_namespace

34   // CPOs:
     static_assert(seqan3::alphabet_size<my_namespace::my_alph> == 2);
36   static_assert(seqan3::char_is_valid_for<my_namespace::my_alph>('B'));
     static_assert(!seqan3::char_is_valid_for<my_namespace::my_alph>('!'));
38   // Concept (seqan3::writable_alphabet subsumes the others)
     static_assert(seqan3::writable_alphabet<my_namespace::my_alph>);
```

Code snippet 6.5: Example of a user-defined alphabet. This snippet is a valid header file. The `static_assert()`s show that the type models `seqan3::writable_alphabet` and that the CPOs can even be evaluated at compile-time

When designing a new alphabet (independent of whether it will become part of SeqAn3 or will reside in user code), the recommended way to satisfy the requirements of the CPOs and concepts is to implement a class type with members. This is the first lookup mechanism described for CPOs on page 149. It is important to note that while the CPOs and concepts make assumptions about the interfaces of the type, it is not specified how the state of the object is stored internally.[11]

Since I established two representations for alphabet objects, it makes sense to store one of these internally as the state of the object. This means that assignment

---

[11] The concepts do require that the state is not dynamically allocated, see above.

from that representation and conversion to that representation is merely passing the respective value in/out. Conversion to/from the other representation then happens via `if`/`switch` statements (less efficient) or via conversion tables (more efficient). Because the rank representation is likely to be used in performance-critical situations, I recommend storing the rank and converting to/from character representation when needed—but this is not a requirement.

Snippet 6.5 shows a full implementation of a user-defined alphabet. It is called `my_namespace::my_alph`, has an alphabet size of 2 and can be either in the "A"-state or in the "B"-state. The rank is stored internally as an 8-bit unsigned integer[12] and the respective member functions for rank representation simply read from/write to this value.

The alphabet size is defined via a `static constexpr` member variable and all comparison operators are defined in the usual way.[13] Together this is sufficient to make the type satisfy all the requirements for `seqan3::writable_semialphabet` (and thereby also `seqan3::semialphabet`). The other concept requirements stated previously (regarding copiability and constructability) are implicit for such simple types if one does not meddle with the constructors.

Finally, the member functions `.to_char()` and `.assign_char()` make the type models `seqan3::alphabet` and `seqan3::writable_alphabet`, respectively. These functions are implemented as conversions of the rank value to/from a character. For an alphabet of size 2, a simple comparison operator/ternary operator is sufficient; larger alphabets would need `switch` statements or table lookups.

All members are marked as `constexpr` which enables them to be called at compile-time. This is illustrated by the `static_assert()`s at the bottom; the usefulness of this feature will become clearer in Sect. 6.5. The usage of `seqan3::char_is_valid_for` highlights that not only the CPO "wrappers" for the defined members work but also that an implementation for `seqan3::char_is_valid_for` is generated in the absence of the respective member function. It correctly identifies `'B'` (and `'A'`, not shown) as *valid* and all other characters as *invalid*.

The example is simple and well suited to explain the design of alphabets because it is self-sufficient. In the context of a large library like SeqAn3, however, it is tedious to have to write even this amount of code for every alphabet. The repeated definition of the comparison operators and small differences between alphabets make copy-and-pasting error-prone. For this reason, SeqAn3 provides `seqan3::alphabet_base`, a CRTP base class that can be used to reduce this overhead. Derivates of `seqan3::alphabet_base` need to pass the size of the

---

[12] Strictly speaking, `bool` could be used here, as well, but as all objects are at least one byte big in memory and certain operations like `++` have been deprecated on `bool`, `uint8_t` is preferable.

[13] It is recommended to define binary operators as `friend`s and not as regular members because of certain advantages in overload resolution and implicit conversion (Brown & Sunderland, 2019).

alphabet as a template parameter and define two *static* member tables: one for mapping rank values to char values and one for the reverse direction. All members and comparison operators will then be defined by the base class; the base class stores the rank as state using the smallest possible unsigned integral type. The character type defaults to `char` unless a different type is given as third template parameter.

An example is given in the appendix (Snippet A.3). It behaves exactly like the example from Snippet 6.5. The decrease in lines of code is modest, but the code expresses the specific nature of the given alphabet much more cleanly. More sophisticated alphabets profit more strongly from this mechanism.

6.2.2 Adapting Existing Types as Alphabets

Being able to extend SeqAn3 with new user-provided types is a part of adaptability. But the second important question is how well existing types can be integrated. Provided that one has access to the namespace that type is defined in, this can be done with free functions that perform the necessary tasks and are picked up by SeqAn3's CPOs via *argument-dependent lookup*. This is the second lookup mechanism described for CPOs on page 149.

An example is shown in Snippet 6.6. The original type is an enumerator type that may have some existing significance in the user's code base (and should therefore not be changed itself). It has three states: `my_alph::ZERO`, `my_alph::ONE` and `my_alph::TWO`. These shall be adapted as an alphabet where the numerical values `0`, `1` and `2` are the rank representation and the characters `'0'`, `'1'` and `'2'` are the char representation. It is an example of where the alphabet's internal state is neither of the representations and both sets of free functions involve conversions.

All free functions take an object of the adapted type as argument. The assignment functions have an additional `_to` in the name to clearly indicate the order of the parameters. `alphabet_size()` is a special case, because it does not operate on the argument, and it only needs the type of the argument so that it is found via *argument-dependent lookup*. Since the size of all alphabets in SeqAn3 is fixed, this function is *required* to be marked `constexpr`; for the other functions, this is optional. It is not necessary to define comparison operators because enumerator types are already ordered.

This second way of meeting the requirements of the CPOs and concepts is more flexible than the first, but I would argue that the first is easier to understand and more *compact*, so I recommend it for any newly created types. A case where the second method is clearly superior is when integrating multiple third party types: if they are specialisations of the same template or can be expressed via a shared concept, one can easily define wrappers as in Snippet 6.6 to adapt them all at once.

It should be noted that neither the example in Snippet 6.5 nor the example in Snippet 6.6 makes use of any declarations or definitions from SeqAn3. No

```
   namespace my_namespace
 2 {
   /* A pre-existing type that shall not be replaced, only adapted */
 4 enum class my_alph { ZERO, ONE, TWO };
   /* Required for seqan3::semialphabet */
 6 constexpr uint8_t alphabet_size(my_alph) noexcept { return 3; }
   constexpr uint8_t to_rank(my_alph const a) noexcept { return static_cast<uint8_t>(a); }
 8 /* Required for seqan3::alphabet */
   constexpr char to_char(my_alph const a) noexcept
10 {
       switch (a)
12     {
           case my_alph::ZERO: return '0';
14         case my_alph::ONE:  return '1';
           default:            return '2';
16     }
   }
18 /* Required for seqan3::writable_semialphabet */
   constexpr my_alph & assign_rank_to(uint8_t const r, my_alph & a) noexcept
20 {
       switch (r)
22     {
           case 0:  a = my_alph::ZERO; return a;
24         case 1:  a = my_alph::ONE;  return a;
           default: a = my_alph::TWO;  return a;
26     }
   }
28 /* Required for seqan3::writable_alphabet */
   constexpr my_alph & assign_char_to(char const c, my_alph & a) noexcept
30 {
       switch (c)
32     {
           case '0': a = my_alph::ZERO; return a;
34         case '1': a = my_alph::ONE;  return a;
           default:  a = my_alph::TWO;  return a;
36     }
   }
38 } // namespace my_namespace

40 #include <seqan3/alphabet/concept.hpp> // only needed for concept check
   static_assert(seqan3::writable_alphabet<my_namespace::my_alph>);
```

Code snippet 6.6: Adapting an existing type as an alphabet. This snippet is a valid header file. Since the original type shall not be changed, the necessary functionality is implemented as free functions in the namespace of the type. SeqAn3's CPOs will find them via argument-dependent lookup

functions inside SeqAn are overloaded and no templates are specialised; in fact, they never open `namespace seqan3` other than in the optional concept check in the end. This is very different from other forms of generic programming.

Finally, if one does not have access to the namespace of the type that one wishes to adapt, e.g. because one is adapting a type from another third party library, one can use the third mechanism that is evaluated by the alphabet CPOs: specialising the "upload" space `seqan3::custom::alphabet<T>` and providing wrappers inside that specialisation. Example code that illustrates this mechanism is omitted here but available in the appendix (Snippet A.4).

SeqAn3 does this for adapting the built-in character types (`char`, `wchar_t` etc.) to behave as alphabets. This means that all the CPOs are defined for them although they are usually implemented as pure assignments or conversions, e.g. `assign_char_to('A', c)` where `c` is of type `char` is literally an assignment.

The adaptations are used internally, and they allow users to explicitly opt out of the type safety normally associated with alphabets (more on this in Sect. 6.5). They might also be useful when developing SeqAn3 interfaces to other programming languages.

6.3   The Nucleotide Submodule

Nucleotides are the elements of DNA and RNA sequences and thereby the foundation of all genetic, genomic and transcriptomic research. Table 5.3 show an overview of the submodule. DNA is composed of the four nucleotides adenine (A), cytosine (C), guanine (G) and thymine (T). RNA uses Uracil (U) in place of Thymine. The relation between Uracil and Thymine is biologically close (DNA T results in RNA U after transcription) and they do not appear in ambiguous contexts (either the sequence is DNA and can contain T but not U, or it is RNA and can contain U but not T). Thus, they are typically treated as the same informational unit in the context of sequence analysis.

DNA appears as a double-stranded polymer where every nucleotide base is part of a base-pair. Adenine is paired with Thymine, and Cytosine is paired with Guanine. This "partner" in base-pairing is called the *complement* of the respective nucleotide and is significant for many applications in bioinformatics. Such a DNA double-strand is displayed in Fig. 6.2. RNA usually appears single-stranded, but if it does appear double-stranded (or if it is paired with DNA in transcription), canonical pairing is identical to that of DNA (with T being replaced by U).

Beyond the symbols defined for the five nucleotides that appear naturally, the International Union of Pure and Applied Chemistry (IUPAC) defines 11 further symbols that each refer to a group of nucleotides (see Table 6.5). These symbols can be used to represent ambiguity in sequencing data. The most common symbol among these is N which indicates that it is unknown which nucleotide is present

Table 6.4 Nucleotide submodule overview	Alphabet: Nucleotide submodule	
	Concepts	seqan3::nucleotide_alphabet
	Class types	seqan3::dna15, seqan3::dna3bs, seqan3::dna4, seqan3::dna5, seqan3::nucleotide_base, seqan3::rna15, seqan3::rna4, seqan3::rna5, seqan3::sam_dna16
	Function objects	seqan3::complement

Fig. 6.2 The DNA double
strand. ©2013 by
Blausen.com, licensed under

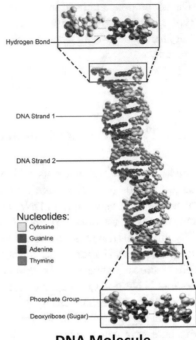

Hydrogen Bond

DNA Strand 1

DNA Strand 2

Nucleotides:
☐ Cytosine
■ Guanine
■ Adenine
■ Thymine

Phosphate Group
Deoxyribose (Sugar)

DNA Molecule

at a given location in the sequence. Common biological file formats like FASTA
and SAM recognise all of these symbols; however, other formats or interfaces only
accept a subset. Furthermore, certain algorithms, including sequence alignment and
indexed search, profit from smaller alphabets due to reduced complexity. Some
formats and conventions require characters for the symbols to be upper case, but
often lower case characters are used interchangeably with upper case letters.

The variety of use-cases and notations suggest that a single type is not sufficient
to address "nucleotides". On the other hand, the interoperability between different
nucleotide types is important and generic algorithms should be able to handle all
nucleotide types and differentiate them from non-nucleotide alphabets.

6.3.1   General Design

The *Nucleotide* submodule defines the `seqan3::complement` CPO which returns
for every nucleotide the respective complementary value. This CPO uses the same
mechanisms as previously discussed, i.e. any types that provide a `.complement()`
member (or for whom a free function is found via ADL...) results in a valid definition of the CPO. Since the behaviour of complementing is unique to nucleotides, the
presence of such a valid CPO definition can be used as an identifier for "nucleotide
types".

Table 6.5 IUPAC nucleotide symbols and their complements

IUPAC	Description	Complement
A	Adenine	T/U
C	Cytosine	G
G	Guanine	C
T/U	Thymine/Uracil	A
M	A or C	K
R	A or G	Y
W	A or T	W
Y	C or T	R
S	C or G	S
K	G or T	M
V	A or C or G	B
H	A or C or T	D
D	A or G or T	H
B	C or G or T	V
N	A or C or G or T	N

```
   /* assignments */
2  dna5 d{};              // -> A by default
   d = 'C'_dna5;          // -> C
4  d.assign_char('X');    // -> N
   d.assign_rank(0);      // -> A
6
   /* conversions */
8  d = static_cast<dna5>('G'_dna15); // -> G
   d = 'U'_rna5;          // -> T (implicit)
10
   /* complement */
12 d = d.complement(); // -> A
```

```
   /* print any alphabet */
2  void print_alph(alphabet auto && a)
   {
4      std::cout << to_char(a) << '\n';
   }
6
   /* for nucleotides, also print complement */
8  void print_alph(nucleotide_alphabet auto && a)
   {
10     std::cout << to_char(a) << ' '
                 << to_char(complement(a)) << '\n';
12 }
```

Code snippet 6.7: Nucleotide alphabet examples. On the left, the various forms of assignment and conversion are shown. Conversion from a differently sized nucleotide alphabet has to be made explicit, while conversion between `seqan3::dna5` ↔ `seqan3::rna5` is implicit. On the right, a generic "algorithm" is shown that has specialised behaviour for nucleotide alphabets. Namespace `seqan3::` is assumed for both

This is what the `seqan3::nucleotide_alphabet` concept does: it is defined as `seqan3::alphabet` plus the requirement that `seqan3::complement` be defined for the type. The concept can be used to constrain generic algorithms to work only on nucleotide types or provide specialised behaviour for them (see the right side of Snippet 6.7). It further suggests that any nucleotide types be explicitly constructible from any other nucleotide types to facilitate simple conversions when necessary (e.g. data is available in one type, but the algorithm prefers another). However, this is not enforceable via concepts.

A new base class is provided by the *Nucleotide* submodule called `seqan3::nucleotide_base`. As with `seqan3::alphabet_base` (which `seqan3::nucleotide_base` derives from), using the base class is entirely optional for the purpose of polymorphism (or meeting the concept requirements), it is simply a way to reduce redundant code. The base class provides the following predefined behaviour:

1. An explicit constructor accepting any other types that model `seqan3::nucleotide_alphabet`. Conversion happens through the character representation. If both alphabets' operations are usable in a `constexpr` context, the usage of this constructor will trigger the generation of a conversion table at compiletime. This means conversion performs only a single table lookup at runtime instead of actually going through both alphabets' character representation.
2. A `.complement()` member function that depends on a complement-table being defined by the derived type (similar to how the derived type provides `char_to_rank` and `rank_to_char` tables, see Snippet A.3).
3. A static member function `.char_is_valid(c)` that is found by the `seqan3::char_is_valid_for` CPO. This behaviour includes declaring `'U'` and `'T'` as valid independent of whether these represent distinct states or not and declaring both the upper case and the lower case versions of all characters in `rank_to_char` valid. Implementation of this check is also done via a table synthesised from a lambda expression.

In essence, the base class provides all the behaviour suggested previously so that actual implementations only need to provide the three tables `char_to_rank`, `rank_to_char` and `complement_table`. All alphabets defined in the *Nucleotide* submodule are derived from `seqan3::nucleotide_base`.

6.3.2 Canonical DNA Alphabets

Three general-purpose DNA alphabets are provided that differ primarily in their size. Whenever SeqAn3 offers differently sized alphabets for common use-cases, this is represented in their name to avoid confusion. Thus, there are the following:

`seqan3::dna15`	All IUPAC symbols other than U.
`seqan3::dna5`	A, C, G, T and N.
`seqan3::dna4`	A, C, G, T.

When assigning through the character representation, all of these alphabets accept lower and upper case characters. Converting back to character representation always yields an upper case letter as this is the more widely accepted notation. To improve interoperability with RNA data, assigning the characters `'U'` or `'u'` to a DNA alphabet will always result in the T state and not the "unknown state" denoted by N. Entirely unknown characters (e.g. `'!'`) are converted to N for `seqan3::dna15` and `seqan3::dna5` (for the latter, this also includes IUPAC symbols other than the five represented ones). For `seqan3::dna4`, it is not possible to indicate any ambiguity, and by default, unknown characters are converted to the letter A. However, for characters representing IUPAC letters, the first letter from the ambiguous set is chosen; e.g. `'B'` (which represents "C, G or T") is converted to C and not to A since it is clearly not A.

Which alphabet to choose depends mainly on space considerations, but it is important to remember that single objects in C++ can never be stored in less than a byte. Any storage advantages of smaller alphabets are only relevant when used in combination with data structures that perform some kind of compression based on the alphabet size. This includes `seqan3::bitcompressed_vector` (Sect. 7.2.2), composite alphabets (Sect. 6.5) and FM-indexes (Sect. 9.1).

For most users, `seqan3::dna5` is likely the best choice because it has a good trade-off in case any of the aforementioned mechanisms is used. I recommend `seqan3::dna4` only in high-performance contexts and only when benchmarks prove that the critical component is indeed slowed down by the additional size of `seqan3::dna5`. `seqan3::dna15` exists for use-cases where distinguishing the ambiguous characters is important or where input data needs to be preserved as is.

6.3.3 Canonical RNA Alphabets

To mirror the behaviour of the DNA alphabets, SeqAn3 also provides `seqan3::rna15`, `seqan3::rna5` and `seqan3::rna4`. These behave exactly as their DNA counterparts with the exception that if they are in the T/U state and converted to their character representation, this will be `'U'` instead of `'T'`.

It should, however, be noted that RNA data can be represented safely in DNA alphabets without loss of information. The only difference is the character printed. So in applications that deal with DNA and RNA data, it is possible to use a single alphabet for most parts of the application and convert on-the-fly when generating user output. This can be done via `seqan3::views::convert` (Sect. 7.3). Since SeqAn3 guarantees that RNA alphabets have the same binary representation as DNA alphabets of the same size, it is even possible to for example `reinterpret_cast<>` containers of `seqan3::dna5` to containers of `seqan3::rna5`.[14]

In addition to the *explicit* conversions that are possible between all nucleotides, RNA and DNA alphabets of the same size are also *implicitly* convertible to each other as they represent essentially the same data.

Finally, it should be noted that there is currently no alphabet in SeqAn3 that differentiates between T and U, i.e. has different states to denote these nucleotides. This follows the principle that features are only added based on use-cases, and in the extensive research done for SeqAn3 and based on the experience with SeqAn1/2,

---

[14] This is supported by all major compilers but undefined behaviour by the standard.

none such use-cases emerged. In fact, most use-cases suggested that it is beneficial to represent both nucleotides in one state. Should an application require different behaviour in the future, such an alphabet could be added easily.

6.3.4   Other Nucleotide Alphabets

The *Nucleotide* submodule offers two more alphabets whose scope is more limited:

`seqan3::dna3bs`   A three-letter DNA alphabet that collapses the letter C onto T. It is used in the context of bisulfite sequencing where the biochemical processes convert unmethylated Cs to U/T.

`seqan3::sam_dna16`   A 16-letter alphabet that includes all IUPAC symbols (except U) but that has an additional state to represent `'='`. It is the alphabet mandated by the SAM/BAM/CRAM file formats; the binary representation in SeqAn3 is guaranteed to correspond to that of BAM/CRAM.[15]

The full definition of `seqan3::sam_dna16`, which exemplifies the ease of defining such an alphabet, is given in the appendix (Snippet A.5).

6.4   The Amino Acid Submodule

Amino acids are the second major elemental types in sequence analysis. They are the molecular building blocks of proteins and the alphabets of protein sequences. See Table 6.6 for an overview of the submodule.

Nucleotides and amino acids are deeply linked through the biochemical process of *translation*. In this process, a sequence of amino acids is synthesised from an RNA sequence by assigning one amino acid to every (non-overlapping) triplet of nucleotides, also called *codons*. The set of all codon-to-amino-acid assignments is

Table 6.6 Nucleotide submodule overview

Alphabet: Amino acid submodule	
Concepts	`seqan3::aminoacid_alphabet`
Class types	`seqan3::aa10li,` `seqan3::aa10murphy,` `seqan3::aa20,` `seqan3::aa27,` `seqan3::aminoacid_base`
Traits	`seqan3::enable_aminoacid`
Functions	`seqan3::translate_triplet()`

---

[15] It is not required to use this alphabet in SeqAn3's alignment I/O since other nucleotide alphabets will be converted, but it can have slight performance advantages.

Fig. 6.3 Standard genetic code displayed as the "codon sun". It includes the 20 canonical amino acids and the RNA-triplets that code for them. Public domain image

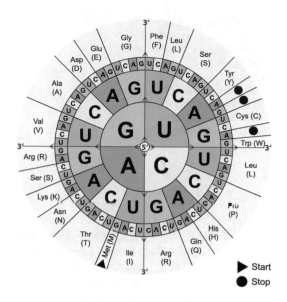

called the genetic code,[16] and codon tables can be used to describe it. A common visualisation of this is called the "codon sun", and it is displayed in Fig. 6.3. The biological process of translation (*in vivo*) is often emulated in bioinformatics (*in silico*) as it can be important to simulate which amino acid sequence a nucleotide sequence would be translated to in the cell.

The codon table displayed in Fig. 6.3 contains the 20 naturally occurring amino acids in the outer ring, each represented by a 3-letter abbreviation and a 1-letter abbreviation. The inner rings represent the three RNA bases that code for the respective amino acid. Some amino acids are coded by more codons than others. As with nucleotides, the single-letter abbreviations are very common in bioinformatics. Beyond the symbols for the aforementioned 20 amino acids, seven more symbols are often considered part of amino acid alphabets (see Table 6.7). These include symbols for real biological (but rarely occurring) amino acids, symbols for ambiguity (similar to those found in nucleotides but not for all combinations) and a symbol for the stop-codon which is never found in a protein *in vivo* but occurs frequently in sequences translated *in silico*.

6.4.1 General Design

The general design of the *Amino acid* submodule is very close to that of the *Nucleotide* submodule. Because there is no alphabet-specific function

---

[16] Variations from the standard genetic code exist but are omitted here for simplicity.

Table 6.7 Non-standard
amino acid symbols with
three-letter codes (3l) and
one-letter codes (1l). They
include symbols for two
rarely found proteinogenic
amino acids, the stop-codon
(not an amino acid) and
multiple symbols for
ambiguity

Description	3l	1l
Stop-Codon		*
Non-standard AAs		
Selenocysteine	Sec	U
Pyrrolysine	Pyl	O
Ambiguity		
Asp or Asn	Asx	B
Glu or Gln	Glx	Z
Leu or Ile	Xle	J
Unknown/any	Xaa	X

that identifies amino acids, types can opt in to being recognised as one via
the `seqan3::enable_aminoacid` trait. The `seqan3::aminoacid_alphabet`
concept requires only that this trait be enabled for a type and the type model
`seqan3::alphabet`. This mechanism ensures that `seqan3::aminoacid_`
`alphabet` is still considered more refined than `seqan3::alphabet` and is used
frequently in the C++ 20 standard library when no other discerning factor exists for
a concept.

The submodule also offers the distinct base class `seqan3::aminoacid_base`.
Analogously to the *Nucleotide* submodule and `seqan3::nucleotide_base`, it
provides a converting constructor for other amino acid alphabets and a validity table
that includes lower case characters of valid letters. All alphabets provided by the
Amino acid submodule are derived from `seqan3::aminoacid_base`.

6.4.2 Amino Acid Alphabets

`seqan3::aa27` is the main amino acid alphabet provided by the submodule. It
encompasses all the 27 symbols described above and is recommended for almost
all use-cases. There is also `seqan3::aa20` which only contains the standard amino
acids, but, since both alphabets require 5 bits in compressed representation, there is
little reason to store data in this alphabet. The only case where it is useful is when
writing to file formats that mandate the 20-letter alphabet and the application needs
to ensure that only valid symbols are written. As with other conversions, this can
happen lazily via `seqan3::views::convert` (see Sect. 7.3).

The submodule further provides two *reduced* amino acid alphabets. These are
smaller alphabets that are intentionally ambiguous by grouping certain amino acids
into single symbols. This clustering is based on certain physical or biochemical
properties (polarity, acidity etc.) or on frequently recorded substitutions/mutations.
It is based on the observation that certain single amino acid mutations happen
much more frequently than others and have little to no functional impact. Thus, it

may be beneficial to not differentiate between them in certain algorithms to reduce computational complexity.

The alphabets currently provided are `seqan3::aa10li` by Li et al. (2003) and `seqan3::aa10murphy` by Murphy et al. (2000). Both are used by Lambda, and Part III covers them in more detail.

6.4.3   Translation

```
template <genetic_code gc = genetic_code::CANONICAL, nucleotide_alphabet nucl_type>
constexpr aa27 translate_triplet(nucl_type const n1, nucl_type const n2, nucl_type const n3) noexcept
{ /*...*/ }
```

Code snippet 6.8: Signature of `seqan3::translate_triplet()`

Translation of three nucleotides to an amino acid is handled by the free function `translate_triplet()` (Snippet 6.8). It takes three nucleotides as individual parameters, looks up the correct amino acid in a table and returns it. The genetic code can be given as an optional template parameter.[17] While it would be possible to generate tables for all combinations of nucleotide types (input), amino acid types (output) and genetic code (mapping), the resulting number of tables and their sizes are daunting. The process would either entail hard-coding many such large tables or re-computing them on every build of the library which considerably increases the compile-time if the tables are large.

The solution is to generate a single hard-coded table for `seqan3::dna15` as input, `seqan3::aa27` as output and the standard genetic code as mapping; it has $15^3 = 3375$ entries. Using the largest available alphabets actually makes a difference for some combinations because ambiguous nucleotides can still lead to unambiguous amino acid symbols due to the redundancy of the genetic code. To be able to handle other nucleotide types as input and other amino acid types as output, the function internally can convert any nucleotide input to `seqan3::dna15` and the user can convert the `seqan3::aa27` output to any other desired amino acid type. This is the only mechanism used to cover different outputs.

The smaller input types `seqan3::dna4` and `seqan3::dna5`, however, generate smaller tables on-demand at compile-time. Due to their smaller alphabet size, the reduction of the table size compared to `seqan3::dna15` is significant (64/125 vs. 3375). This strongly affects cache efficiency at runtime, and since the tables are small, their specific impact on compile-times is low. The RNA alphabets provided by SeqAn use the respective tables of their equally sized DNA counterparts. Finally,

---

[17] Support for this is available although most non-standard tables have yet to be added to the SeqAn3 master branch.

tables for other genetic codes could be generated by a lambda expression that copies the original table of the standard genetic code and changes only the affected cells (these tables would still be large, but the compile-time needed to create them would be small). Any such tables are only generated at compile-time when there is a code-path that uses them.

This design is *generic* and *simple*; there is only one function interface that is called by the user. It provides correct behaviour on all combinations of nucleotide types and amino acid types, even those that may be supplied by the user. It is also *compact* and *simple* from the perspective of the maintainer, because only one table is hard-coded and all other ones are expressed as functions on this one table—instead of multiple huge, hard-coded tables that differ only in single values as frequently found in SeqAn2. The *performance* at runtime should be optimal (a single table lookup in a suitably sized table) and the impact on compile-times is kept to a minimum.

6.5   Composite Alphabets

Nucleotides and amino acids are alphabets that directly represent biochemical compounds, but many alphabets useful to bioinformatics are more abstract. Often it is helpful to combine existing alphabets into new alphabets and/or hide multiple alphabets behind a single type. The *Composite* submodule provides such facilities which are introduced in this subsection. See Table 6.8 for an overview concrete applications of these are also presented here although they appear in their own submodules in the library (gap and mask).

6.5.1   *Alphabet Variants*

Like `std::variant` (see Sect. 3.10) or a C `union`, `seqan3::alphabet_variant` is a type that can hold the values of different types alternatively, i.e. it always holds only either the value of one type **or** another type, not two values at once. It requires that its types model `seqan3::writable_alphabet` and share the same

Table 6.8 Composite submodule overview

Alphabet: Composite submodule	
Class types	`seqan3::alphabet_tuple_base`, `seqan3::alphabet_variant`, `seqan3::semialphabet_any`

`seqan3::alphabet_char_t`. For implementation reasons, it also requires that all operations on the alphabet are callable in a `constexpr`-context.

Recall that variants perform a kind of *type-erasure*, i.e. objects of different types can be stored in an object of a single type, and information that is previously encoded in the type (a compile-time property) is now encoded in the value (a runtime property). See also Sect. 3.10.

What makes `seqan3::alphabet_variant` special (compared to just using a union or `std::variant` over two alphabets) are the following properties:

1. It is itself an alphabet and it is possible to call all the usual operations on it.
2. It "knows" in which type's state it is in and allows for safe conversion to that type.
3. It can be constructed from, assigned from and compared with any of its alternative types and also those types that they are constructible from/assignable from/comparable with;[18] essentially it behaves as all of its alternative types.
4. Its alphabet size is the sum of the alphabet sizes of its alternative types, and it compresses to the smallest rank type that can represent this.

```
   gapped<dna15> v;                        // gapped<T> is a typedef for alphabet_variant<T, gap>
2
   static_assert(alphabet_size<gapped<dna15>> == 16);  // alphabet size is 15 + 1
4  static_assert(sizeof(gapped<dna15>)       == 1);   // memory size is 1 byte

6  v = 'C'_dna15;                          // assign from one alternative type
   std::cout << v.to_char();               // prints 'C'
8  v = 'A'_rna15;                          // implicitly converts to 'A'_dna15 and assigns that
   std::cout << v.to_char();               // prints 'A'
10 v = gap{};                              // assign from the other alternative type
   std::cout << v.to_char();               // prints '-'
12
   v.assign_char('G');                     // character results in state belonging to dna15
14 std::cout << v.to_char();               // prints 'G'
   std::cout << (v == gap{});              // false, because not in gap-state
16
   dna15 d = v.convert_to<dna15>();        // successful
18 //gap g = v.convert_to<gap>();          // would throw an exception because v is in dna15-state
   v.assign_char('-');                     // in gap state
20 gap   g = v.convert_to<gap>();          // successful
```

Code snippet 6.9: Alphabet variants and the gap alphabet. `seqan3::gap` is an alphabet with only a single state, the gap state. It can be combined with nucleotide or amino acid alphabets via `seqan3::alphabet_variant` to form an alphabet for aligned sequences

Snippet 6.9 illustrates this behaviour. It combines the single-letter `seqan3::gap` alphabet from the *Gap* submodule with `seqan3::dna15` to form a new 16-letter alphabet that contains all the states of `seqan3::dna15` as well as a state for the character "-". An object of the resulting type behaves as either of its alternative types depending on how it is used. While the example may suggest that classification

---

[18] It even replicates the exact behaviour in regard to implicit and explicit conversions.

into the states of different alternative types happens via the character representation, this is not the case and "overlapping" characters are allowed. Snippet 6.10 demonstrates this: `seqan3::dna4` and `seqan3::rna4` represent unique states (ranks) in this example, and they do not compare equal (there is also no implicit conversion since both types are valid alternatives).

```
     alphabet_variant<dna4, rna4> v;
2
     static_assert(alphabet_size<alphabet_variant<dna4, rna4>>  == 8);  // alphabet size is 4 + 4
4    static_assert(sizeof(alphabet_variant<dna4, rna4>)         == 1);  // memory size is 1 byte

6    v = 'C'_dna4;                  // assign from one alternative type
     std::cout << v.to_char();      // prints 'C'
8    v = 'C'_rna4;                  // assign from the other
     std::cout << v.to_char();      // prints 'C'
10   std::cout << (v == 'C'_dna4);  // false, because 'C'_dna4 and 'C'_rna4 are distinct states!

12   v.assign_char('A');            // ambiguous in theory, results in state of first matching alternative
```

Code snippet 6.10: Advanced alphabet variants. Arbitrary alphabets can be combined in the `seqan3::alphabet_variant` and the full range of the states of each is preserved

All the shown behaviour is achieved without `virtual` functions or any other kind of polymorphism. **There is zero runtime or memory overhead for using such a combined alphabet compared to writing an alphabet with the described properties from hand.** This is achieved through modern metaprogramming: the `seqan3::alphabet_variant` is designed in a way that makes the compiler generate all states of the involved alphabets and their converting behaviour (in regard to characters) and then integrate that into the respective tables for the combined alphabet.[19] In the end, it simply supplies these to `seqan3::alphabet_base` like any other alphabet, and access to character or rank happens accordingly. It does provide additional constructors, assignment operators and comparison operators, but these all make use of the given tables, so any runtime operation on the `seqan3::alphabet_variant` is at most one table lookup—independent of how many alphabets are joined or their sizes.

Beyond the regular alphabet interface, the variant offers three members:

`bool is_alternative<T>()`   Check whether the variant is in a state belonging to the given type.

`T convert_to<T>()`   Convert to the alternative type if it is in one of its states, otherwise throw an exception. This is the only member that involves an `if`-check and might throw.

---

[19] This involves no access to the internal states of the involved alphabets; the compiler uses the public API described by the concepts. That is why `constexpr` is important on the entire interface.

`T convert_unsafely_to<T>()`   Convert to the alternative type if it is in one of its states, otherwise the behaviour is undefined. Faster than the previous function (marked `noexcept`, no `if`-check), but potentially unsafe.

6.5.2   Alphabet Tuples

While the `seqan3::alphabet_variant` combines multiple alphabets in an "either-or" fashion, it is often also useful to be able to combine multiple alphabets into an object that can hold a state for each of them, similar to a `std::tuple`. This can be achieved with `seqan3::alphabet_tuple_base`. Similar to `seqan3::alphabet_variant`, it provides:

1. Constructors, assignment operators, conversion operator and comparison operators for the individual component types
2. Computation and storage of a rank value from the rank values of the represented component values
3. Access to the rank representation
4. The alphabet size which is the product of the alphabet sizes of the component types[20] because all combinations of values are possible

However, it differs fundamentally in that it is not a directly usable class template but a (CRTP) base class that needs to be derived from. This is strongly linked to it not having a character representation, only a rank representation. The character representation needs to be provided by the derived class, because it is not possible to determine what the character representation would be for arbitrary combinations of alphabets (e.g. an alphabet tuple that contains both `'A'_dna4` and `'P'_aa27`).

An example of how `seqan3::alphabet_tuple_base` is used can be seen in Snippet 6.11. While I previously explained that lower and upper case letters have no semantic meaning for nucleotide and amino acid alphabets (and are thus both accepted and silently converted to the same state), there is one popular application where case is used to denote different states: the masking of sequences. Masking happens as a result of filters being applied on the sequences, typically repeat filters or *low complexity region* (LCR) filters. These are used to indicate that certain parts of the sequence likely do not contain what is being searched for (e.g. a gene or a certain protein domain) and that they should be excluded from the more expensive parts of the algorithm (Altschul et al., 1994). Initially such regions were just replaced with N/X symbols to prevent them from being searched, but later lower case was also used to mask the regions (Wootton & Federhen, 1996). Alignment algorithms for example may choose not to start an alignment in an LCR but can still perform a correct alignment *through* such a region if the original base pair information is preserved.

---

[20] In contrast to the sum of alphabet sizes for `seqan3::alphabet_variant`.

```
   mask m;                            // essentially a boolean semi-alphabet
2  m.assign_rank(false);              // indicates that position is unmasked
   m = mask::MASKED;                  // enum-like interface identical to assigning true
4
   masked<dna15> md;                  // masked<T> derives from alphabet_tuple_base<masked, T, mask>
6                                      // "version of T that differentiates lower and upper case"

8  static_assert(alphabet_size<masked<dna15>>  == 30);  // alphabet size is 2 * 15
   static_assert(sizeof(masked<dna15>)         == 1);   // memory size is 1 byte
10
   /* character representation */
12 md.assign_char('C');               // sets dna15-component to C state and mask-component to "unmasked"
   std::cout << md.to_char();         // prints 'C'
14 md.assign_char('c');               // sets dna15-component to C state and mask-component to "masked"
   std::cout << md.to_char();         // prints 'c'
16
   /* assigning individual components */
18 md = 'N'_dna15;                    // only changes dna15-component, mask-state still "masked"
   std::cout << md.to_char();         // prints 'n'
20 md = mask::UNMASKED;               // only changes mask-component, dna15-component still N
   std::cout << md.to_char();         // prints 'N'
22
   /* conversion to component types */
24 dna15 d = md;                      // implicitly convertible to its component types
   mask m2 = get<1>(md);              // explicit access via tuple-like interface possible
26 std::cout << (md == 'N'_dna15);    // true, only the respective components are compared
```

Code snippet 6.11: Alphabet tuples and the mask alphabet. `seqan3::mask` is a semialphabet with the two states masked/unmasked. `seqan3::masked<T>` is a derived type of `seqan3::alphabet_tuple_base` that combines `seqan3::mask` with a nucleotide or amino acid alphabet to represent data that uses the case of the character to mark ("mask") filtered regions. Namespace `seqan3::` is assumed

SeqAn3 offers the `seqan3::mask` semialphabet that represents the per-character information of whether a single given sequence position belongs to a masked region or not. It is only a semialphabet and not a full alphabet, because it cannot be visually displayed without the accompanying sequence data, essentially it is a `bool` with an "alphabet-ised" interface. To become useful to users, it needs to be combined with a sequence alphabet which can be done via `seqan3::masked<T>`. The latter derives from `seqan3::alphabet_tuple_base` and creates from its template argument `T` a new alphabet that behaves exactly like `T` except that lower case and upper case character inputs lead to distinct states (the total number of states/the alphabet size is doubled).

Similar to `seqan3::alphabet_variant`, `seqan3::alphabet_tuple_base` performs the bulk of the metaprogramming that is generic for any kind of "alphabet tuple". Beyond the points listed above, it provides a true tuple interface similar (and compatible) to that of `std::tuple`, including `get<index>()` and `get<type>()` functions to retrieve the components and a specialisation of `std::tuple_size` that holds the number of contained components. Since the alphabet tuple does not contain distinct member objects of the components (an important point is compressing into a single object), the `get()`-functions cannot return references to the components. To still be able to assign through the return value, they return *proxy* objects that appear as objects of the component type but update the tuple on assignment.[21]

```
template <typename sequence_alphabet_t>
    requires writable_alphabet<sequence_alphabet_t>
class masked : public alphabet_tuple_base<masked<sequence_alphabet_t>, sequence_alphabet_t, mask>
{
private:
    using base_type = alphabet_tuple_base<masked<sequence_alphabet_t>, sequence_alphabet_t, mask>;
    using char_type = alphabet_char_t<sequence_alphabet_type>;

public:
    constexpr masked() noexcept = default;    // Explicitly defaulted default constructor.
    using base_type::base_type;               // Inherited constructors.

    using base_type::alphabet_size;           // Inherit alphabt_size static member...
    using base_type::operator=;               // ... and assignment operators.

    /* Required to model seqan3::writable_alphabet */
    constexpr masked & assign_char(char_type const c) noexcept
    {
        assign_char_to(c,           get<0>(*this));
        assign_rank_to(is_lower(c), get<1>(*this));
        return *this;
    }
    constexpr char_type to_char() const noexcept
    {
        return seqan3::to_rank(get<1>(*this))
            ? to_lower(seqan3::to_char(get<0>(*this)))
            : seqan3::to_char(get<0>(*this));
    }
};
```

Code snippet 6.12: A possible definition of `seqan3::masked`. The actual definition in SeqAn3 uses predefined tables to avoid going through the proxies and writing twice in character assignment

To underline the usefulness of the abstractions and the simplicity of implementing such a derived class, see Snippet 6.12. Only the implementations for reading and writing character representation are required to complete the definition of an alphabet tuple. The given implementations reflect the semantics very closely: when assigning a character, it is first assigned to the sequence component of the tuple (this "stores" the letter but ignores case), and then the case of the character is separately stored to the mask-component. Both proxy objects update the respective parts of the alphabet tuple, so the combined state is stored. For reading the character representation, the reverse happens: the character representation of the sequence component is retrieved and, depending on the value of the mask component, returned as is or transformed to lower case.

While the specific way these functions are implemented in Snippet 6.12 is easy to understand and highlights the flexibility of accessing different aspects of the alphabet tuple in an object-oriented way, it should be clear that creating two proxy

---

[21] This is a common pattern in C++, it is for example used in bit-compressed vectors that do not hold their elements as distinct objects.

Table 6.9 Alphabet micro-benchmark. The first row is the base line, the second row is the implementation given in Snippet 6.12 and the third row is the actual implementation of `seqan3::masked` (Snippet A.6)

Micro-benchmark	`to_rank`	`assign_rank_to`	`to_char`	`assign_char_to`
`dna4`	138 ns	139 ns	139 ns	138 ns
`masked_obj<dna4>`	139 ns	140 ns	255 ns	580 ns
`masked<dna4>`	140 ns	140 ns	142 ns	145 ns

objects during assignment and writing to the base object twice is not the best solution from a performance perspective. The actual implementation in SeqAn3 therefore also pre-computes tables for character-to-rank conversion and vice versa. Such expressions for pre-computing tables are only marginally longer than the function definitions. The full implementation of `seqan3::masked` is given in the appendix (Snippet A.6). Table 6.9 shows the cost associated with not using such tables.

6.5.3 Alphabet "any" Types

As discussed above, `seqan3::alphabet_variant` can hold values of different types and perform a particular kind of type erasure. It offers type-safe access to the contained values, because it "knows" which type it currently represents. And for most use-cases that involve storing one type's values or another type's values in the same object, this is the best approach.

However, it has two drawbacks:

1. The types that are to be encoded need to be given as template parameters, and they cannot be arbitrary alphabets.
2. The alphabet size of the variant is the sum of its alternatives.

SeqAn3 offers a different kind of class template that also performs type erasure for alphabets: `seqan3::semialphabet_any<size>`. As the name suggests, it is modelled (vaguely) after `std::any` (see Sect. 3.10). Compared to `seqan3::alphabet_variant`, it has the following important differences:

- It is only a semialphabet and not an alphabet.
- It has a fixed size given as the only template parameter; possible types are not encoded.
- It is (explicitly) constructible from and convertible to *any* other (semi)alphabet of the *same size*.
- It does not "remember" which type it was assigned from, and this needs to be encoded separately if required.

- This also means no type safety; it is possible to convert it to a different type than it was assigned from.

Its main use-case is avoiding template instantiations in contexts where the type information is not needed, e.g. in algorithms that only work on the rank representation.

```
   template <typename alph_t>
2  void pre_algo(std::vector<alph_t> const & in)
   {
4      std::vector<semialphabet_any<10>> vec;
       std::ranges::copy(in,
6                        std::back_inserter(vec));

8      /* encode type in value */
       bool is_murphy =
10        std::same_as<alph_t, aa10murphy>;

12     algo(vec, is_murphy);
   }
```

```
                                                Type erasure
```

```
                                                    ↓
```

```
   void algo( // <- not a template
15    std::vector<semialphabet_any<10>> const & in,
      bool is_murphy)
17 [
      /*... algorithm implementation ...*/

      if (is_murphy)
21        post_algo(in | views::convert<aa10murphy>);
      else
23        post_algo(in | views::convert<aa10li>);
   }
```

```
   template <typename range_t>
26 void post_algo(range_t & in)
   {
28    // e.g. print the range:
      debug_stream << in << '\n';
30 }
```

```
                                                    ↓
```

```
                                                Reification
```

Code snippet 6.13: Type erasure and reification. This snippet shows how the alphabet type (`seqan3::aa10li` or `seqan3::aa10murphy`) can be *erased* via `seqan3::semialphabet_any<10>` (on the right). It can later be *reified* when needed again (on the left)

One such example is shown in Snippet 6.13. The left side contains two function templates that are each instantiated for `seqan3::aa10li` and `seqan3::aa10murphy`. On the right is an algorithm as a function (no template) that takes a vector over `seqan3::semialphabet_any<10>`. The algorithm could for example measure the distribution of alphabet values in the input; since it only needs the rank to perform this operation, it is not necessary to instantiate two instances of the algorithm. A practical example from SeqAn3 is the indexed search which can operate on semialphabets and does not need the character representation.

It should be noted that this mechanism is powerful but not very well known. The benefits in compile-time need to strongly outweigh the added complexity from such an approach.[22] While it is not likely to appear in many user applications, it is used internally in various places and might be useful when developing pre-compiled libraries or interfaces to other programming languages.

---

[22] In the example shown in Snippet 6.13, there is even runtime overhead due to the initial conversion; however, this need not be the case in all applications.

Why There Is No `seqan3::alphabet_any<char_type>`
A different kind of "any type" that type-erases to the character representation would be possible. It would allow assignment from any alphabet with the specified character representation and could be converted to any other with the same character type, e.g. also nonsense conversions like `'A'_dna4` ↔ `seqan3::alphabet_any<char>` ↔ `'A'_aa27`. The obvious drawback (beside the type-unsafety) is a strong increase of the alphabet size (to 256 or more)—but this may be irrelevant for the affected algorithm. However, since the built-in character types like `char` are already adapted as alphabets, one can use `to_char()`/`assign_char_to()` for the purpose, e.g. `'A'_dna4` ↔ `'a'` ↔ `'A'_aa27` (but this never happens implicitly).

6.6   The Quality Submodule

Most next-generation DNA sequencers emit a *quality* sequence together with the DNA sequence. This quality sequence has the same length as the DNA sequence and encodes for every nucleobase the probability that an error occurred in the sequencing of that base (Ewing et al., 1998; Ewing & Green, 1998). The quality submodule helps represent such quality values; see Table 6.10 for an overview. There are two quality definitions of which the first is the de facto standard now (P is the error probability):

Phred quality score   defined as $Q = -log_{10} P$; this score is always ≥ 0.
Solexa quality score   defined as $Q = -log_{10} \frac{P}{1-P}$; can be negative.

Quality sequences are typically found in FASTQ files (Cock et al., 2010) where the individual quality scores are stored as characters and not as numbers to achieve a 1-to-1 correspondence to the DNA sequence characters. The range of valid quality scores and the corresponding range of characters depend on the sequence technology and generation.

Table 6.10 Quality submodule overview

Alphabet: Quality submodule	
Concepts	`seqan3::quality_alphabet`, `seqan3::writable_quality_alphabet`
Class types	`seqan3::phred42`, `seqan3::phred63`, `seqan3::phred68legacy`, `seqan3::qualified`, `seqan3::quality_base`
Function objects	`seqan3::assign_phred_to`, `seqan3::to_phred`
Traits	`seqan3::alphabet_phred_t`

6.6.1 General Design

Since the alphabet design already covers character representations, the main design task for the *Quality* submodule is to handle access to the quality scores. These are modelled as their own independent representation. To this end, the submodule provides the `seqan3::assign_phred_to` and `seqan3::to_phred` CPOs that perform conversion from/to the phred representation. They are accompanied by the (non-customisable) `seqan3::alphabet_phred_t` trait which is always the return type of `seqan3::to_phred` (and typically `int8_t`).

On a design level, SeqAn3 only works with phred quality scores and not Solexa quality scores, because the latter have long been deprecated. There are, however, facilities to store Solexa values and convert to/from them, see below.

The alphabet interface additions specific to qualities contain both a read-only aspect and a writable aspect, so the submodule also defines two new concepts:

`seqan3::quality_alphabet`   Requires `seqan3::alphabet` and additionally a valid implementation of the `seqan3::to_phred` CPO.

`seqan3::writable_quality_alphabet`   Requires `seqan3::quality_alphabet`, `seqan3::writable_alphabet` and a valid definition of the `seqan3::assign_phred_to` CPO.

The *Quality* submodule also provides `seqan3::quality_base` which derives from `seqan3::alphabet_base`. Due to all the value ranges (rank, phred, character) being contiguous for quality alphabets, the base class can generate all required conversion tables for derived classes based on the first respective character/phred value (which that derived class needs to define). It also provides a constructor from the numeric quality values and a converting constructor from other quality alphabets (conversion is based on the quality values).

6.6.2 Quality Alphabets

The predefined alphabets and their representations are shown in Table 6.11. The most commonly used quality score notation is the original Sanger phred notation which is also used by Illumina since version 1.8 of their pipeline[23] and formats like SAM (Li et al., 2009). Since phred scores are ≥ 0 (see the definition at beginning of this section), the phred representation can be stored internally and is identical to the rank. The character offset for this notation is 33, i.e. the character range begins at the character ′!′.

Originally the Sanger phred notation was meant to hold quality values from 0 to 93, but any sequencing technology since the Illumina 1.3 pipeline only

---

[23] https://www.illumina.com/Documents/seminars/presentations/2011_09_smith.pdf.

Table 6.11 Quality alphabets and their representations. The first two types only differ in the alphabet size

Format	Type	Phred repr.	Rank repr.	Char repr.
Illumina ≥ 1.8 & Sanger	`seqan3::phred42`	[0 .. 41]	[0 .. 41]	[`'!'` .. `'J'`]
Illumina ≥ 1.8 & Sanger	`seqan3::phred63`	[0 .. 62]	[0 .. 62]	[`'!'` .. `'_'`]
Illumina < 1.8 (& Solexa)	`seqan3::phred68legacy`	[-5 .. 62]	[0 .. 67]	[`';'` .. `'~'`]

produced values up to 41 (Cock et al., 2010). Higher values were reserved for confidence achievable through post-processing; typically values only up to 62 are used. `seqan3::phred42` and `seqan3::phred63` both implement this notation, one with a maximum representable quality value of 41 and the other with 62. Both accept quality values outside their range which will result in a conversion to the respective maximum (or minimum for negative values). As with other alphabets, the main difference is the size of the alphabet.

Before version 1.8, versions 1.3 and 1.5 of the Illumina pipeline were popular. Their quality notation was also phred-based, but the character offset was 64 (character `'@'`) instead of 33 (character `'!'`). A type for these notations is provided for backwards-compatibility: the `seqan3::phred68legacy` alphabet. However, a second type with a smaller value range is not provided since these formats are rarely used today.

To be able to also represent Solexa quality values, `seqan3::phred68legacy` has a minimum quality value of −5. This means that for `seqan3::phred68legacy`, rank and phred value are not identical, and phred values are shifted by −5. Solexa quality notation has the same character offset as legacy Illumina, so this is compatible—both have quality value 0 at character `'@'`, and however the valid range now starts at −5 (character `';'`).

It should be noted that SeqAn3 in general assumes phred scores. If data is read from a file in the `seqan3::phred68legacy` alphabet and it indeed contains negative quality values, this indicates that the underlying score is Solexa and not phred. The *Quality* submodule will offer functions to correctly convert these scores based on the previously defined formulas, and however, for higher quality values, their difference is negligible.

6.6.3 Quality Tuples

Quality alphabets can be combined with nucleotide alphabets via `seqan3::qualified`, a derived type of `seqan3::alphabet_tuple_base`. The main advantage is compression, and the following combinations all still fit into a single byte:

- `seqan3::qualified<seqan3::dna4, seqan3::phred42>` (alphabet size 168)

- `seqan3::qualified<seqan3::dna4, seqan3::phred63>` (alphabet size 252)
- `seqan3::qualified<seqan3::dna5, seqan3::phred42>` (alphabet size 210)

A vector over such a combined alphabet can thus save 50% of storage space compared to two separate vectors (of course the same combinations with RNA alphabets are possible).

`seqan3::qualified` exposes the quality component's phred interface as its own, so it models the quality concept and can be used in any context that relies on respective interfaces. In contrast to `seqan3::masked` which combines an alphabet with a semialphabet and can generate a new unique character representation, `seqan3::qualified` has two existing character representations to handle. Since the phred interface already exposes the quality component and the nucleotide component is usually the "primary" information if both are combined, I chose to implement the character representation of `seqan3::qualified` to simply be that of the nucleotide component. This not only allows using `seqan3::qualified` as if it were a nucleotide but also allows access to the phred scores when for example wanting to trim the sequence. Some practical use-cases will be discussed in Sect. 7.3.2.

An important implication of this approach is that the state of a `seqan3::qualified` object is not fully determined by the character representation and assignment from `char` cannot rely on a fixed char-to-rank conversion table.[24] Thus, an optimisation, as performed for `seqan3::mask`, is not possible and writing the character representation is bound to be slower than for other alphabets. The current implementation of `seqan3::alphabet_tuple_base` generates a new bit representation for the value range, so a char assignment of `seqan3::qualified` leads to multiple arithmetic operations on the current rank. Changing the internal design of `seqan3::alphabet_tuple_base` to store the components' rank values in independent bit representations adjacent to each other would make this update operation faster (although still slower than for other alphabets), because the arithmetic operations could be replaced by bit operations. However, this would come at the cost of compressibility as bits would be wasted, e.g. with the current design `seqan3::qualified<seqan3::dna5, seqan3::phred42>` can be represented in a byte, because the combined alphabet size is $5 * 42 = 210 \leq 256$, but with the changed design, the type would require two bytes ($\lceil log_2(5) \rceil + \lceil log_2(42) \rceil = 3 + 6 = 9$ bits).

Since compression is a main reason for using `seqan3::alphabet_tuple_base` and data is likely to be read more often than written, this compromise is deemed

---

[24] The quality component contributes to the combined object's rank and is not represented in the character being assigned. For example, `seqan3::qualified{'A'_dna4, 34}.assign_char('C')` should only change the dna4 aspect of the object and not quality information.

acceptable. It should be noted that such a performance overhead is inherent to updating only part of the compressed representation, and also any manually implemented alphabet with the same interfaces and storage properties will exhibit it.

6.7   Discussion

Table 6.12 shows an overview over alphabets in SeqAn3, SeqAn2 and BioPython. On SeqAn3's side further quality alphabets, `seqan3::aa10li`, the RNA structure alphabets and the CIGAR alphabet are not shown—all of which are not available in SeqAn2 and BioPython (except `Dna5Q` in SeqAn2). BioPython provides only few more reduced amino acid alphabets (but also lacks `seqan3::aa10li`). SeqAn3's alphabets are clearly a super-set of those available SeqAn2 and BioPython.

Table 6.12   Alphabets in SeqAn3, SeqAn2 and BioPython

SeqAn3 (seqan3::)	SeqAn2 (seqan::)	BioPython
dna4	Dna	IUPACUnambiguousDNA
dna5	Dna5	
dna15	Iupac[a]	IUPACAmbiguousDNA
rna4	Rna	IUPACUnambiguousRNA
rna5	Rna5	
rna15	Iupac[a]	IUPACAmbiguousRNA
aa20		IUPACProtein
aa27	AminoAcid	ExtendedIUPACProtein
aa10murphy	ReducedAminoAcid<Murphy10>	Murphy10
gap		
gapped<dna4>	GappedValueType<Dna>	Gapped(...)
alphabet_variant		AlphabetEncoder
phred63		
mask, masked<>		
qualified<dna4, phred63>	DnaQ	
dssp9[b]		SecondaryStructure

[a] This has size 16 (distinct U) in SeqAn2 but size 15 in SeqAn3 and BioPython
[b] Protein structure alphabet; BioPython only supports a subset of the valid symbols

6.7.1 Performance

Table 6.13 shows micro-benchmarks for a large subset of SeqAn3's alphabets. As can been seen, almost all operations on the alphabets are as fast as assigning a value to a `char`.[25] This shows that performance-wise the alphabet interface is a *zero-cost abstraction*. Beyond those alphabets implemented directly, SeqAn3 also provides composite alphabets which are one abstraction level higher, because they create new alphabets from existing ones. The table shows that `seqan3::alphabet_variant` (including the example `gapped<dna4>`) is also always a *zero-cost abstraction*. For composite alphabets derived from `seqan3::alphabet_tuple_base`, the abstraction is also free in the sense that it is not possible to implement better behaviour by hand.

However, the design of the latter comes with an inherent theoretical overhead for some cases: those derived types where the mapping from character representation onto ranks is *surjective* perform as well as other alphabets (e.g. `masked<T>`). But for those types where this does not hold, a higher cost is associated with assigning

Table 6.13 Alphabet micro-benchmarks. Alphabets marked with [a] are more than one byte in size

SeqAn3: Regular alphabets	to_rank	assign_ rank_to	to_char	assign_ char_to
gap	144 ns	140 ns	145 ns	146 ns
dna4	146 ns	145 ns	138 ns	149 ns
dna5	146 ns	146 ns	138 ns	148 ns
dna15	146 ns	146 ns	138 ns	148 ns
aa20	146 ns	146 ns	141 ns	147 ns
aa27	145 ns	145 ns	139 ns	146 ns
SeqAn3: Composites				
gapped<dna4>	146 ns	146 ns	139 ns	148 ns
alphabet_variant<dna4, char>[a]	146 ns	146 ns	148 ns	148 ns
masked<dna4>	146 ns	146 ns	142 ns	148 ns
qualified<dna4, phred42>	146 ns	146 ns	148 ns	709 ns
qualified<dna5, phred63>[a]	146 ns	146 ns	148 ns	814 ns
SeqAn3: Adaptations				
char	146 ns	146 ns	145 ns	146 ns
char32_t[a]	146 ns	146 ns	146 ns	146 ns
SeqAn2	ordValue	=	(char)	=
Dna	147 ns	148 ns	138 ns	149 ns
AminoAcid	146 ns	150 ns	140 ns	148 ns
GappedValueType<Dna>::Type	143 ns	239 ns	142 ns	241 ns
Dna5Q	150 ns	200 ns	151 ns	200 ns

---

[25] This is exactly what happens for `seqan3::assign_char_to` on `char`.

from the character representation, because essentially one component needs to be extracted, updated and then merged with the other components' states. This is described on page 181 in more detail and is visible for `seqan3::qualified` in Table 6.13.

SeqAn2's `Dna5Q` does not suffer from the same overhead as SeqAn3's `qualified` types, because it does not truly treat the states as independent, i.e. assignment of a new character resets also the quality information. However, both forms of assignment (even by rank) have a notable overhead, because they involve a conversion plus bit-shifting. `seqan::GappedValueType` is the only true composite in SeqAn2, and its assignment functions (including by rank) require notably more time, because an `if`-check is used instead of conversion tables. Regular SeqAn2 alphabets have the same performance as SeqAn3 alphabets.

6.7.2   Simplicity

On a very fundamental level, the design of alphabets in SeqAn3 is rooted in that of prior SeqAn versions. SeqAn2 already offered for all of its alphabets the possibility to implicitly cast to `char` which would return what is now considered the character representation. It also allowed casting to any other integral type which would return what is now considered the rank representation. To explicitly request the latter, it also offered the free function `ordValue()`.

```
   seqan::Dna d{'A'};        // assigns character representation -> A
2  d = 1;                    // assigns rank representation       -> C

4  unsigned char c = 'G';
   d = c;                    // oops, assigns rank 71, undefined behaviour!
6
   seqan::Iupac i;
8  i = 'A' + 1;              // does not assign 'B' because integral promotion to 'int'
                             // assigns rank 66, undefined behaviour!
```

Code snippet 6.14: SeqAn2's alphabet interface. For the most obvious use-cases, the behaviour was as expected, but there was also large potential for silent error

As I discussed initially, the design left room for ambiguity and unexpected (as well as undefined) behaviour. Snippet 6.14 shows some examples. These cannot happen in SeqAn3, because access to either representation is always explicit. Since SeqAn2 defined implicit conversions to and from all integral types for its alphabets, there was also no way to detect unwanted conversions via compiler warnings. It also did not offer assertions for invalid rank values and it offered no interface to detect assignment of invalid character values. Both of these shortcomings have been addressed in SeqAn3's design leading to much safer code that is easier to use and debug.

Altogether the alphabet interfaces in SeqAn3 are a little more explicit but not more complicated than before. Understanding the design background given in this chapter (concepts, CPOs, CRTP-bases etc.) is not a prerequisite for *using* SeqAn3 alphabets.

6.7.3 Integration

Generic interfaces that take alphabets in SeqAn3 are constrained by alphabet concepts (Sect. 6.1.3) that use customisation point objects (CPOs; Sect. 6.1.2). When new user defined types are designed, these can be directly modelled to satisfy the requirements of the CPOs/concepts. Existing user-defined types can be integrated via wrappers as explained in Sect. 6.2.2. For most types, this can be done without ever specialising templates in SeqAn3 or even opening namespace `seqan3::`.

Due to the presented design, the ability of SeqAn3 to integrate third party types is far superior to SeqAn2's. Multiple ways exist to satisfy the requirements of the CPOs/concepts which reflect different use-cases. The more common use-cases are covered by easier solutions and the more exotic use-cases have slightly more elaborate solutions. Entire groups of third party types can be integrated without code duplication (see Snippet A.3); this was not possible in SeqAn2 (without changing SeqAn2's library code).

6.7.4 Adaptability

The existing set of alphabets can easily be *extended* by adding new alphabets; multiple examples were given. Generic use-cases have been formalised as concepts. Care has been taken to codify the respective minimum requirements so that interfaces are not over-constrained and users need not implement alphabet functionality that they do not use. Using concepts makes the interfaces much more *generic*, and the way the concepts are defined allows to adapt and integrate other types much more easily. Multiple *refinements* of the main alphabet concepts were presented (e.g. `seqan3::nucleotide_alphabet`), and, with some practice, users should have no problems defining their own refinements if desired.

6.7.5 Compactness

An overview of the provided alphabets is given at the beginning of this section (Table 6.12). The feature set of SeqAn3 is greater than that of SeqAn2 while at the same being much more compact: the SeqAn3 alphabets (including base classes and detail code) amount to 1271 lines of code (loc), while SeqAn2 (including alphabet code in basic) requires 2070 loc.

A closer look at an example alphabet reveals the details. The "definition" of `seqan::Dna5` consists of:

- Ten specialisations of SeqAn2 metafunctions (`ValueSize<>`, `BitsPerValue<>`, ...)
- Six free function (template) overloads (`unknownValueImpl()`, `assign(char, Dna5)`, ...)
- Four manually defined conversion tables—two of which are 256 hard-coded values

The code is shown in the appendix (Snippet A.7) and highlights how complex and exotic a "type definition" in SeqAn2 is. It should also be noted that this is only the "local" part of `seqan::Dna5`, e.g. `seqan::Dna5Q` has additional tables and functions that are specific to interaction with `seqan::Dna5`.

In contrast, the definition of `seqan3::dna5` encompasses a single type definition with:

- A converting constructor for `seqan3::rna5`.
- Three conversion tables: `rank_to_char` (5 fixed values), `complement_table` (5 fixed values) and `char_to_rank` (generated by short lambda expression).
- Remaining members are inherited from `seqan3::nucleotide_base`.

SeqAn3's alphabets are built up much more modularly by making use of base classes and composite alphabets. SeqAn2 has no equivalent for `seqan3::alphabet_tuple_base` and consequently implements `seqan::DnaQ` and `seqan::Dna5Q` as their own fixed types. It does provide the template `ModifiedAlphabet<T, ModExpand<CHAR, TSpec>>` which behaves a little like `seqan3::alphabet_variant` and is the basis for `GappedValueType<T>`, but it only allows expanding an alphabet by exactly one character and does not provide type-safe access. The metaprogramming behind this alphabet is a lot more extensive, with the cloc tool reporting 857 loc vs. only 326 for `seqan3::alphabet_variant` (which is arguably a lot more generic and powerful).

One can summarise that SeqAn3 achieves more features with simpler and smaller interfaces and notably less code.

Chapter 7
The Range Module

I have extensively written about C++ Ranges in Sect. 3.6. SeqAn3 uses and recommends using the containers and views provided by the standard library (or indirectly through SeqAn3's *STD* module). In contrast to SeqAn2, it only offers custom types if the functionality is not available in the standard and there is actual demand for a new data structure.

Of course this is still often the case and many general-purpose ("non-biological") ranges that are not (yet) part of the standard library are provided by SeqAn3's range module. The *Range* module also provides ("biological" or "bioinformatical") ranges specific to SeqAn3 alphabets. See Table 7.1 for an overview.

Ranges, on the other hand, that are only useful in the context of another module (e.g. alignment or search) are found in that respective module. This follows the principle that the library is divided up by use-cases. The *Alphabet* module and the *Range* module together address the use-case of sequence storage and manipulation.

The *Range* module provides the following submodules:

Container   Provides container concepts, general-purpose containers and alphabet-specific containers; see Sect. 7.2.

Views   Provides general-purpose and alphabet-specific view adaptors as well as utilities to simplify the definition of such; see Sect. 7.3.

7.1 General Design

Almost all the general range machinery (concepts, traits etc.) is provided by the standard library/*STD* module. The only larger addition by SeqAn3 to the design space of C++ Ranges is *decorators*. Decorators are ranges that differ in their storage behaviour from containers (which own all their elements) and views (which do not own their elements) by taking an intermediate position: they depend on an existing range (typically a container) but annotate ("decorate") it with additional data. This

© The Author(s), under exclusive license to Springer Nature Switzerland AG 2022
H. Hauswedell, *Sequence Analysis and Modern C++*, Computational Biology 33,
https://doi.org/10.1007/978-3-030-90990-1_7

Table 7.1 Range module overview

Range module	
Submodules	Container, views

Concepts	`seqan3::const_iterable_range,`
	`seqan3::pseudo_random_access_iterator,`
	`seqan3::pseudo_random_access_range`

is fundamentally different from views which may also adapt existing ranges but provide an *algorithm* on top of the existing data and not additional data; destructing and possibly copying a view must be in $O(1)$. In contrast, decorators may contain arbitrary additional data that may result in a runtimes of $O(n)$.

There is currently no dedicated submodule for decorators, because there is only a single one (`seqan3::gap_decorator`) which is found in the *Alignment* module (discussed in Chap. 10). A port of SeqAn2's journaled string under the name of `seqan3::journal_decorator` is planned for a later release of SeqAn3. It will allow adapting a container and recording changes (substitutions, insertions, deletions) inside the decorator and without updating the underlying container. This is useful in modelling mutations in sequence data and is the basis for the *journaled string tree* (Rahn et al., 2014), a data structure that models a container of sequences where all sequences but the first are expressed as modifications of the first. The method is also known as reference-compression and is also planned for a later release of SeqAn3.

The only entities the *Range* module provides at top level (not in a submodule) are the following concepts:

`seqan3::const_iterable_range`   This concept verifies whether for a given input range type `rng_t` the type `rng_t const` is also a range—and whether that range has identical "strength" (input or forward or ...) as the original. While the recommended way of accepting ranges in generic code is via forwarding references (`rng_t && r`, see Sect. 3.2.2), it is still useful to know whether it is possible to also accept a given range via `rng_t const & r`. In particular this is not true for certain views that have an internal state that changes when iterating over the view.

`seqan3::pseudo_random_access_iterator` and `seqan3::pseudo_random_ access_range`   The standard concept `std::random_access_iterator` has certain syntactical requirements on the iterator, e.g. `operator+` and `operator[]` for "jumping". But the concept also has the semantic requirement that random access is in $O(1)$. SeqAn3 provides certain iterators/ranges that satisfy the syntactic requirements of `std::random_access_iterator`/ `std::ranges::random_access_range` but where the runtime behaviour is between constant and linear (typically $O(log(n))$). These are covered by SeqAn3's pseudo-random access concepts, an intermediate "level" between the bidirectional and random access concepts (see Fig. 3.1 on p. 65 for a visualisation of the concept hierarchy). Subsection 10.1.2 illustrates how these concepts are used in practice.

7.2 Container

Containers are ranges that own all their elements and do not "depend" on other ranges, e.g. `std::vector`. Since SeqAn3 primarily works with standard library containers, it is important to handle and classify these correctly. As discussed in Sect. 3.6.2, C++ 20 adds range concepts. Different containers meet different range concepts, but containers typically allow many more actions (e.g. being able to insert elements). Requirements for these actions are described as standard language (ISO/IEC 14882:2017, 26.2) but have not (yet) been turned in C++ concepts. Whether to add such concepts and how to design them are described below.

Those standard containers that allocate dynamic memory (all but `std::array`) do so via *allocators*. The allocator type can be given as a template argument to these containers and can be used to for example allocate from a predefined memory pool or perform better in highly parallel environments. For most cases, the default allocator provides a very good performance, but in the context of SIMD operations this is not always true. SIMD operations rely on elemental types that are larger than the machine word size (up to 512 bits), and transforming existing data into compatible data is significantly faster when this data is already aligned to memory blocks of the target size. To guarantee this, SeqAn3 offers `seqan3::aligned_allocator` that can be plugged into all allocator-aware standard library containers.

Finally, the container module provides several container templates that fulfil roles not covered by the standard library types. These are discussed in their own sections below. See Table 7.2 for an overview.

It should be noted that while the standard discusses *associative containers* (maps, sets) together with sequence containers, these are not handled here as there was no need to extend these or define concepts for them, yet.

7.2.1 Concepts

As already elaborated on in the context of designing the alphabet concepts, one often has the false intuition to derive concept requirements from the interfaces of existing types instead of the other way around. The first design for SeqAn3 container

Table 7.2 Container submodule overview

Range: container submodule	
Concepts	`seqan3::back_insertable_with`
Class types	`seqan3::aligned_allocator`, `seqan3::bitcompressed_vector`, `seqan3::concatenated_sequences`, `seqan3::dynamic_bitset`, `seqan3::small_string`, `seqan3::small_vector`

concepts followed this approach, essentially dividing up the requirements discussed in the standard (ISO/IEC 14882:2017, 26.2) into four concepts that vaguely map onto the interfaces of the container types and formalising these with no regard to how they relate to each other or the range concepts. An immediate problem with this approach already was that the interfaces are not subsets of each other and it is difficult to define a concept hierarchy.

Instead, one should design the concepts based on use-cases. Analysing the use-cases leads to the first important observation: **Whenever data is just being read, the range concepts are sufficient to interact with containers.** And if one only writes to individual elements, `std::ranges::output_range` is still sufficient. The nature of a container comes into play only if elements are to be inserted or removed. A comprehensive analysis of the SeqAn3 codebase revealed that surprisingly few places perform such modifications on objects whose type is generic—and those places that do only ever use `.push_back()`.[1]

As such, it was decided to only define the following concept:

`seqan3::back_insertable_with<cont_t, element_t>` This concept checks that the type `cont_t` offers the member function[2] `.push_back()` and that this accepts objects of type `element_t`. Calling the function is assumed to be in amortised $O(1)$ with respect to the current container size.

This is cognate with `std::back_insert_iterator`, a standard library iterator that can be created on exactly those containers. If use-cases arise for further container concepts, e.g. `front_insertable_with` (requires `.push_front()`), `insertable_with` (requires `.insert()`) or `reservable` (requires `.reserve()`), these can be added later. In theory, it would be possible to subsume (some of) these into container concepts, but based on the aforementioned study of use-cases and the inconclusive overlap of features between standard library containers, I anticipate little need for such concepts.

The container types defined by SeqAn3 still try to offer the full standard library interfaces (not just `.push_back()`). This is done to enable them to be used in contexts not anticipated by SeqAn3 or even as drop-in replacements for standard library containers. Auxiliary concepts and integration tests are used internally to ensure this compatibility, but it is not relevant for SeqAn3's own generic interfaces.

---

[1] It is important to stress *generic contexts* here, because, of course, `std::vector` is used in many places including its other member functions. But in those cases, the type is fixed and not interchangeable by the user.

[2] It would be possible to add a CPO instead, however, there are currently only few places this concept is used and the specific member function is well-established. A CPO can still be added later if greater extensibility is required.

7.2.2 Bit-Compressed Container

In the context of alphabet sizes, I have already explained that a single object in C++ always requires at least one byte of storage in memory if stored individually. Since almost all alphabets can be represented by fewer bits, significant storage space can be saved by *bit-compressing* the storage, i.e. not storing the elements individually but in a compressed form.

```
  std::vector<seqan3::dna4>              v0{"ACGT"_dna4}; // data occupies 4 bytes in memory
2 seqan3::bitcompressed_vector<seqan3::dna4> v1{"ACGT"_dna4}; // data occupies 1 byte in memory
```

Code snippet 7.1: A bit-compressed vector. Note that the sizes given are conceptual and that, due to over-provisioning, both types have likely pre-allocated more memory

This can be achieved via `seqan3::bitcompressed_vector` which has an interface almost identical to `std::vector` and is displayed in Snippet 7.1. But since elements are not stored *as is* adjacently (they are compressed), the type models "only" `std::ranges::random_access_range` and not the stronger `std::ranges::contiguous_range` concept. The bit-compressed vector requires that its elements be alphabets and be default-constructible (the latter is required by all containers).

A similar type was available in SeqAn2 under the name `seqan::String<T, seqan::Packed<THostSpec>>`. The implementation in SeqAn3 uses `sdsl::int_vector` from the SDSL library (Gog et al., 2014) internally but provides a full container interface and performs conversion to/from rank (which is then stored in the `sdsl::int_vector`). Calling `operator[]` on the container or dereferencing its iterators yields a *proxy type* that is derived from the actual alphabet type (and convertible to it) but provides custom assignment operators to update the underlying data in the correct way. This is similar to how accessing `seqan3::alphabet_tuple_base` works; in fact, both have reference types that derive from `seqan3::alphabet_proxy`, an auxiliary type that exists exactly for the purpose.

7.2.3 Containers of Containers

In bioinformatics, one often does not only handle single sequences but also collections of sequences, e.g. a set of reads or a protein database. These can easily be stored in a container of containers, e.g. `std::vector<std::vector<seqan3::dna4>>` or `std::vector<seqan3::bitcompressed_vector<seqan3::dna4>>` to reduce storage size. However, experience with SeqAn2 has shown that often the cache behaviour of such two-dimensional containers is bad, because subsequent elements of the outer container (the individual sequences) may be allocated in entirely different memory regions.

```
2   seqan3::concatenated_sequences v1{"ACGT"_dna4, "GATA"_dna4};

    /* access concatenation directly */
4   std::cout << v1.concat_size();      // == 8 (cumulative size in constant time)
    v1.concat_reserve(10000);           // pre-allocate if future cumulative size can be anticipated
```

Code snippet 7.2: Concatenated sequences. The full type of `v1` is automatically deduced via CTAD (Sect. 3.1.2) to `seqan3::concatenated_sequences<std::vector<seqan3::dna4>>`

SeqAn2 offered `seqan::StringSet<TString, seqan::Owner<seqan::ConcatDirect>>` to address this issue, a data structure that appears like a container of containers but internally stores all sequences concatenated into a single one plus a vector of sequence begin-positions. Accessing individual sequences then returns an "infix" on the underlying container. This design is only useful for data that is read often and rarely updated, because inserting new containers into the outer container becomes very expensive (the data of all containers might have to be copied where before it could have been moved).

SeqAn3 offers a similar type called `seqan3::concatenated_sequences`. Useful applications can be seen in Snippet 7.2. It should be noted that this type (the "outer" container) is also not a contiguous range, because it does not store values of its value type at all; i.e. `v1` in Snippet 7.2 has a value type of `std::vector<seqan3::dna4>`, but the type returned by `operator[]` will be a view type similar to a `std::span<seqan3::dna4>` (this "inner" type may model `std::ranges::contiguous_range`).

7.2.4 Fixed-Capacity Containers

All containers other than `std::array` use dynamic memory allocation due to their size being dependent on runtime choices. But dynamic memory allocation is considerably more expensive than stack storage, and often it is known at compile-time that memory requirements will never exceed a certain threshold. This includes local buffers in algorithms, fixed-width identifier strings and any storage that is guaranteed to be very small. While `std::array` can be used in these situations, it is not very convenient, because its size is fixed.

The solution to this problem is a container that is internally implemented as an array but behaves as a vector; it has dynamic size but fixed maximum size/capacity. A huge advantage of this type is that it can also be used at compile-time and that it is trivially copyable (and thus types that hold a member of this type can also be trivially copyable).

```
    constexpr small_string s1{"foo"};      // small_string<3>
2   constexpr small_string s2{"bar"};      // small_string<3>

4   constexpr small_string s3{"foobar"};   // small_string<6>
    constexpr small_string cm = s1 + s2;   // small_string<3 + 3>
6
    static_assert(cm == s3);
```

Code snippet 7.3: Small strings. Note that concatenation and comparison happen at compile-time. Notice also how helpful automatic deduction of template arguments is in this context and that the null-terminator is handled without exposing it to the user

SeqAn3 provides two types that are based on `std::array` and behave in such a way:

`seqan3::small_vector<T, cap>` Behaves exactly like `std::vector<T>` but has a `max_size()` and fixed `capacity()` of `cap`. It models `std::ranges::contiguous_range`.

`seqan3::small_string<cap>` Like `seqan3::small_vector<char, cap>` but adds a null-terminator and is convertible to `std::string` and C-Strings. It is also printable and can be concatenated with `operator+`. It is shown in Snippet 7.3.

```
    constexpr dynamic_bitset b0{"010010101"};      // can be initialised from string literals!
2
    /* query properties at compile-time */
4   static_assert(b0.size()    == 9);      // size
    static_assert(b0.count()   == 4);      // number of set bits (popcount)
6   static_assert(b0[0]        == 1);      // right-most element
    static_assert(b0[1]        == 0);      // next element
8
    /* binary operations at compile-time */
10  constexpr dynamic_bitset b1{"010000101"};
    constexpr dynamic_bitset b2 = b0 ^ b1;         // XOR two bitsets
12  constexpr dynamic_bitset b3 = b0 << 1;         // bit shift by 1
14  static_assert(b2 == dynamic_bitset{"000010000"});
    static_assert(b3 == dynamic_bitset{"100101010"});
```

Code snippet 7.4: Dynamic bitsets. The bitset can be treated as either a container of bits or an unsigned integer. All operations on it can be performed at compile-time and the size is dynamic up to the given maximum capacity (56 by default)

The *Container* submodule also provides `seqan3::dynamic_bitset<bit_cap>` which is a very interesting data structure. It combines the properties of `seqan3::small_vector<bool, bit_cap>` and `std::bitset`, i.e. it has a full vector-like interface with access to individual bits via `operator[]` and the ability to insert, push and pop elements. But it also has an integer-like interface with typical binary operators (AND, OR, XOR), bit-shift operators and convenience functions to retrieve the number of set/unset bits. And all of this is `constexpr`-safe; examples are shown in Snippet 7.4.

Storage of this type is a plain `uint64_t` and the dynamic size is encoded within this integer, as well. The data structure is very compact. The maximum size is currently limited to 56 (8bits reserved for size), but it is planned to expand the data structure, so it can rely on an array of integers for higher desired capacities. It enables encoding bit patterns, and it is the basis for defining the shapes of gapped k-mers (see Sect. 9.2.1). Algorithms can be optimised based on certain properties of bit-patterns (e.g. whether any 0s are contained at all), and if the shape is fixed at compile-time (and can be queried), this allows for even greater optimisation. It highlights how one can use the same data structure and very similar interfaces for storing compile-time and runtime arguments in Modern C++.

7.3   Views

Views play a central role in SeqAn3. In contrast to containers, they generate their elements via an algorithm or transformation, usually from an existing range (either a container or another view). Because they can be combined easily and their design is well suited to the design goals of the library (see Sect. 4.1), views are the preferred way of implementing all algorithms on ranges. See Table 7.3 for an overview of the submodule. Table 7.4 shows that SeqAn3 provides many such views with very different purposes. This subsection will introduce alphabet-specific views and some general-purpose views but cannot cover all of them for space reasons. The background of views has been discussed extensively in Sect. 3.6.3, but this subsection will give some details on the implementation.

7.3.1   General Design

As I explained in Sect. 3.6.4, the view type (template) is usually associated with a function object that can create objects of the actual view type (and facilitate the piping behaviour). These function objects are called a "range/view factory" if they do not depend on another range as input; they can only be placed at

Table 7.3 Views submodule overview

Range: views submodule	
Class types	`seqan3::views::deep`
Function objects	Various, see Table 7.4

Table 7.4 Views provided by the *Views* submodule. *A/F/T* denotes an adaptor/factory/terminator. The implicit range parameter is omitted. Namespace `seqan3::views::` assumed for the left column

Subrange-y views	↔	Description
`slice(n, m)`	A	= `views::take(m)` | `views::drop(n)`.
`take_exactly(n)`[a]	A	Like `std::views::take` but always returns sized range.
`take_until(fn)`[a]	A	Like `std::views::take_while` but inverted predicate.
`take_until_and_consume(fn)`[a]	A	Like `views::take_until` but skips final element(s).
`take_line`[a]	A	Like `views::take_until_and_consume` with EOL functor.
Element-wise conversion		
`convert<T>`	A	Convert type to T (implicitly or via `static_cast<T>`).
`as_const`	A	Convert type to `const` type.
`move`	A	Convert type to `&&` type.
`to_lower`	A	Convert value to lower-case (`char` input only).
`to_upper`	A	Convert value to upper-case (`char` input only).
Alphabet-related		
`to_char`	A	Convert from alphabet to its character representation.
`to_rank`	A	Convert from alphabet to its rank representation.
`char_to<T>`	A	Convert to alphabet T from its character representation.
`rank_to<T>`	A	Convert to alphabet T from its rank representation.
`complement`	A	Generate complements from range of nucleotides.
`translate_single(frame)`	A	Translate nucleotides to amino acids.
`translate(frames)`	A	Translate nucleotides to multiple frames of amino acids.
`translate_join(frames)`	A	Optimised version of `views::translate` | `std::views::join`.
Tuple/pair related		
`get<i>`	A	Calls `get<i>()` on every element.
`zip(rng1, rng2)`	F	Creates a range-of-tuples from a multiple ranges.
`unzip`	T	Creates a tuple of ranges from a range-of-tuples.
`all_pairs(rng1[, rng2])`	F	Creates tuples of all elements between ranges.

(continued)

Table 7.4 (continued)

Subrange-y views	↔	Description
Miscellaneous		
`async_input_buffer(size)`	A	Cache elements from underlying range in concurrent queue.
`enforce_random_access`	A	Makes a random-access-range from a pseudo-random-access-r.
`interleave(str, i)`	A	Interleave `str` into underlying range every `i` elements.
`persist`	A/F	Makes a view from a non-viewable range.
`repeat(v)`	F	The value `v` repeated infinitely.
`repeat_n(v, n)`	F	The value `v` repeated `n` times.
`simplify_type`	A/F	Perform type-erasure if possible.
`single_pass_input`	A	Turn every range into single-pass input range.
`to<container_t>`	T	Converts the whole view into a container (by copying).

[a] For each of these adaptors, another one called `*_or_throw` exists that throws an exception if the end is reached without satisfying the condition.

the beginning of a series of pipe operations. The function objects are called a "range/view adaptor" if they require another range as input; these can only appear after another range in a series of pipe operations. SeqAn3 also introduces the term "range/view terminator" for objects that are used at the end of a series of pipe operations but that do not actually return a view. The latter are not very common, but the `seqan3::views::to<T>` terminator is important because it allows converting a view (back) into a container.

Snippet 7.5 shows examples of the objects in use. The respective objects in the standard library are located in the namespace `std::views::`, so their counterparts in SeqAn3 are located in `seqan3::views::`. Since the function objects are the recommended way of creating views (instead of the view type's constructor), it was decided to not expose the actual view type in SeqAn3 at all. This has the advantage that the interface is easier to learn, because there is only one way to get a view. It also gives future SeqAn3 developers more room to change the actual returned type, because the documentation does not guarantee a specific type at all, it only promises that the type returned behaves in a certain way. One implication is that `auto` needs to be used liberally, but this is already strongly recommended for views because the resulting types grow very complex quickly.[3]

If the actual view types are not publicly specified in SeqAn3, it is very important to specify in documentation what users can expect of these types that are

---

[3] The original designers of "C++ Ranges" stated in my presence that they would have preferred such a design for the standard library, as well, but that it was not possible because the standard library requires ABI stability which implies fixed types. SeqAn3 deliberately makes no promises in regard to ABI (Subsection 4.3.4), so it can choose a different design here.

```
   /* views::repeat is a "factory", must come at beginning of pipe */
2  auto v0 = seqan3::views::repeat(false) | /*...*/ | /*...*/; // repeats the value 'false' infinitely

4  /* views::take is an "adaptor", must come after another range */
   std::vector vec0{1, 2, 3, 4, 5};
6  auto v1 = vec0 | seqan3::views::take(2);                    // view over {1, 2}

8  /* views::to is a "terminator", must come at end (does not return a view) */
   std::vector vec1 = v1 | seqan3::views::to<std::vector>;     // vector over {1, 2}
```

Code snippet 7.5: View factories, adaptors and terminators. Objects that create views are differentiated by whether they adapt an existing range or not. There are also very few "terminator" objects—they do not actually return a view but appear at the end of view pipes

hidden behind `auto`. This is especially important in the context of view *adaptors* where the properties of the returned view depend not only on the view adaptor itself but also on the properties of the underlying range. The properties that are essential to know are whether and which range concepts are satisfied and what the `std::ranges::range_reference_t` of the range is. This is the "element-type", i.e. the type returned by dereferencing an iterator of the range or calling `operator[]` on a random access range.

An example of such a documentation entry can be seen in Table 7.5. `seqan3::views::to_char` transforms a range of alphabet letters into their respective character representation. It is a range adaptor, i.e. it works on an existing range whose type is denoted by `urng_t` in the table. It has certain requirements on this range type: that it be at least a `std::ranges::input_range` ("readable elements") and a `std::ranges::viewable_range` ("not a temporary container"). These requirements are shared with almost all other view adaptors, although many have stricter requirements. The adaptor also requires that the range reference type of the underlying range models the `seqan3::alphabet` concept, because it calls `seqan3::to_char` on the elements of the underlying range and this is only guaranteed to work if that type is an alphabet.

The third column in Table 7.5 represents the promises regarding the properties of the type of the returned range (`rrng_t`) which is the actual type of the view that is intentionally not exposed. The properties of this type are expressed in relation to the properties of `urng_t`, because they strongly depend on them. In particular the documentation promises that the range reference type of the returned range is exactly the type returned by `seqan3::alphabet_char_t<uref_t>` where `uref_t` is the range reference type of the underlying range. This is expected, because the entire purpose of this view is to transform from an alphabet to that alphabet's character representation. The table further states that the properties `std::ranges::contiguous_range` and `std::ranges::output_range` are *lost*, which means that even if the underlying range type models these concepts, the returned type will not.

Table 7.5 View documentation of `seqan3::views::to_char`. The second column expresses requirements on the underlying range (`urng_t`) and the third column expresses promises regarding the returned type (`rrng_t`). `uref_t` refers to `std::ranges::range_reference_t<urng_t>`

Input range concepts	urng_t	rrng_t
`std::ranges::input_range`	Required	Preserved
`std::ranges::forward_range`		Preserved
`std::ranges::bidirectional_range`		Preserved
`std::ranges::random_access_range`		Preserved
`std::ranges::contiguous_range`		Lost
Other concepts		
`std::ranges::viewable_range`	Required	Guaranteed
`std::ranges::view`		Guaranteed
`std::ranges::sized_range`		Preserved
`std::ranges::common_range`		Preserved
`std::ranges::output_range`		Lost
`seqan3::const_iterable_range`		Preserved
`std::semiregular`		Preserved
Traits		
`std::ranges::range_reference_t`	`seqan3::alphabet`	`seqan3::alphabet_char_t<uref_t>`

```
  /* vec is an output range */
2 std::vector vec = "ACGT"_dna4;
  vec[0] = 'G'_dna4;                  // type of "vec[0]" is "dna4 &", can be assigned to
4
  /* v is not an output range */
6 auto v = vec | views::to_char;      // behaves like std::string{"ACGT"}, but:
  std::cout << v[0];                  // type of "vec[0]" is "char", not "char &", cannot be assigned to
```

Code snippet 7.6: Preservation of range concepts. This example shows that the `std::ranges::random_access_range` concept is preserved (vector and view have `[]`), but that the range reference type has changed, because temporary values are created on-the-fly when the view is read. This means that the elements are not assignable (`std::ranges::output_range` is lost). Since they are not stored adjacently in memory (they are not stored at all), `std::ranges::contiguous_range` is lost, too

See Snippet 7.6 for a detailed example. Most other (range) concepts in the table are marked as being *preserved*, i.e. the returned range models the respective concept if (and only if) the underlying range models it. The concept `std::ranges::view` is marked as *guaranteed* which indicates that the returned type is always a view, independent of whether the underlying range was a view or not; this is true of all view factories and view adaptors. `std::ranges::viewable_range` is also *guaranteed*, because all views are viewable ranges by definition (see Subsection 3.6.2). Except for these two concepts, it is very rare for a view to "gain" properties that the underlying range does not have, e.g. if the underlying range is `std::list<dna4>`, the returned range will be at most a bidirectional range, because it is impossible for a view to implement random access on top of a (pure) bidirectional range without violating its space or time complexity constraints.

Tables like Table 7.5 are provided for all SeqAn3 view adaptors in the API documentation. View factories have only the `rrng_t` column (there is no underlying range), and terminators typically only provide the `urng_t` column (they do not return a view). I hope that similar documentation will appear for the standard library views.

```
  std::vector foo{"AAATTT"_dna5, "CCCGGG"_dna5};
2 auto v1 = foo | std::views::reverse;                                // == [ [C,C,C,G,G,G], [A,A,A,T,T,T] ]
  auto v2 = foo | seqan3::views::deep{std::views::reverse}; // == [ [T,T,T,A,A,A], [G,G,G,C,C,C] ]
4
  // auto v3 = foo | std::views::transform(seqan3::complement);                  // doesn't work
6 auto v4 = foo | seqan3::views::deep{std::views::transform(seqan3::complement)}; // works!
  auto v5 = foo | seqan3::views::complement;                                      // implicitly deep
```

Code snippet 7.7: Deep views. This example shows the effect of wrapping a view adaptor in `seqan3::views::deep`. For `std::views::reverse`, this leads to distinct (and possibly unexpected) behaviour. But for views that normally only work on one-dimensional ranges, it only adds usability. Note that `seqan3::complement` is the CPO described in Sect. 6.3 and `seqan3::views::complement` is a view adaptor

When dealing with biological sequences, one often deals with collections of sequences—or "ranges of ranges" in Modern C++ vocabulary. Applying a view to a range of ranges applies the view to the "outer range" which is the obvious and usually correct behaviour. In particular when working with alphabet-specific views, however, it is clear that these can only ever apply to a range whose element type is an alphabet. So, for example, `seqan3::views::complement` (which complements the values in a range of nucleotides; equivalent to a transform view that applies the `seqan3::complement` CPO) can never be applied to a range of ranges. Because it may still be useful to complement all the ranges in a range of ranges, SeqAn3 offers `seqan3::views::deep`, which could be described as an "adaptor adaptor". It can be used to construct a "deep adaptor" from an existing adaptor; this behaves like the original with the only difference that if it is passed a range of ranges, it applies to the innermost range, not the outermost one. See Snippet 7.7 for two examples. Only view adaptors where the behaviour is unambiguous are *deep* by default in SeqAn3. This includes most alphabet-specific views.

7.3.2 Alphabet-Specific Views

```
2   std::string s{"GATTACA"};                          // plain string

    std::ranges::copy(s | seqan3::views::char_to<dna5> // converts char to dna5
4                       | seqan3::views::complement    // complements the values
                        | seqan3::views::to_char,      // converts dna5 back to char
6               s.begin());
```

Code snippet 7.8: Alphabet-related views. The snippet demonstrates how various views can be used to update a plain `std::string` in-place by using the complement semantics of the `seqan3::dna5` alphabet but without creating extra storage (the values from `s` are transformed and copied back into `s`)

Several of the view adaptors provided by the views submodule are specific to (biological) alphabets. This includes the following which are all based on `std::views::transform` and perform a per-element conversion: `seqan3::views::to_rank` (semialphabet → rank repr.), `seqan3::views::to_char` (alphabet → character repr.), `seqan3::views::rank_to<T>` (rank repr. → semialphabet T), `seqan3::views::char_to<T>` (character repr. → alphabet T) and `seqan3::views::complement` (nucleotide value → complemented nucleotide value). An application of some of these is shown in Snippet 7.8. All of these adaptors are *deep*, i.e. they can be applied to multi-dimensional ranges and apply to the innermost ranges.

In contrast to the previous views that are element-wise transformations, `seqan3::views::trim` is a conditional take view, i.e. it represents elements from the beginning of the underlying range until (excluding) the first element

```
  // combined nucleotide and quality storage in a single vector:
2 std::vector<dna5q> vec{{'A'_dna5, 'I'_phred42}, {'G'_dna5, 'I'_phred42}, {'G'_dna5, '?'_phred42},
                        {'A'_dna5, '5'_phred42}, {'T'_dna5, '+'_phred42}};
4
  // trim by phred score
6 auto v1 = vec | views::trim(20);  // [{'A'_dna5, 'I'_phred42}, ..., {'A'_dna5, '5'_phred42}]

8 // trim by phred object, then 'drop' quality representation
  auto v2 = vec | views::trim('A'_phred42) | views::get<0>; // [ 'A'_dna5, 'G'_dna5 ]
```

Code snippet 7.9: Using `seqan3::views::trim`. The view shown is based on `seqan3::views::take_until` and the abort-condition is encountering a letter with a quality below the given threshold. The threshold can be given as a quality alphabet value or numeric phred score. Namespace `seqan3::` is assumed

that meets the "abort-condition". Specifically, this is that an element has a phred score below the given minimum threshold (the element type is required to model `seqan3::quality_alphabet`). Snippet 7.9 shows two applications of this view. It shows that this works also on combined nucleotide/quality alphabets and that "after" trimming has happened, and the quality information can even be dropped from the view. The alphabet type of `v2` is thus simply `seqan3::dna5`, but the content of `v2` of course still depends on the quality values in the underlying data (whether the end is reached is determined lazily).[4]

Finally, there are various views to perform nucleotide to amino acid translation, each optionally takes a `tf` argument which is an `enum` that specifies the frame(s). The view adaptors differ in how they handle range of ranges as input (namespace `seqan3::views::` is assumed):

`translate_single(tf)` Translates a range of nucleotides to a range of amino acids; the frame can be selected, but only a single frame is supported. It is a deep view; range is transformed to range; range of ranges is transformed to range of ranges (same dimension).

`translate(tf)` Translates a range of nucleotides to multiple amino acid frames; the exact frame configuration can be given (defaults to `SIX_FRAME`). It is a deep view; it always increases the dimension by 1, e.g. single range to range of frames.

`translate_join(tf)` Appears like `seqan3::views::translate(tf) | std::views::join`, i.e. a flattened version of the previous adaptor. But in contrast to applying `std::views::join`, it preserves random access and sized-ness. It only accepts exactly two dimensions of input (a range of ranges) and always returns exactly two dimensions.

---

[4] This kind of view does not preserve `std::ranges::sized_range`, because it is not known a priori when/if the termination condition will be met.

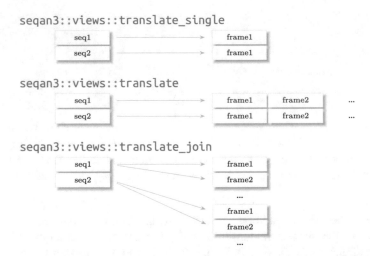

Fig. 7.1 The translate views. `seqan3::views::translate_single` always returns a single frame per input range; `seqan3::views::translate` always returns a range of ranges for every input range (i.e. multiple frames); `seqan3::views::translate_join` also returns multiple frames but flattened consecutively into a single range of ranges

Figure 7.1 illustrates this visually. The different adaptors are useful, because the dimension of a range (of ranges) is a hard parameter, i.e. fixed at compile-time and not abstractable. The least generic adaptor is `seqan3::views::translate_join`, but it provides an optimised implementation for the very common use-case of translating a set of DNA reads into a set of protein sequences where the frames are stored consecutively.

7.3.3   Some General-Purpose Views

It is not in the scope of this chapter to cover all the general-purpose views, but I will briefly introduce some interesting and/or representative ones. Some view adaptors with more specialised use-cases will be introduced in later sections, e.g. `seqan3::views::async_input_buffer` (Sect. 8.4.4) and `seqan3::views::enforce_random_access` (Sect. 10.1.2).

SeqAn3 offers multiple views that deal with "taking" a certain amount of elements from the beginning of the underlying range. The standard already provides `std::views:take(n)` (which takes the first `n` elements) and `std::views::take_while(cond)` (which takes elements from the beginning as long as the condition `cond` is satisfied). Both of these implicitly also have the termination criterion that the end of the underlying range is reached, so taking for example the first 10 elements of a range is valid even if the range has only

5 elements (the returned view will also only have five elements). For many use-cases, this is the desired behaviour, but especially in the context of *tokenisation* and parsing of formatted input, it is often required that the given termination criterion is met—and reaching the end of the underlying range before such a condition is met indicates an error. To still be able to use views in this context, SeqAn3 provides views that throw an exception if the underlying range ends before their termination condition is satisfied. These are `seqan3::views::take_exactly_or_throw(n)` and `seqan3::views::take_until_or_throw(cond)`. Note that to improve readability, SeqAn3 has inverted termination criterion of the latter ("until" instead of "while").

```
  //        key_type        value_type
2 std::map<std::string, uint64_t> name_to_id{{"Tom", 135454},  {"Jane", 768445},  /* ... */};

4 auto keys   = name_to_id | seqan3::views::get<0>;   // [ "Tom", "Jane", ... ]
  auto values = name_to_id | seqan3::views::get<1>;   // [ 135454, 768445, ... ]
```

Code snippet 7.10: Using `seqan3::views::get<i>`. Since the range reference type ("element type") of a map is a pair over the key and value, one can use a view adaptor to get a range over either keys or values

Various adaptors relating to tuples are provided. For a range whose element type is a tuple, `seqan3::views::get<i>` returns a view over a single element of that tuple. Iterating over such a view does not copy any elements because references to the original items are returned.

`seqan3::views::zip(l, r)` does the reverse; it can zip multiple individual ranges into a single range of tuples. `views::zip(l, r)` is also part of the range-v3 library but has not yet been standardised, so SeqAn3 provides an implementation based on the draft specification. This view also creates tuples of references if possible.

`seqan3::views::all_pairs(l, r)` behaves like the zip view but instead of creating a 1-to-1 pairing of the elements, it creates an n-to-m mapping, i.e. all pairwise combinations between the elements of both ranges. If only one range is provided, all pairs of elements from that range are created.

`seqan3::views::unzip` is a view terminator. It turns range of tuples into a tuple of ranges, i.e. it returns an object that behaves like `std::tuple<seqan3::::views::get<0>, seqan3::views::get<1>, ... >`.

7.3.4   Implementation Notes

The division between *view* and range/view *adaptor* has been discussed in the previous paragraphs and also in Sect. 3.6.4. That subsection also discusses a further differentiation between *adaptor* and *adaptor closure* objects which is important for adaptors that take further arguments beyond the underlying range. It should be clear

Table 7.6 Adaptors and adaptor closures. For adaptors that take arguments other than the underlying range, "range adaptor" and "range adaptor closure" are distinct entities—only the closure can be combined with a range. For adaptors that do not take further arguments the adaptor is always a closure. The expressions in the third column are each equivalent; `Vec` is for example `std::vector<int>` and `vec` an object of that type

"Take view"	Range adaptor	Range adaptor closure	View
Expression	`std::views::take`	`std::views::take(3)`	`std::views::take(3)(vec)`
			`vec | std::views::take(3)`
			`std::views::take(vec, 3)`
Type	Impl. defined	Impl. defined	`std::ranges::take_view<Vec>`
"Reverse view"		Range adaptor closure	View
Expression		`std::views::reverse`	`std::views::reverse(vec)`
			`vec | std::views::reverse`
Type		Impl. defined	`std::ranges::reverse_view<Vec>`

that `auto v = vec | std::views::take;` is not a valid expression, because it is not specified how many elements should be taken. `std::views::take` by itself cannot adapt a range, and it is not a range adaptor *closure*. The expression `std::views::take(3)`, on the other hand, returns an entirely different type that stores `3` and can adapt a range. Table 7.6 illustrates this.

```cpp
   /* using views::take */
2  auto const views_take_3 = std::views::take(3);  // a new closure object is created

4  std::vector vec{1, 2, 3, 4, 5};
   auto v = vec | views_take_3;        // [0, 1, 2] -> equivalent to "vec | std::views::take(3)"
6
   /* using views::transform */
8  auto const views_to_rank = std::views::transform(seqan3::to_rank);
   // pass CPO to adaptor -> returns adaptor closure object
10 // This is sufficient to implement "seqan3::views::to_rank"!

12 std::vector avec = "ACGT"_dna4;
   auto v2 = avec | views_to_rank;     // [0, 1, 2, 3]
```

Code snippet 7.11: Defining "new views" from existing adaptors

While this may seem like an unimportant detail of the implementation, it is actually very useful, because it enables defining "new views" simply by providing arguments to existing (non-closure) adaptors. Snippet 7.11 highlights how this can be used in practice: no types need to be defined, no templates need be specialised explicitly and a one-liner is sufficient to define `seqan3::views::to_rank`. Many of the views offered by SeqAn3 are defined in such a way with the majority being implemented as a closure object attained by passing a lambda expression or function object to `std::views::transform`.

```cpp
   std::vector vec = "TCAT"_dna4;
2  // use two adaptors to create a view
   auto v0 = vec | std::views::reverse | seqan3::views::complement; // [A, T, G, A]
4
   // define new view adaptor by combining two adaptors into one
6  auto views_rev_comp = std::views::reverse | seqan3::views::complement;
   // use the new view adaptor
8  auto v1 = vec | views_rev_comp; // equivalent to above
```

Code snippet 7.12: Defining "new views" by combining existing adaptor closure objects

Another simple way of defining a "new view" is by combining two existing adaptor closure objects into a new one. This can be seen in Snippet 7.12. Note that the pipe operators used in line 3 and in line 6 are fundamentally different: the first takes a 'range + adaptor' and returns a range, and the other one takes 'adaptor + adaptor' and returns a new adaptor.

Only if these two simple methods are not sufficient for defining a new adaptor, one should consider implementing a "full view", i.e. a class template for the actual view definition, the adaptor type, possibly the adaptor closure type and the adaptor

```
     namespace seqan3::detail
 2   {

 4   template <std::ranges::view urng_t>
     class kmer_hash_view                                          // <- Actual view type
 6   {
         /* full implementation: begin(), end(), iterator type, ... */
 8   };

10   struct kmer_hash_fn                          // <- Type of range adaptor object (non-closure)
     {
12       template <std::ranges::range urng_t>
         constexpr auto operator()(urng_t && urange, shape const & shape_) const
14       {
             return kmer_hash_view{std::forward<urng_t>(urange), shape_};      // returns actual view
16       }

18       constexpr auto operator()(shape const & shape_) const
         {
20           return adaptor_from_functor{*this, shape_};               // returns range adaptor closure
         }
22   };

24   } // namespace seqan3::detail

26   namespace seqan3::views
     {
28   inline constexpr auto kmer_hash = detail::kmer_hash_fn{}; // <- range adaptor object (non-closure)
     } // namespace seqan3::views
```

Code snippet 7.13: Skeleton of `seqan3::views::kmer_hash` and related types.
The first `operatator()` of the adaptor delegates to the view's constructor (use-
case: `seqan3::views::kmer_hash(vec, 0b101_shape)`). The adaptor's second
`operatator()` returns the closure object which stores the shape (use-case:
`seqan3::views::kmer_hash(0b101_shape)`). The range adaptor closure object
is defined through the generic `seqan3::detail::adaptor_from_functor` (it
provides all the pipe operators, definition not shown)

object. An example is shown in Snippet 7.13. Implementing the different features
of range adaptor closure objects (combination with other closure object through
`operator|`, combination with range through `operator()` and combination with
range through `operator|`) is quite sophisticated, so I developed several class
templates that automate these tasks via elaborate metaprogramming (in namespace
`seqan3::detail::`):

`adaptor_for_view_without_args<view_type>`   Can be used to define the
    entire adaptor closure object for a view that does not take further arguments.

`adaptor_from_functor<non_closure_type, ... stored_args_ts>`   Can
    be used to create the adaptor closure object from the adaptor non-closure object
    and the to-be-stored arguments; see Snippet 7.13.

`combined_adaptor<left_t, right_t>`   Defines a new adaptor from two given
    adaptors. This is used by the other two templates to define one of the two pipe
    operators.

While it appears that the internals of views are not without complexity, I would like to summarise these implementation notes as follows:

- Most new views can be defined in terms of existing views in a few lines.
- If a full-fledged definition is necessary, SeqAn3 provides the developer with the tools to get the adaptor machinery with little to no extra code.

Although I expect most developers to be content with the views defined in the standard library and those provided by SeqAn3, there is exhaustive documentation on defining a custom view in the form of a How-To in SeqAn3's online documentation.[5] I expect this to be helpful to future SeqAn developers but also any third parties interested in learning about the details of C++ ranges and views.

7.4 Discussion

This section introduced the Range module of SeqAn3. It provides containers and views that are entirely generic or specific SeqAn3's alphabets. Also found in this module is machinery to help users define their own views.

SeqAn3 currently provides only a few custom containers, but there are no SeqAn2 containers whose use-case is not covered by SeqAn3 or the standard library. SeqAn3 provides more than 30 view adaptors in the *Range* module and several more in other modules. These have a vastly greater scope than the "view"-like types in

Table 7.7 Ranges in SeqAn3 and SeqAn2. Comparison of similar data structures in SeqAn3 and in SeqAn2

SeqAn3 containers	SeqAn2 containers
`bitcompressed vector<T>`	`String<T, Packed<>>`
`small_vector<T, i>`	`String<T, Array<i>>`
`concatenated_sequences<TStr>`	`StringSet<TStr, ConcatDirect<>>`
+ `small_string`, `dynamic_bitset`	+ various types that duplicate `std::*`
SeqAn3 view adaptors	SeqAn2 "equivalent"
`views::slice`	`Segment<THost, InfixSegment>`
`views::zip`	`ContainerView<T, ZipContainer<TSpec> >`
`views::convert`	`ModView<FunctorConvert<TIn, TOut>>`
`views::complement`	`ModView<FunctorComplement<TAlph>>`
`views::to_lower`	`ModView<FunctorLowcase<TAlph>>`
`views::to_upper`	`ModView<FunctorUpcase<TAlph>>`
`views::reverse`	`ModifiedString<THost, ModReverse>`
+ 30 other view adaptors	+ 1 more "view" not present in SeqAn2

---

[5] https://docs.seqan.de/seqan/3-master-user/howto_write_a_view.html.

SeqAn2 (see Table 7.7). Currently, there is only a single type of the latter that does not have SeqAn3 counterpart: one that enables iterating over the gapped k-mers of a sequence based on a given shape. This will likely be added together with k-mer-indexing in a future release of SeqAn3. In any case, SeqAn3 offers much more than SeqAn2 in the domain of ranges, containers and views.

Python has various different features that are each comparable to certain aspects of C++ views:

"Views"   This term is used for different non-owning, container-like abstractions.[6] These seem to be mostly just reference types (like `std::ranges::ref_view`) or slices (like `std::ranges::subrange`). They are not very common.

"List comprehensions"   They are very common and can be used to filter and/or transform lists (like `std::ranges::filter` and `std::ranges::transform`) or retrieve members/sub-items (like `seqan3::ranges::get<i>`).[7] However, they imply returning a new list (with copies of the elements).

"Generators"   These are more comparable to C++ views, and they lazily produce objects and can be iterated over.[8] However, they are always "single-pass ranges" in C++ terminology, i.e. they are empty after they have been iterated over once, and they never allow random access.

While I would argue that none of these approaches can achieve the full flexibility of C++ views and that it is *simpler* to have a uniform abstraction for all of these use-cases, I am by no means an expert in Python and have no empiric data that compares the usability of these programming techniques. In any case, none of these methods seem to be employed by BioPython—although I have been told that it would certainly be possible to implement something like a reverse complement view as a generator in Python. I will compare with regular Python in some places and with BioPython when considering operations that copy anyway.

7.4.1   Performance

Container

Table 7.8 shows benchmarks of the most important container-alphabet-combinations. `char` and `uint32_t` are given as "baseline" alphabets and the less common standard containers `std::deque` and `std::list` are given to highlight how structural differences affect micro-benchmark performance independent of asymptotic complexity—which is constant for the read and write operations and amortised constant for the push-back operation **on all the shown containers**. For

---

[6] Python3 dictionary: https://docs.python.org/3/library/stdtypes.html#dictionary-view-objects
NumPy: https://docs.scipy.org/doc/numpy/glossary.html#term-view.

[7] https://docs.python.org/3/tutorial/datastructures.html#list-comprehensions.

[8] https://wiki.python.org/moin/Generators.

Table 7.8 Container benchmarks. The runtime is shown for reading/writing 10,000 elements from/to a pre-allocated container; and appending 10,000 random pre-constructed elements to a newly created, empty container

Container benchmarks	Read	Write	Push-back
Regular storage			
`std::vector<char>`	4.6 μs	9.1 μs	10.9 μs
`std::vector<uint32_t>`	4.6 μs	9.3 μs	14.7 μs
`std::vector<seqan3::dna4>`	4.6 μs	9.1 μs	10.7 μs
`std::vector<seqan3::aa27>`	4.6 μs	9.1 μs	10.7 μs
`seqan::String<seqan::Dna>`	4.6 μs	9.1 μs	20.1 μs
`seqan::String<seqan::AminoAcid>`	4.6 μs	9.1 μs	20.1 μs
`std::deque<char>`	5.0 μs	9.2 μs	21.0 μs
`std::deque<uint32_t>`	6.5 μs	9.8 μs	31.7 μs
`std::deque<seqan3::dna4>`	5.0 μs	9.2 μs	20.9 μs
`std::deque<seqan3::aa27>`	5.0 μs	9.2 μs	20.9 μs
`std::list<char>`	18.5 μs	23.7 μs	597.4 μs
`std::list<uint32_t>`	18.5 μs	24.1 μs	615.1 μs
`std::list<seqan3::dna4>`	18.6 μs	23.8 μs	619.2 μs
`std::list<seqan3::aa27>`	18.5 μs	23.7 μs	600.8 μs
Stack storage			
`seqan3::small_vector<char, 10000>`	4.6 μs	9.1 μs	44.6 μs
`seqan3::small_vector<uint32_t, 10000>`	4.6 μs	9.1 μs	38.1 μs
`seqan3::small_vector<seqan3::dna4, 10000>`	4.6 μs	9.1 μs	38.5 μs
`seqan3::small_vector<seqan3::aa27, 10000>`	4.6 μs	9.1 μs	38.4 μs
`seqan::String<seqan::Dna, seqan::Array<10000>>`	4.6 μs	9.1 μs	37.9 μs
`seqan::String<seqan::AminoAcid, seqan::Array<10000>>`	4.6 μs	9.1 μs	41.9 μs
Compressed storage			
`seqan3::bitcompressed_vector<char>`	4.6 μs	6.9 μs	38.9 μs
`seqan3::bitcompressed_vector<uint32_t>`	4.6 μs	9.2 μs	42.9 μs
`seqan3::bitcompressed_vector<seqan3::dna4>`	18.4 μs	35.6 μs	54.9 μs
`seqan3::bitcompressed_vector<seqan3::aa27>`	20.7 μs	32.1 μs	52.9 μs
`seqan::String<seqan::Dna, seqan::Packed<>>`	16.8 μs	27.6 μs	45.5 μs
`seqan::String<seqan::AminoAcid, seqan::Packed<>>`	15.2 μs	27.6 μs	55.8 μs
Compressed storage (const-access)			
`seqan3::bitcompressed_vector<char> const`	4.6 μs		
`seqan3::bitcompressed_vector<uint32_t> const`	4.6 μs		
`seqan3::bitcompressed_vector<seqan3::dna4> const`	10.8 μs		
`seqan3::bitcompressed_vector<seqan3::aa27> const`	15.9 μs		
`seqan::String<seqan::Dna, seqan::Packed<>> const`	16.7 μs		
`seqan::String<seqan::AminoAcid, seqan::Packed<>> const`	14.1 μs		

SeqAn3, the alphabets `seqan3::dna4` and `seqan3::aa27` are chosen, because they represent a small and a rather large alphabet and because they are used frequently. The equivalent alphabets `seqan::Dna` and `seqan::AminoAcid` were selected for SeqAn2. All the results given here are produced by benchmarks that are part of the performance regression suite—which also contains many more alphabet and container combinations.

The read benchmark creates a random sequence of size 10,000 and then repeatedly loops over the entire sequence copying the current element into a buffer variable. Times given cover a full loop over the entire sequence. The write benchmark is set up similarly but copies a local buffer value into the current element of the container. The push-back benchmark starts with a new empty container per loop and appends 10,000 elements in one iteration. After every element that is appended, the size of the current container is read to prevent the operation from being optimised out. Since a new container is created for every loop, this implies regular dynamic memory allocation.

The first important observation is that reads, writes and back-insertions are fairly consistent between all alphabets in standard library containers and SeqAn2's `AllocString` ("Regular storage" in Table 7.8). This shows that there is no overhead associated with using for example `seqan3::dna4` instead of `char` in one of these data structures. A slight exception to this is `uint32_t`, which incurs a higher runtime in some metrics, likely due to a worse cache behaviour because of its larger size. The read performance on `std::deque`, the "double-ended queue", is about 10% lower compared to a vector which is expected:

> typical implementations use a sequence of individually allocated fixed-size arrays, with additional bookkeeping, which means indexed access to [a] deque must perform two pointer dereferences, compared to vector's indexed access which performs only one. (https://en.cppreference.com/w/cpp/container/deque)

Back-inserting into the deque is twice as slow as into the vector. This is perhaps surprising since the deque never needs to reallocate previous storage (if elements are not deleted), but the element types are quite small in these benchmarks, possibly favouring the very cache efficient vector. Moreover, the number of memory allocations decreases exponentially in the vector, while the deque needs to allocate new storage in fixed intervals. The performance of the `std::list` is notably worse, especially the back-insertion (30x slower) which dynamically allocates every item individually.

The only unexpected result for "regular storage" is that back-insertions into SeqAn2's `AllocString` are twice as slow as into `std::vector` (and more comparable to insertions into `std::deque`). It is not obvious why this is the case, I surmise that compilers are particularly good at optimising for standard library containers as back-insertions are indeed very fast compared with reads/writes (see also below). This would strengthen the argument that one should choose `std::vector` over custom types unless one has very good reasons.

Fixed-capacity containers allocated on the stack ("Stack storage" in Table 7.8) perform exactly like `std::vector` in regard to reads and writes. Curiously, there is a notable overhead to performing back-insertions. This is consistent between SeqAn2's stack-based string and `seqan3::small_vector`, and it happens also with `char` as element type. One would expect these operations to be faster (no dynamic memory allocations happen at all), but as mentioned above, the performance of the `std::vector` is also suspiciously good. Cursory analysis with a profiler suggested that, even though I took precautions, `push_back()` on `std::vector` in the micro-benchmark may not be entirely serial.

While SeqAn2 had its own custom implementation of bit-compressed storage, SeqAn3 builds its vector on the `sdsl::int_vector`. They seem to have similar performance in general with SeqAn2's implementation being slightly faster in reads and writes. When comparing the performance of accessing the bit-compressed vector in a `const` context, SeqAn3's implementation performs better.[9] Back-inserting into compressed storage performs similar between SeqAn2 and SeqAn3. Based on the element-wise write speed, the back-insert performance appears reasonable. It should be noted that there is no read/write overhead to storing incompressible alphabets like `char` in compressed storage. However, back-insertion has the same "baseline" performance as when inserting into other custom containers like stack storage (see above).

Views

Table 7.9 shows benchmarks of some use-cases implemented with views in SeqAn3 and respective data structures in SeqAn2. In each benchmark, 10,000 elements are parsed, although not all views have size 10,000, e.g. the filter view will only "present" on average 7,500 elements, because it filters out the letter C.

Table 7.9 View benchmarks. Comparing the performance of those use-cases that could be achieved with "views" in both SeqAn3 and SeqAn2. The views/view combinations are applied to a `std::vector`/`seqan::String` of size 10,000 (or adjusted accordingly to reach 10,000 processed elements). Alphabet is `seqan3::dna4`/`seqan::Dna` except "to-upper" which works on `char`

View benchmarks	SeqAn3	SeqAn2
Baseline (no view)	4.6 µs	4.6 µs
Slice (10k out of 30k)	4.6 µs	4.6 µs
Convert to complement	5.0 µs	7.3 µs
Reverse	4.6 µs	4.6 µs
Reverse-complement	5.0 µs	9.6 µs
Convert to upper-case	4.6 µs	9.1 µs
Convert to char. representation	4.6 µs	4.6 µs

---

[9] In a `const` context, the compressed vectors return actual values, whereas in a non-`const` context they need to create and return proxy types that provide semantics for updating the position in the vector. It is expected that the latter is more expensive.

The most important observation is that, in SeqAn3, the shown views have no measurable impact on the performance. This is not true for all views on all containers (see the following paragraphs), but slices and transform-based views are very fast, especially if the underlying range is a random access range. `seqan3::views::complement` is a small exception with an overhead of ~10% over the baseline. The reason for this is not obvious, but the overall impact is very low. Combining two views (as is the case for the "Reverse-complement" use-case) also does not result in a combined higher runtime, it only reflects the more expensive complement operation in the runtime.

For SeqAn2, the picture is entirely different. Already, some simple transform-like "views" ("Convert to upper case") are 2x slower than the baseline. Other transform-like views ("Convert to char. representation") are not affected by this slowdown. It is particularly noteworthy that combining multiple "views" causes incremental slowdown in SeqAn2, e.g. the "Reverse-complement" use-case is noticeably slower than either of its individual views.

In Table 7.9, I showed use-cases that were solved with view-like abstractions in SeqAn2 and SeqAn3, but since views are much more powerful, now, they are used for many things that they could not previously be used for. It is important to analyse closely if introducing this abstraction is indeed free or associated with runtime overhead. Comparisons are either given with plain C++ code (loops, if-checks) or against functions of for example SeqAn2. It should be noted that many such comparisons are "unfair" towards views in the regard that they process the view completely and copy all its elements—which could be considered the worst-case usage scenario for a view, because it ignores many of a view's benefits like lazy evaluation and being able to represent a modified range without copying all its elements. Comparisons against plain Python and BioPython are given for broad reference although it is clear that C++ and Python operate under very different constraints.

Table 7.10 shows the use-case of copying certain elements from one range to another, based on a filter condition. The kernel of the benchmark code is shown in Snippet 7.16 on p. 215. A first glance reveals that runtime strongly depends on the condition itself and the content of the range. The CPU's branch prediction seems to

Table 7.10 Filter benchmarks. This table compares various solutions to the problem of filtering a container and copying all elements that meet a criterion into a new container. Runtimes are given for processing 10,000 DNA elements. "Happy" numbers represent a criterion that is always true, "sad" number represent a criterion that is always false and the "random" column shows the results for a criterion that is on average true in 50% of the cases

Filter benchmarks	Happy	Random	Sad
C++ baseline (copy elements, no check)	19.3 μs	–	–
C++ for-loop, if-check, copy	19.3 μs	36.9 μs	4.6 μs
C++ view (`std::views::filter` + copy)	19.3 μs	40.9 μs	4.6 μs
Python list-comprehension	9094.2 μs	10,163.6 μs	9078.4 μs

Table 7.11 Translate benchmarks. These benchmarks show the use-case of translating a DNA sequence of length 1000 into the respective amino acid sequence (single frame) which is stored in a new container.

Translate benchmarks	Runtime
SeqAn3 translate view + copy	0.51 μs
SeqAn2 translate function	0.79 μs
BioPython translate function	656.57 μs

have significant influence, so the runtime of a "happy" path (one where all elements pass the filter) is identical to the baseline where all elements are copied (the `if`-check is free). The "sad" path on the other hand (no elements pass the filter) is even faster, because no push-back operations happen at all. When the outcome of the condition is not predictable (50% chance per item randomly on the sequence), runtime is almost twice as high as the baseline—even though only half of the push-back operations happen. This shows (again) that performing an (unpredictable) `if`-check is very expensive. The interesting part is how the view performs against the loop: runtimes are identical in the happy and sad paths, but in the random case the "traditional" C++ approach is almost 10% faster than when copying through the view. Apparently, compilers cannot (yet) optimise the view machinery as reliably as the for-loop/if-check. Numbers for Python highlight that while the syntax for performing a filter-and-copy operation is remarkably compact, the performance is entirely in a different realm.

An example that is more closely based on biological applications is the translate benchmark of which the results are shown in Table 7.11 and the code is shown in Snippet 7.17 on p. 216. The translate view is not directly based on `std::views::transform`, because it is more complex, but it is implemented similarly. As the benchmark results show, it is very fast, even outperforming SeqAn2's implementation by 50%. This is impressive considering that (as mentioned above) the view is much more versatile and it is not primarily designed to be copied en bloc.

I would summarise that SeqAn2's performance in this area has been superb and SeqAn3's performance is very comparable with few micro-benchmarks favouring one or the other. As has been the case with SeqAn2, speed-optimisation is an ongoing process for the library developers. In contrast to earlier versions, I expect performance of some use-cases to also increase "automatically" with future compiler versions as views become used more widely and compilers receive improvements tuned to them. Compared with BioPython, SeqAn3 leads in the tested cases by factors of up to 1000x which shows the importance of C++ -based software for large computational workloads.

7.4.2   *Simplicity*

SeqAn2 primarily used the `seqan::String<TAlph, TSpec>` template for containers. Specialisations included `seqan::Alloc<>` for `std::vector`/`std::string`-like behaviour and `seqan::Packed<>` for bit-compression. SeqAn3 does not rely on template specialisation for polymorphism, so all its containers are distinct classes/class templates. It uses standard library containers where possible and only provides definitions of new container types where important use-cases demand it, e.g. `seqan3::bitcompressed_vector`. Relying on well-established standard types should make using SeqAn3 *simpler* than before.

```
   /* SeqAn2 */
 2 template <typename rng_t>
   void print_first3_revcomp(rng_t && range)
 4 {
       using InfixT = seqan::Segment<std::remove_reference_t<rng_t>, seqan::InfixSegment>;
 6     using AlphT  = typename seqan::Value<InfixT>::Type;
       using ComplT = seqan::ModifiedString<InfixT, seqan::ModView<seqan::FunctorComplement<AlphT>>>;
 8     using RevT   = seqan::ModifiedString<ComplT, seqan::ModReverse>;

10     RevT v(seqan::infix(range, 0, 3));
       std::cout << v;
12 }
```

```
   /* SeqAn3 */
 2 void print_first3_revcomp(auto && range)
   {
 4     auto v = range | std::views::take(3) | std::views::reverse | seqan3::views::complement;
       seqan3::debug_stream << v;
 6 }
```

Code snippet 7.14: "View" usability SeqAn2 vs. SeqAn3. This example illustrates how to print the reverse complement of the 3-letter prefix of a nucleotide range in SeqAn2 and in SeqAn3. It shows that SeqAn2 indeed offered similar functionality as some SeqAn3 views, however, using this required complicated "assembly" of the type beforehand

Views in their Modern C++ definition were not present in SeqAn2, but SeqAn2 provided various abstractions that behaved similarly. Some but not all of these were implemented as specialisations of the `ModifiedString` template. Snippet 7.14 illustrates how difficult it was in SeqAn2 to combine these abstractions; SeqAn2 was notorious for these large blocks of type definitions at the beginning of every function. Small typos in these sections would lead to very complex error messages and user frustration. In SeqAn3, on the other hand, the same semantics can be expressed in a one-liner and all usage of templates is invisible. It should also be noted that SeqAn3's adaptors are much more flexible: they would handle a pure bidirectional range (e.g. a `std::list`) as input to the above function, whereas SeqAn2's function would fail because it requires random access.

The example in Snippet 7.15 shows BioPython code that has the same result as Snippet 7.14 for SeqAn2/SeqAn3. It is very short and fairly easy to read, and

```
    def print_first3_revcomp(seq):
2       v = seq[0:3][::-1].complement()
        print(v)
```

Code snippet 7.15: Absence of views in BioPython

however, all three operations each involve a potential copy of the sequence making such a composition computationally expensive on real-world data. A shortcut for `.reverse_complement()` exists, but it does not change the fundamental problem of not having "view"-like abstractions in BioPython.

This is not true for Python in general, as I pointed out at the beginning of this section. Especially, if the final data is copied, comparisons might make sense. Two examples are shown in Snippets 7.16 and 7.17. Note that these are not typical use-cases of C++ views, because copying the entire "content" of a view into a container is usually not necessary—the view itself can be used in place of the container. However, sometimes it might be desirable and the given examples help to visualise such use-cases. They also help create comparability with SeqAn2 and BioPython which always generate "owning output".

```
std::ranges::copy(inpt | std::views::filter(fun), std::ranges::back_inserter(outpt));   // C++20

for (auto && c : inpt) if (fun(c)) outpt.push_back(c);                                   // C++11

outpt = [ i for i in inpt if fun(i) ]                                                    # Python
```

Code snippet 7.16: Filtered copy operation. The code shows copying those elements from `inpt` to `outpt` that evaluate the function `fun` (not shown) to `true`. Modern C++ (algorithm and view-based), traditional C++ and Python list comprehension are shown. Benchmark results are given in Table 7.10 on p. 212

Snippet 7.16 shows a conditional copy in the respective languages/styles. C++20 uses a view to generate the input data and then performs a copy to the output using an algorithm and a back-inserter object that internally calls `push_back()` when being invoked. An alternative is the purely functional style in Snippet 7.17. If one wanted only the view, `auto outpt = inpt | std::views::filter(fun)` would suffice. The C++11 code works in a classic imperative style by iterating over the input, doing an explicit conditional check and calling the respective function on the target object. Python, on the other hand, uses a functional-style list comprehension to express the filter. To get a view-like generator object with the same semantics (although single pass), one would only need to replace the brackets with parentheses.

A more "biological" application is given in Snippet 7.17, which shows single frame translation from nucleotide to amino acid (see also Sect. 7.3.2). In this example, SeqAn3 uses the purely functional/declarative style and conversion to a container happens via the view terminator `seqan3::views::to<>` and not

```
auto outpt = inpt | seqan3::views::translate_single | seqan3::views::to<std::vector>;   // SeqAn3

seqan::String<seqan::AminoAcid> outpt;     seqan::translate(outpt, inpt);               // SeqAn2

outpt = inpt.translate()                                                                # BioPython
```

Code snippet 7.17: Translate examples. The code shows nucleotide to protein translation (single frame) in SeqAn3, SeqAn2 and BioPython. Benchmark results are given in Table 7.11 on p. 213

via `std::ranges::copy()`—although they can be used interchangeably here.[10] Simply omitting the terminator will make the object a light-weight view instead of a container. SeqAn2 needs to be used in the notorious imperative/procedural style that includes out-parameters. BioPython, on the other hand, models translation as part of the object interface of its sequence type. Neither SeqAn2 nor BioPython can perform translation lazily (as view or generator). Examples of actual library code simplified through the use of views will be shown later (e.g. Snippet 8.13 on p. 241).

I conclude that SeqAn3's design is based on well-founded Modern C++ principles and strongly reflects the C++ 20 standard library instead of "homegrown" project-specific styles. Users can use and combine views without metaprogramming skills—often without even realising that C++ templates are involved.

7.4.3   Integration

Central to integrating well with the standard library is using standard library types and functions—which SeqAn3 does very consistently with respect to ranges. This includes standard library containers, standard library views/view adaptors and "range-ified" algorithms. When the latter are not provided by the compiler, they are emulated by the *STD* module (see Sect. 5.2.4).

Using standard library range concepts (Sect. 3.6.2) within SeqAn3 means that any third party range types can be used in conjunction with SeqAn3 if they satisfy the same concepts. And it means that third party libraries that constrain their algorithms with these concepts can consume SeqAn3 range types. This compatibility comes without the need for common base classes or base templates. Alphabet-specific ranges in SeqAn3 use the previously introduced alphabet concepts (see Sect. 6.1.3), so the users who wish to use for example `seqan3::bitcompressed_vector` simply need to make their element types satisfy the respective alphabet concept.

---

[10] The view terminator always allocates a new object, so when repeatedly writing to for example a buffer, it is more efficient to clear the buffer and use `std::ranges::copy`, which does not result in new memory allocation.

Another angle to integration is the combinability of the view adaptors. Although a mechanism is not yet defined by the standard library for integrating user-defined view adaptor's with the standard library's, SeqAn3 provides its adaptors with operators that provide full compatibility. Many examples in this chapter have demonstrated how easy it is to combine them.

7.4.4 Adaptability

As elaborated upon in the previous subsection, SeqAn3 is more generic than SeqAn2 by relying on concepts. Since these concepts are mostly defined by the standard library, this increases the chance that third party types will already satisfy the requirements.

I also demonstrated how one can refine these concepts when I introduced pseudo-random access ranges (Sect. 7.1). Such ranges will be recognised as bidirectional ranges by interfaces unaware of the refinement but can receive specialised treatment if desired.

I have also shown how average users can extend existing adaptors to perform refined or specialised tasks (Sect. 7.3.4). Furthermore, library maintainers and expert-users of the library can rely on the provided base classes and templates to create their own views with as little extra work as possible.

7.4.5 Compactness

The high number of views available in SeqAn3 reflects how simple it is to define new views and adapt existing ones. At the beginning of this section, Table 7.7 presents those views that have counterparts in SeqAn2, but SeqAn3 provides a total of 30 more view adaptors. The entire source code for these, including auxiliary entities like iterator templates, base classes, concept definitions etc. (the whole *Range* module), amounts to only 5837 lines of code.[11] SeqAn2, on the other hand, offers only a fraction of the functionality but requires 19,022 lines of code.[12] This difference is in part due to SeqAn2 duplicating functionality of the standard library but in large part also through the compactness of view definitions in SeqAn3 (Sect. 7.3.4).

---

[11] This does not include code provided by the C++20 standard library / the *STD* module.

[12] This includes the *Sequence* module, the parts of the *Modifier* module that relate to sequences and the iterator headers of the *basic* module. Notably it does not include functionality not implemented as ranges like translation—which is included in SeqAn3' count.

Examples like Snippet 7.14 on page 214 further illustrate that the syntax and interfaces are much less verbose and much more compact than SeqAn2's—although not always as concise as (Bio)Python's.

Chapter 8
The Input/Output Module

The reading and writing of files is a crucial part in almost all bioinformatics pipelines. In contrast to other computer-aided sciences that often deal with computationally expensive problems on a small set of input data (e.g. molecular dynamics), sequence analysis in bioinformatics is especially data-intensive. This means that the amount of data is very large compared to the problem that is being solved, and moving such data between different forms of memory and storage can constitute a large part of all program runtime or even be the bottleneck (Buffalo, 2015; Kosar, 2012).

The *I/O* module in SeqAn3 covers low-level streaming utilities and high-level formatted files. See Table 8.1 for an overview. A great variety of file formats exist in bioinformatics and the lack of standardisation makes solid, reusable implementations all the more necessary. SeqAn2 played an important role in this regard and SeqAn3 continues this effort.

I will first introduce the *Stream* submodule as the core of SeqAn3 I/O (Sect. 8.1). Although serialisation is not implemented within the *I/O* module, I will briefly show how it works in Sect. 8.2. Then I will explain the general design of formatted files (Sect. 8.3) and subsequently introduce the *Sequence file* submodule as an application of this design (Sect. 8.4). The *Sequence file* was the first formatted file introduced in SeqAn3, and I was not only responsible for its conception but also for most of the implementation. The *Alignment file* and the *Structure file* have a different composition of fields but are otherwise almost identical to the *Sequence file*, so these submodules are not discussed individually. Section 8.5 will wrap up the chapter and explain in how far the design goals have been reached for this module.

© The Author(s), under exclusive license to Springer Nature Switzerland AG 2022   219
H. Hauswedell, *Sequence Analysis and Modern C++*, Computational Biology 33,
https://doi.org/10.1007/978-3-030-90990-1_8

Table 8.1 I/O module overview

I/O Module	
Submodules	Alignment file, Sequence file, Stream, Structure file

Exception types	`seqan3::file_open_error,`
	`seqan3::format_error,`
	`seqan3::io_error, seqan3::parse_error,`
	`seqan3::unexpected_end_of_input,`
	`seqan3::unhandled_extension_error`

Enum types	`seqan3::field`
Other class types	`seqan3::fields, seqan3::record`

Table 8.2 Stream submodule overview

I/O: stream submodule	
Class types	`seqan3::detail::fast_istreambuf_iterator,` `seqan3::detail::fast_ostreambuf_iterator`
Function objects	`seqan3::views::istreambuf`

8.1 The Stream Submodule

Most of the *Stream* submodule's content is auxiliary functionality in the namespace `seqan3::detail::`. Some important parts are introduced here. See Table 8.2 for an overview.

```
2   std::istringstream is{"foo bar"};                                    // could be ifstream, too

4   for (char c : seqan3::views::istreambuf(is))                         // create view from stream
        std::cout << c;                                                  // prints "foo bar"

6   is.seekg(0);                                                         // rewind stream

8   for (char c : seqan3::views::istreambuf(is) | seqan3::views::to_upper)   // combine with other views
        std::cout << c;                                                  // prints "FOO BAR"
```

Code snippet 8.1: Views over streams. `seqan3::views::istreambuf` allows using a stream like a range. The view is single pass; the stream needs to be reset and a new view created to parse it again

Since many of the transformations used on strings and other ranges are implemented as C++ views, it makes sense to also provide a range abstraction for input streams. This allows plugging view adaptors directly onto an input stream. The C++20 standard library provides `std::ranges::istream_view` for this purpose. But to access the underlying stream, it uses `std::istream_iterator` (or a comparable type) which has a poor performance and results in undesired character processing like skipping whitespace. `std::istreambuf_iterator` is more efficient and reads directly from the buffer, but experience with SeqAn2 and further research showed that its performance is still far from optimal.

For this reason, I developed `seqan3::detail::fast_istreambuf_iterator` which provides an interface like `std::istreambuf_iterator` but is notably faster (see Sect. 8.5.1). Performance gains are achieved by only performing virtual function calls when absolutely needed, i.e. only inside `operator++` and only if the underlying buffer is exhausted. Importantly, `operator*` and `operator==` never involve a virtual function call.[1] This iterator is used heavily within I/O and also as part of `seqan3::views::istreambuf`, a fast replacement for `std::ranges::istream_view`.

The corresponding output operator works similarly and provides a dedicated function for writing a range of characters to the stream. If the range is a `std::ranges::sized_range`, it can be written "en-bloc" or in chunks directly into the output buffer which performs no intermittent virtual function calls. If the range is a `std::ranges::contiguous_range`, the compiler may also use `std::memcpy()`. These optimisations make output much faster (see Sect. 8.5.1). Both iterators can be used as drop-in replacements for the standard library streambuf iterators and may be moved out of `seqan3::detail::` in the future.

Conceptionally, the *Stream* submodule also encompasses the compression and decompression streams, although they are currently found in the *Contrib* module. This includes `*streambuf`, `*streambase` and `*stream` types, each for input and output and for the compression formats currently supported: GZip, BZip2 and BGZF. It is likely that these will be moved to the *Stream* submodule in the future.

8.2 Serialisation

```
   struct storage_t
2  {
       std::vector<seqan3::dna4> data;          13    /* Serialisation */
4      bool option1;                            
       bool option2;                            15    std::ofstream os{"outfile", std::ios::binary};
6                                                      cereal::BinaryOutputArchive archive{os};

       void serialize(auto & ar)                17    storage_t storage;
8      {                                                /* ... fill storage ... */
           ar(data, option1, option2);          19
10     }                                              archive(storage);          // stores to disk
   };
```

Code snippet 8.2: Serialisation with Cereal/SeqAn3. De-serialisation is not shown but works very similar to serialisation (input stream/archive instead of output stream/archive)

---

[1] This design assumes that the stream is only accessed through this iterator, which is valid for SeqAn3's use-cases but may not always be.

Serialisation is the process of transforming arbitrary objects into a binary or text representation that is streamable. It can also be considered a part of input/output, although, as discussed in Sect. 4.4.1, the infrastructure for this is not provided by SeqAn3 but by the *Cereal library*. Wherever class-specific hooks are required to enable serialisation, these are provided directly within the respective data structure.

Snippet 8.2 shows an example of this: the class `storage_t` provides a `serialize()` member that passes all data members to the archive. Note that this method is not called directly. The recommended way of serialising objects is passing them to `operator()` of the archive as shown on the right side of Snippet 8.2; this then recurses through all members of the object using said `serialize()` function. Cereal can serialise to/from any C++ stream including SeqAn3's compression/decompression streams. For more information, see the documentation of Cereal.[2]

A storage class can include data and/or options that an application needs to access later. It is especially useful to store a pre-computed index for efficient search (see Snippet 9.1). But it is also helpful in networking and distributed computing.

8.3 Formatted Files

The following terms are used to describe different levels of abstraction within the *I/O* module:

File   A *file* describes a common use-case. Files typically support multiple *formats* and each file is either an *input* or an *output* file. They are implemented as class templates and model `std::ranges::input_range` or `std::ranges::output_range`, i.e. **a file is range over** *records*. An example of a file is `seqan3::sequence_file_input`.

Format   A *format* describes a specific file format; it can generate a *record* from a *stream*. Most formats are specific to one *file*, but some can be used by multiple files. Formats in SeqAn3 are class types that are not used directly by most users (they are used internally by files). An example is `seqan3::format_fasta`.

Record   The element type of a *file* is a *record*. It consists of multiple *fields* and provides a tuple-like interface. The exact composition of the record is specific to the *file* but can be configured within certain file-specific constraints. All record types are specialisations of `seqan3::record` and the default record of `seqan3::sequence_file_input` encompasses sequence, ID and qualities.

Field   The elements of a record are called *fields*. Fields have a fixed purpose that is given through a field ID (`seqan3::field`), but the exact type is configurable. An example is the sequence field in a sequence file's *record* that has the ID `seqan3::field::seq` and can be a `std::vector<seqan3::dna4>` or of a different alphabet (or just `std::string`).

---

[2] http://uscilab.github.io/cereal/index.html.

8.3.1 Files and Formats

Differentiating between *file*[3] and *format* allows for reusing more code and it helps to structure program interfaces. Furthermore, formats of the same *file* are easily convertible to each other. It is important to note that deciding which *file* to use is up to the programmer, it is a compile-time decision based on the use-case.

Files can be created from filenames or stream objects. In the former case (which I expect to be the more common one), the *file* creates its own stream object from the filename[4] and automatically determines the format based on the extension. This is a runtime decision that depends on user input. In the case where the file is constructed with an existing stream object, the format needs to be given as a "tag"-parameter and no auto-detection takes place. This is still a runtime choice because the parameter is accepted as a `std::variant` over the possible formats (and not as a template parameter); see Sect. 3.10 on how variants can be used to "store types as values". This is also how the choice of format is stored internally in the *file*, and the visitor mechanism is the way of dispatching from *file* to format.

Like in SeqAn2, the *file* provides transparent (de-)compression of input/output depending on file extension and/or magic header, but in contrast to SeqAn2 this only happens if compression is independent of the format (e.g. for `foo.fasta.gz`). If (de-)compression is format-specific (e.g. `foo.bam`), it happens inside the format. This design change allows for more sophisticated format-specific stream handling as required by the CRAM format.

Table 8.3 shows the currently available and planned *files* with the respective formats. Some formats like `seqan3::format_sam` are supported by multiple *files*. Which combinations are possible depends on the interface of the format. `seqan3::sequence_file_input` requires that formats model `seqan3::sequence_file_input_format` which is a concept that in turn requires:

1. The static member `::file_extensions` that holds the valid extensions. This is required by all format concepts.
2. The member function `.read_sequence_record()` that takes the fields as out-parameters. This is specific to `seqan3::sequence_file_input`, and the function required for `seqan3::alignment_file_output` would be `.write_alignment_record()`.

While average users are not expected to ever deal with formats or their interfaces (they only use *files*!), this mechanism clearly allows differentiating which *files* a format supports. It also permits the users to implement their own format if they wish;

---

[3] In this chapter, I am using *file* in italics to denote SeqAn's abstraction (versus a file on the disk/as visible by the operating system).

[4] This is a regular `std::ifstream`/`std::ofstream` except that the underlying buffer is exchanged for a larger one (1 MB).

Table 8.3 Files and formats. Namespace `seqan3::` is assumed

File	Format
`sequence_file_input`	`format_embl`, `format_fasta`, `format_fastq`, `format_genbank`, `format_sam`, `format_bam`[a]
`sequence_file_output`	Same as input
`alignment_file_input`	`format_sam`, `format_bam`, `format_cram`[a]
`alignment_file_output`	Same as input + `format_blast_tabular`[a], `format_blast_xml`[a]
`structure_file_input`	`format_vienna`, `format_stockholm`[a], `format_connect`[a]
`structure_file_output`	Same as input
`annotation_file_input`[a]	`format_gff`[a], `format_gtf`[a], `format_vcf`[a], `format_bcf`[a]
`annotation_file_output`[a]	Same as input

[a] These files/formats are work in progress or planned (but not yet available)

the list of formats considered by a file can be specified as a template parameter to the file.

Most formats support both read and write interfaces, but conceptually this is entirely separate. It is likely that certain future formats (like the Blast formats) will only support writing as there has been no demand for reading these formats in SeqAn2 (although support exists), and they are too different from SAM/BAM/CRAM to provide inter-convertibility under the current abstraction.

8.3.2   Records and Fields

`seqan3::record<ts, fs>` behaves very similar to a tuple, and the number and types of the elements can be configured via template parameters. The first template parameter is a `seqan3::type_list<...>` with the element types, and the second template parameter is `seqan3::fields<...>` with an equal number of `seqan3::field::*` identifiers. `seqan3::field` is an `enum` that provides IDs for all possible fields of all *files*. In contrast to `std::tuple`, the record then provides not only access via `get<pos>(tup)` (position-based) and `get<T>(tup)` (type-based—if unique) but also access via `get<seqan3::field::x>(tup)` (by field ID).

```
/*                                    1st element      2nd element    3rd element    */
using record_type = record<type_list<std::vector<dna5>, std::string,   std::vector<phred42>>,
                           fields<field::seq,           field::id,     field::qual>>;
record_type r;

/* these are eqivalent: */
auto & seqA = get<field::seq>(r);
auto & seqB = get<0>(r);
auto & seqC = get<std::vector<dna5>>(r);

/* structured bindings work like for regular tuples */
auto & [ s, id, q ] = r;
debug_stream << s << '\n';  // prints the sequence
```

Code snippet 8.3: Records and fields. The record type shown is the type also returned by `seqan3::sequence_file_input` by default; typically these need not be defined "by hand". Namespace `seqan3::` is assumed

Files differ by which fields they support. A certain subset of these fields is usually read/written by default, and the files can be configured via a template parameter to produce/accept other supported fields. See the following section for examples.

8.4   The Sequence File Submodule

Sequence I/O is the most widespread and the most basic form of input/output in sequence analysis. It encompasses reading sequence data (nucleotides or amino acids) and usually also at least an identifier string and sometimes sequence quality information. This is represented by the following fields supported by sequence files:

`seqan3::field::seq`   The sequence, typically a range over nucleotides or amino acids.

`seqan3::field::id`   The identifier string.

`seqan3::field::qual`   The quality values, typically in phred notation.

`seqan3::field::seq_qual`   The fields `::seq` and `::qual` encoded in a single range.

By default, the sequence files' record contains the first three fields, but the *file* can be configured to provide/accept a different combination or order of fields. `::seq_qual` is mutually exclusive with `::seq` and `::qual`, and it results in sequence and quality data being represented by a single field.

See Table 8.4 for an overview of the submodule. SAM and BAM are actually alignment formats but are frequently (mis)used for sequence storage, so SeqAn3 allows using them in this context, as well.[5]

---

[5] One reason for doing this is that the formats are well designed and standardised compared with popular "pure" sequence formats. Another reason is that BAM is a compressed format by specification—and while "pure" sequence formats can be compressed with e.g. GZip, support for handling this varies.

Table 8.4 Sequence file submodule overview

I/O: Sequence file submodule	
File types	`seqan3::sequence_file_input,` `seqan3::sequence_file_output`
Concepts	`seqan3::sequence_file_input_format,` `seqan3::sequence_file_output_format`

Format types	`seqan3::format_embl,`
	`seqan3::format_fasta,`
	`seqan3::format_fastq,`
	`seqan3::format_genbank`

Traits types	`seqan3::sequence_file_input_default_traits_aa,`
	`seqan3::sequence_file_input_default_traits_char,`
	`seqan3::sequence_file_input_default_traits_dna`

8.4.1   Input

```
    // Passing no template arguments defaults to sequence_file_input_default_traits_dna:
2   sequence_file_input                                      f0{"example.fastq"};

4   // To read amino acids, pass a different traits class:
    sequence_file_input<sequence_file_input_default_traits_aa>    f1{"example.fasta"};
6
    // To read alls fields as plain std::strings:
8   sequence_file_input<sequence_file_input_default_traits_char> f2{"example.fastq"};

10  // Reading from an existing stream instead of opening a file:
    sequence_file_input                                      f3{std::cin, format_fasta{}};
```

Code snippet 8.4: Constructing a sequence file. Different traits classes can be passed to the sequence file to modify the storage types. Files can be constructed from filenames or stream objects. Namespace `seqan3::` is assumed

Snippet 8.4 shows how to create an input file. Since SeqAn3 uses distinct types for its alphabets, the user needs to tell the sequence file which kind of alphabets they expect and which containers they want to use for storage. By default, DNA data is assumed and the respective traits class (Sect. 3.3.2) is chosen. To read data into vectors of amino acids, the user can pass the traits class `sequence_file_input_default_traits_aa`.

For users of dynamically typed languages, this may be surprising, but since information is encoded in types, this is inherently a compile-time decision. But whether a FASTA-file provided by the user actually contains DNA or protein data is only detectable at runtime. While this may lead to conflicts, most use-cases (applications) clearly only work with one type of data or the other, so it makes sense for the developer to restrict this. Type-safety is augmented in a user-friendly way by throwing a descriptive exception if incompatible data is read.

In those cases where an application is truly agnostic of the alphabet or wishes to perform conversion of the data at a later point, the traits class `sequence_file_input_default_traits_char` can be selected.[6] This prevents any kind of "interpretation" of the sequence/quality data and simply stores it in `std::string`s.

Users may also provide their own traits classes to set a different alphabet, e.g. `seqan3::dna15` instead of `seqan3::dna5` (which is the default). Furthermore, this allows changing the container type from `std::vector` to e.g. `seqan3::bitcompressed_vector`.

`f3` in Snippet 8.4 shows how one can use existing streams including `std::cin` to construct sequence files. Since there is no file extension to use for classification, the file format has to be specified manually, although it can also be provided as a `std::variant` whose value is determined by a runtime decision (e.g. an extra command line argument).

```
2   sequence_file_input f0{"example.fastq"};

    // Option1: read record-by-record, access fields via get:
4   for (record & r : f0)
        debug_stream << get<field::id>(r) << '\n' << get<field::seq>(r) << '\n';
6
    // Option2: read record-by-record, decompose record immediately:
8   for (auto & [ s, id, q] : f0)
        debug_stream << id << '\n' << s << '\n';
10
    /* Custom fields */
12  sequence_file_input f1{"example.fastq", fields<field::id, field::seq_qual>{}};

14  // Composition of record is different now:
    for (auto & [ id, sq] : f1)
16      debug_stream << id << '\n' << sq << '\n';
```

Code snippet 8.5: Simple sequence file parsing. The file is read record by record; ID and sequence are printed (qualities ignored). The record is of type `seqan3::record`, and it can be accessed as is or decomposed. Namespace `seqan3::` is assumed

Snippet 8.4 illustrates iterating over a sequence input file. The value type of the file is a specialisation of `seqan3::record`. As described in Sect. 8.3.2, it can be accessed via `get<>()` and/or be decomposed with structured bindings. The latter is much less verbose, but it requires that the developer knows the order of the fields in the record. This composition can be changed by passing a second argument with a different fields-configuration to the constructor (`f1` in Snippet 8.4). Note how the

---

[6] This traits class has been implemented but not yet merged.

ID is the second element of `f0`'s records but the first element of `f1`'s records.[7] This configuration change also changed sequence and qualities to be returned as a single vector over `seqan3::qualified<dna5, phred42>` instead of two vectors.

If a field is selected that cannot be provided by some format (e.g. qualities and FASTA), the field is simply left empty. While I expect average users to not need to change field configurations, they allow for a very high flexibility. It is possible to omit fields entirely if one is for example interested only in a single field. This may speed up reading the file, because more efficient parsing can be performed.

Using `begin()` and `end()` directly on the file is not shown, because the range-based `for`-loop is much simpler. However, using the iterators manually is absolutely possible. It is safe to call `begin()` multiple times, but it will always return an iterator pointing to the current record of the file (it does not "rewind" and go back to the beginning). The same is true for the member function `front()` which returns a reference to the current record. It is important to remember that (similar to generators in Python) the files are single-pass, i.e. once all records have been iterated over, the range is empty and `begin() == end()`.

The current record is buffered inside the file, but it is safe to `std::move()` this out into different storage if desired. This prevents needless copies if the contents of the record are to be stored and not processed on-the-fly.

8.4.2   Output

```
      sequence_file_output f{"example.fastq"};
 2
      // Option1: Write individual fields:
 4    f.emplace_back("ACGT"_dna5, "SeqNo1", "IIHH"_phred42);

 6    // Option2: Write an object called r of type seqan3::record:
      f.push_back(r);
 8
      // Option3: Write by assigning to output iterator:
10    *f.begin() = r;
```

Code snippet 8.6: Simple sequence file writing. The file is written record by record, typically via a `for` or a `while` loop. There are multiple ways to write records

Writing files is similar to reading files, but it does not require any traits classes, because all the respective functions are templates that simply deduce the alphabets/range types from the input provided by the application developer. Snippet 8.6 shows how to create an output file and how to write records to it. Since output files are output ranges, one can write to them by assigning to the

---

[7] As noted in the beginning of the section, the default fields' configuration is `fields<field::seq, field::id, field::qual>` (namespace `seqan3::`).

dereferenced output iterator returned by `begin()` (Option3). This is, however, not very intuitive and the two member functions `emplace_back()` and `push_back()` (inspired by `std::vector`) are much easier to use. The emplace function takes individual fields as arguments (Option1 in Snippet 8.6). The parameter types are generic although certain constraints are enforced (e.g. sequence data could be given as a plain `std::string`, but passing an `int` would result a constraint failure).

```
   /* Writing to custom stream */
2  sequence_file_output f0{std::cout, format_fasta{}};

4  /* Custom fields */
   sequence_file_output f1{"example.fastq", fields<field::id, field::seq_qual>{}};
6
   // Composition of arguments is different now:
8  f1.emplace_back("SeqNo1", std::vector<dna5q>{{'A'_dna5, 'I'_phred42}, /*...*/});
```

Code snippet 8.7: Advanced sequence file writing. Output files can also be created over existing streams, and they can also be configured with custom field configurations

Similar to how the input file assumes a field order, the emplace function also expects that its arguments correspond to the fields that the file was configured with. This can be changed very similar to how it is done for input files (`f1` in Snippet 8.7). Note that `push_back()` with a `seqan3::record` as parameter is not affected by any changes to the fields configuration, because the file can always map the record's elements to the correct IDs.

When writing files, formats ignore fields that cannot be written (e.g. passing quality data to a FASTA file). This allows for seamless conversion of formats that are a super-set of another one, e.g. FASTQ-to-FASTA. The reverse, however, is not possible since FASTQ files require that a quality string is written. Formats will throw an exception if one attempts to write a record that does not contain the required fields, or they are mismatched (e.g. the qualities-string is shorter/empty).

8.4.3   Combined Input and Output

Snippet 8.8 shows how one can use the functions previously introduced to combine input and output files. But it also demonstrates how one can use the input file like a view and the output file like a view terminator. Note that like other view terminators, adding the output file with `operator|` will result in the whole input file being processed and piped to the output file (i.e. this is an eager expression, and it is not lazy-evaluated). This allows for very concise expressions like the following:

```
     /* Style 1: traditional loops */              /* Style 2: view-like / functional */
  2  sequence_file_input  fin{"in.fastq"};      2  sequence_file_input  fin{"in.fastq"};
     sequence_file_output fout{"out.fastq"};       sequence_file_output fout{"out.fastq"};
  4                                              4
     /* copy everything */                         /* copy everything */
  6  for (record & r : fin)                      6  fin | fout;
       fout.push_back(r);
  8                                              8  /* filter by length */
     /* filter by length */                        auto min_length = [] (auto & r)
 10  for (record & r : fin)                     10  {
       if (std::ranges::size(get<field::seq>(r))      return std::ranges::size(get<field::seq>(r))
 12        > 10)                                12        > 10;
         fout.push_back(r);                        };
 14                                             14
                                                   fin | std::views::filter(min_length) | fout;
```

Code snippet 8.8: Combining input and output files. Two styles for combining input and output files are shown: the first is based on the iterative approach (similar to previous examples), and the second approach uses view-based transformations and the fact that both files are ranges. The top examples simply copy input to output (format conversion would be possible). The bottom examples add a length-based filter

```
seqan3::sequence_file_input{"in.fastq"} | seqan3::sequence_file_output{"out.fasta"};
```

This immediately converts in.fastq into out.fasta without creating any variables or leaving any state. It is debatable whether using this style is superior for the filter example (right side of Snippet 8.8), but in more complex use-cases it can increase readability immensely.

8.4.4   Asynchronous Input/Output

I already established that the reading of files can be a bottleneck in bioinformatics applications. It is thus very helpful to program applications in a way that avoids waiting for I/O operations to complete by performing multiple tasks in parallel. This is called asynchronous I/O. There are dedicated routines defined by e.g. POSIX (Austin Common Standards Revision Group, 2014) to perform async I/O, but regular concurrency can also be used. Besides being platform-independent, this has the advantage that the format-logic for parsing the input can also be parallelised (not just "physically" reading from the disk).

SeqAn3 enables this kind of async I/O via `seqan3::views::async_input` `_buffer(n)`. The view is entirely independent of actual input/output, and it simply allocates space for n elements and spawns a thread on construction that moves elements from the underlying range into this buffer. Whenever elements of the view are read, they are in turn moved out of the buffer and the view refills the buffer via its background thread. This has the effect that all the work related to reading the file now happens outside the main application thread.

```
   auto v = seqan3::sequence_file_input{"in.fastq"}
2            | seqan3::views::async_input_buffer(100);

4  #pragma omp parallel
   {
6      for (seqan3::record & r : v)
           /* ... */
8  }
```

```
   seqan3::sequence_file_input{"in.fastq"}
2        | seqan3::views::async_input_buffer(100)
         | seqan3::sequence_file_output{"out.fasta"};
```

Code snippet 8.9: Async I/O and parallel processing. The left shows how input is asynchronously buffered while being written/converted. The right shows parallel processing of input via a buffer

If a combination of input file and async buffer is combined with an output file (left side of Snippet 8.9), the main thread can perform output while the view's thread performs input at the same time. The benefit of this is higher for complex formats, in combination with (de-)compression, if the records are large (chromosomes) and/or if physical read throughput is low (spinning disks or network storage). It is likely less pronounced for short read processing.

Since I designed the interfaces of `seqan3::views::async_input_buffer` to be (mostly) thread-safe, such a view can also be processed by multiple threads. This is helpful when e.g. "streaming" over the reads in a file to perform per-read processing of some kind (search, statistics, alignment etc.). The right side of Snippet 8.9 shows such an example. An OpenMP parallel region is created that launches multiple threads that each execute the following block of code.[8] Each of the threads now iterates over the buffer simultaneously until the buffer appears as empty—which is the case when the underlying range (the file) has reached its end.[9]

Key aspect of this design is that the programmer does not need to perform any explicit synchronisation. It appears as though each thread is processing the full range—and in case only one thread is started, this is true. The view appears as a simple range of elements but caches the data, waiting for the individual threads to "steal" the items. Of course, this means that the view always only models `std::ranges::input_range`; it is single-pass by nature.

This is all realised through the ranges/views design, and, due to its generic nature, the buffer could even be used to concurrently process integers in a vector. However, when the elements are cheaply copyable and only little work is associated with producing them, a more light-weight design with less moving of objects would likely perform better.

---

[8] Note that the code does not create an OpenMP `for`-loop; it just uses OpenMP to create threads. A lambda expression together with `std::async()` (Sect. 3.8) could also have been used here but would have resulted in notably more code. It also highlights the compatibility of the different threading frameworks.

[9] The view is not at end if the buffer runs empty because processing is faster than input. In that case, it simply blocks until more data is available.

8.5   Discussion

This chapter introduced the functionality and terminology surrounding the reading and writing files in SeqAn3. In this section the performance is analysed and SeqAn3's code-style is compared with other libraries. This includes not only SeqAn2 but also several Python-based libraries and two C-libraries. SeqAn2 provides custom low-level I/O but can also work with standard library streams/stream iterators. When comparing against the first, "SeqAn2" is given as the library name; the latter is denoted by "SeqAn2 (`std::ios`)".

8.5.1   Performance

There are many aspects to performance in input/output. The following benchmarks cover use-cases of increasing complexity.

Table 8.5 shows a benchmark of the different stream iterators. Regarding input, SeqAn3's iterator is more than 3x faster than the standard library's streambuf iterator which is the reason it was implemented. Although the implementations are very similar, SeqAn2's iterator is even faster which may be due to the underlying custom stream implementation. Curiously, SeqAn3's input performance equals SeqAn2's if the benchmark is compiled with machine-specific optimisations (`-march=native`; the other benchmarks are unaffected). Regardless, the throughput attainable with either is well beyond what the hardware typically used in large-scale storage systems can offer.[10]

When writing every character individually, SeqAn2's and SeqAn3's iterators provide no benefit over the standard streambuf iterator. However, both provide functions to write ranges "en-bloc" (see Sect. 8.1). This is the mechanism usually used by the format, and the performance gains are quite significant. SeqAn3 is

Table 8.5 Low-level stream iterator performance. Throughput is computed by repeatedly reading/writing a 1 MB file character by character from/to disk. The "en-bloc" benchmarks pass the 1 MB sequence as is to the stream, allowing for more low-level optimisations

Library	Input	OUTPUT
STD (stream iterator)	33.4 MB/s	15.9 MB/s
STD (streambuf iterator)	156.4 MB/s	91.4 MB/s
SeqAn2	610.4 MB/s	91.1 MB/s
SeqAn2 (en-bloc)	–	805.2 MB/s
SeqAn3	497.8 MB/s	91.3 MB/s
SeqAn3 (en-bloc)	–	873.7 MB/s

---

[10] The benchmark system uses high-speed NVME storage that reads/writes multiple GB/s. Cost-efficient spinning disks usually read/write < 200 MB/s.

Table 8.6 In-memory FASTA benchmark. This benchmark does not operate on files but on string-streams, so only the "format logic" is benchmarked. Throughput in MB/s is given, more is better

Library	Input	Output
SeqAn2 (`std::ios`)	226 MB/s	72 MB/s
SeqAn3	225 MB/s	869 MB/s
SeqAn3 (views)	215 MB/s	868 MB/s

almost 10% faster than SeqAn2 here. It should also be noted that while the specific function accepts a `std::vector` in SeqAn2, the performance gains are only available when writing `seqan::String`s. This exemplifies the mistakes and missed optimisations easily made with *template subclassing* (see Sect. 2.4.4).

Table 8.6 illustrates the performance of format handling. To eliminate the influence of low-level software I/O (buffering, paging) and the effects of hardware on this benchmark, it was performed on standard library string-streams, i.e. data is read from an in-memory `std::string` via `std::istringstream` and written to an in-memory `std::string` via `std::ostringstream`. This benchmark indicates the best possible performance attainable on the given CPU/memory assuming that the software and hardware stack below incurred no overhead at all.

Although not chosen when opening a file by filename, SeqAn2 does offer input/output interfaces that take standard library streams. These were chosen here to achieve the desired in-memory behaviour. They are less flexible and require the format to be specified at compile-time. This means that theoretically SeqAn2 has a slight advantage in this benchmark, because the routines do not have the runtime-overhead of having to dispatch to the correct format on every call.

With regard to input, SeqAn2 and SeqAn3 perform almost identically. But the output performance of SeqAn2 is very poor in this benchmark. I suspect that when given a standard library output stream, SeqAn2 does not use its optimised iterator at all and instead resorts to the standard library's streambuf iterator which is slower (see Table 8.5).

In SeqAn3, views are internally used frequently for tokenisation and parsing of input data. Based on the results of Sect. 7.4.1, I was curious if this would result in a performance overhead, so I implemented FASTA and FASTQ format logic with views and without. Table 8.6 shows that there is indeed a difference—although it is only ~5%. This correlates with previous results for stand-alone benchmarks of the filter view and conditional take views. Again, running the benchmark with machine-specific optimisations narrows this gap to ~2%. Right now the implementation without views is chosen by default, but it is not yet decided which codepaths will be shipped in the next SeqAn3 release; the view-based code is much *simpler* and more *compact* (see Snippet 8.13 on p. 241). I also anticipate that once views ship as part of the standard library, compilers will pick up optimisations that could close the performance gap entirely.

The final benchmarks in I/O are shown in Table 8.7 and cover a real-world "application use-case", converting a FASTQ file to FASTA format. In addition to

Table 8.7 FASTQ to FASTA conversion. This benchmark reads a 430 MB FASTQ file with 100bp reads from disk and writes it back out into ∼ 260 MB FASTA file. Times are given (lower is better), because input and output happen at the same time and throughput is different for each. The third column contains numbers for gzip compression (on both input and output files)

Library	Uncompressed	Gzipped
SeqAn2	4.7 s	43.0 s
SeqAn2 (std::ios)	6.3 s	–
SeqAn3	3.0 s	6.2 s
SeqAn3 (views)	3.4 s	6.4 s
LibGenomeTools	12.9 s	72.9 s
BioPython	99.8 s	250.0 s
BioPython (convert)	18.4 s	173.6 s

SeqAn programs, minimal applications written with BioPython and LibGenome-Tools (Gremme et al., 2013) are included. The latter is a performance-focused C library for sequence analysis.

Except for "BioPython (convert)", the conversion is based on regular iteration over the files, i.e. records are created from the input file and are immediately sent to the output file (only the current record is cached). This represents the typical use-case of "streaming" over an input file and writing out a new file with the possibility of performing operations in between. "BioPython (convert)" uses a specific conversion function of BioPython which does not allow intermediate processing. For SeqAn2 (not std::ios), SeqAn3 and LibGenomeTools, the detection of input and output format as well as compression happens fully automatically. For "SeqAn2 (std::ios)", this has to be hard-coded, and for BioPython, it has to be given explicitly, too (although auxiliary functions could get/set this at runtime). Input/output is performed on a real disk—although high-speed NVME storage and likely in-cache.

When performing uncompressed I/O, **SeqAn3 is the fastest** with a factor of 1.5x over SeqAn2, 4x over LibGenomeTools and 33x over generic BioPython. Using standard iostreams slows down SeqAn2 by another ∼30% and using the view-based codepaths for tokenisation slows down SeqAn3 by ∼10% (see above). This performance-drop is not mitigated by building with machine-specific optimisations (they have no notable effect on any of the results in Table 8.7). Nevertheless, the runtime with views is still better than that of any other library.

Operating system profiles of SeqAn2's program showed that it performs more output disk operations than the SeqAn3 application. Apparently SeqAn2 uses a regular-size buffer for I/O (while SeqAn3 uses a 1 MB buffer, see Sect. 8.1).[11] This could explain the difference as removing this optimisation from SeqAn3 almost doubles the runtime. BioPython expectedly has a much higher runtime, although

---

[11] SeqAn2 does use larger buffers when using compression and the performance difference there is lower if the same type of compression is used.

it does offer a special conversion function whose performance is also shown and is much better. In contrast to the regular BioPython benchmark and the numbers for SeqAn3, however, this performance is only indicative of the narrow use-case of conversion and cannot be generalised to other reading and writing tasks. And it still places BioPython behind the others.

If it is known at compile-time that the output format is going to be FASTA, SeqAn3 can easily be instructed to not read the quality field as it is not written anyway (see Sect. 8.4.2). This further reduces the runtime of SeqAn3's program from 3 to 2.5 s.

As shown in Table 4.3 on page 112, the computational cost of compression is expected to be measurable, the impact of decompression likely less so.[12] SeqAn2 uses its own GZip compressor automatically, while SeqAn3 uses a parallel version by default. This is in line with SeqAn3's design goals of making parallelisation easily accessible and on-by-default, and the performance difference is striking. A similar final runtime can be achieved in SeqAn2 by choosing .bgzf as file extension which will trigger parallelised blocked GZip compression. LibGenomeTools uses a single-threaded compressor, too, and it is notably slower than SeqAn2. BioPython does not offer transparent (de-)compression, at all. The respective benchmarks were performed by manually adding calls through Python's GZip module. Although Python uses C code internally to perform (de-)compression, the impact on runtimes is very severe compared to the native C/C++ programs. It should however be noted that LibGenomeTools and BioPython create compressed files that are ~10% smaller than those of SeqAn2 and SeqAn3; likely a higher compression level is set by default.

The *Alignment file* and *Structure file* are not benchmarked here, because I was personally involved only little their implementation. The design is almost identical to the *Sequence file* and the speed of the (de-)compression layer (relevant for BAM files) has been verified independently, so there is no reason why a similar performance should not be attainable.

8.5.2 Simplicity

I hope that the examples in Sect. 8.4 demonstrate how easy it is to use SeqAn3's formatted files in applications. A simple example implemented with different libraries is shown in Snippet 8.10. For sequence I/O, SeqAn3 comes very close in style to BioPython. I would argue that decomposing the record on iteration makes it even simpler to access the individual fields and reduces syntactic overhead.

Also, SeqAn3 detects the format automatically from the file extension and it supports transparent (de-)compression of sequence files. In this regard, SeqAn3 is simpler than even BioPython.

---

[12] Individual benchmarks where only input or output is (de-)compressed confirm this.

```
   /* SeqAn2 (C++) */
2  CharString id;
   Dna5String seq;
4  CharString qual;

6  size_t id_len = 0;
   size_t seq_len = 0;
8
   SeqFileIn f{"example.fastq"};
10
   while (!atEnd(f))
12 {
       clear(id);
14     clear(seq);
       clear(qual);
16
       readRecord(id, seq, qual, f);
18     id_len += length(id);
       seq_len += length(seq);
20 }
```

```
   /* libgenometools (C) */
2  gt_ma_init(false); gt_fa_init();
   GtError        *err      = gt_error_new();
4  GtStrArray     *filenames = gt_str_array_new();
   gt_str_array_add_cstr(filenames, "example.fastq");
6  GtSeqIterator  *seqit    =
       gt_seq_iterator_sequence_buffer_new(filenames, err);
8  const GtUchar *sequence = NULL;
   char          *desc     = NULL;
10 GtUword       len       = 0;

12 uint64_t      id_len    = 0;
   uint64_t      seq_len   = 0;
14
   while (gt_seq_iterator_next(seqit, &sequence,
16                              &len, &desc, err) == 1)
   {
18     seq_len += len;
       id_len += strlen(desc);
20 }

22 gt_seq_iterator_delete(seqit);
   gt_str_array_delete(filenames);
24 gt_error_delete(err);
```

```
   /* SeqAn3 (C++) */
2  size_t id_len = 0;
   size_t seq_len = 0;
4
   sequence_file_input f{"example.fastq"};
6
   for (auto & [seq, id, qual] : f)
8  {
       id_len += id.size();
10     seq_len += seq.size();
   }
```

```
   # BioPython (Python)
2  id_len = 0
   seq_len = 0
4
   f = SeqIO.parse("example.fastq", "fastq")
6
   for record in f:
8      id_len += len(record.id)
       seq_len += len(record.seq)
```

Code snippet 8.10: Usability comparison in sequence I/O. The snippets demonstrate how to calculate the total lengths of IDs and sequences in FASTA/FASTQ files. Namespaces are assumed for SeqAn2 and SeqAn3

SeqAn2 is much more verbose, uses out-parameters and requires that the users clear the sequence buffers on every iteration. However, in this example, it already does not have any visible templates, and compared to LibGenomeTools it seems rather trivial. The latter is not primarily a library design issue, but simply the result of using a C library for such a task.

Snippet 8.11 compares the SeqAn3 alignment file with SeqAn2, HTSlib and two Python libraries (BioPython does not have SAM/BAM support). I have chosen this example to show that the alignment file is used just like the sequence file and to evaluate the overall design of formatted files in the greater context of other libraries and frameworks.

HTSlib (Li et al., 2009) is the reference implementation of the SAM/BAM/CRAM formats and written in C. It is very low level and does not even provide online API documentation. The snippet shown in Snippet 8.11 demonstrates that quite a few functions and data structures have to be known to implement this simple use-case. Also, memory management is explicit, forcing the user to manually free resources to avoid memory leaks.

```
/* SeqAn2 (C++) */
2  BamFileIn f{"data.bam"};

4  BamHeader header;
   BamAlignmentRecord rec;
6
   readHeader(header, f);
8  while (!atEnd(f))
   {
10     readRecord(rec, f);
       if (rec.mapQ > 10)
12         std::cout << rec.seq << '\n';
   }
```

```
/* HTSlib (C) */
2  samFile   *f      = hts_open("data.bam", "r");
   bam_hdr_t *header = sam_hdr_read(f);
4  bam1_t    *rec    = bam_init1();

6  while (sam_read1(f, header, rec) > 0)
   {
8      if (rec->core.qual > 10)
       {
10         uint8_t *q = bam_get_seq(rec);
           for (int i = 0; i < rec->core.l_qseq; ++i)
12             printf("%c", seq_nt16_str[bam_seqi(q,i)]);
           printf("\n");
14     }
   }
16
   bam_destroy1(rec); sam_close(f);
```

```
# PySam (Python)
2  f = pysam.AlignmentFile('data.bam', "rb")

4  for rec in f.fetch(until_eof=True):
       if rec.mapping_quality > 10:
6          print(rec.query_sequence)
```

```
/* SeqAn3 (C++) */
2  alignment_file_input f{"data.bam",
       fields<field::seq, field::mapq>{}};
4
   for (auto & [seq, mapq] : f)
6      if (mapq > 10)
           debug_stream << seq << '\n';
```

```
# PyBam (Python)
2  f = pybam.read('data.bam', \
       ['sam_seq','sam_mapq'])
4
   for seq,mapq in f:
6      if mapq > 10:
           print(seq)
```

Code snippet 8.11: Usability comparison in alignment I/O. The snippets demonstrate how to parse a SAM/BAM file and print the sequence of every record whose mapping quality is greater than 10. Various languages, libraries and styles are represented. Note how similar the usage of SeqAn3 and PyBam is. Namespaces are assumed for SeqAn2 and SeqAn3

SeqAn2 is already considerably simpler with a data structure for the entire record and without manual memory management and pointer semantics. However, it still requires reading the header separately, and records also have to be explicitly read via function calls. All interfaces are free function interfaces with out-parameters, and only the record data structure has members.

PySam[13] is Python wrapper around HTSlib. It is already much more compact with the header being stored implicitly in the file. The records can be iterated over, but to do so one needs to explicitly "fetch" them and provide an extra option to indicate that one wants to iterate in the order they appear and that unmapped records should be included.

PyBam[14] is a different Python library from PySam; it is implemented in pure Python and does not wrap HTSlib. Both support SAM and BAM. PyBam and

---

[13] https://github.com/pysam-developers/pysam.

[14] https://github.com/cmeesters/pybam/.

SeqAn3 have very similar interfaces, and they both support iterating over the file directly which returns a record by default. But both also support configuring the file to return only a subset of the available fields.[15] It is then very easy to decompose the record and work only with these selected fields.

Although SeqAn3 has a little "syntactical overhead" over the Python modules, this is intrinsic to C++. The usage patterns are very comparable to BioPython for sequence I/O (Sect. 8.4) and PyBam for alignment I/O (Snippet 8.11). Since these are promoted as role models in usability and have a huge user base compared to SeqAn,[16] it is fair to say that SeqAn3 has improved strongly in this domain. The formatted file interfaces shown here are simple to use while still offering a great amount of flexibility with regard to configuration. Combining files in pipelines with view adaptors provides an entirely different style than for-loop-based iteration. It is very concise and expressive, but novel to most C++ developers. Both styles are the immediate result of implementing files as ranges, and they are not the result of separate interfaces.

Files in SeqAn3 can be created from filenames and the manner of low-level file-access is hidden from the user by default. If users do want to influence low-level access, they can provide standard library I/O-streams which are very common and well-understood.

8.5.3 Integration

By using standard library streams and providing public interfaces that accept these, the *I/O* module offers a good integration with the standard library. And any other streams that are derived from the standard I/O-streams can be used, e.g. different (de-)compression streams provided by third party libraries.

Serialisation in SeqAn3 (which is based on Cereal) is much more generic than in SeqAn2 and allows to easily combine SeqAn3 types and third party types into a single binary (or text) archive. Since Cereal comes with a wide support of the standard library, most standard types like containers and tuples are already well-supported.

SeqAn3's files are ranges, so all standard library algorithms and view adaptors that work on ranges can be applied to them. See Snippet 8.8 on p. 230 for an example. This is also true for future third party libraries that choose to follow the C++ 20 standard library designs.

---

[15] For SeqAn3, this has the added benefit that complex parsing of deselected fields can be avoided, and the area in the file will simply be skipped.

[16] Based on cursory analysis of publicly available download counts and weak indicators like GitHub followers/stars.

8.5.4 Adaptability

Compared to the *Alphabet* module or the *Range* module, the I/O module does not prominently define extension points for third party software, and it primarily provides a self-contained set of features. It does use the concepts defined by the *Alphabet* module and the *Range* module, though, so any data structures adapted to behave as alphabets or ranges can be used with SeqAn3's I/O.

Output files simply verify these concepts on any objects that are being written, while the input files have traits classes that define which types they produce. The latter are necessary because the input file interfaces do not have out-parameters, instead the file returns records during iteration. However, the traits classes are no less flexible than template parameters, e.g. container types can be set independently of alphabet types.

The interfaces that *files* use to communicate with *formats* are currently being finalised for SeqAn-3.1. Subsequently, it will be possible to define new formats or custom versions of existing formats in user code and supply these to the file as a template parameter, so they are selected automatically based on extension.

```
  format_fasta::extensions.push_back("foosta");        std::vector<std::vector<dna4>>      seqs  = /*...*/;
2                                                     2 std::vector<std::string>           ids   = /*...*/;
                                                        std::vector<std::vector<phred42>> quals = /*...*/;
4 sequence_file_input f{"example.foosta"};            4

  for (auto & [seq, id, qual] : f)                      views::zip(seqs, ids, quals) |
6     /*...*/                                         6     sequence_file_output{"example.fasta"};
```

Code snippet 8.12: Adaptability of sequence I/O. The left shows how to extend the list of file extensions associated with the FASTA format (not changing the extensions would have resulted in an exception). The right shows how data that is not in "record-form" can be easily adapted. Namespace `seqan3::` is assumed

A criticism of SeqAn2's formatted files was the strong reliance on file extensions to detect file formats and apparently projects like BioPython explicitly decided against this (see Sect. 8.5.2). On the one hand, this is understandable because the lack of standardisation in bioinformatics has resulted in the proliferation of file extensions for popular formats.[17] On the other hand, automatic detection of the format is a usability improvement for almost all "regular" use-cases. SeqAn3 still relies on extensions for this reason, but the design now allows to change the list of associated extensions easily. It is a `static` member of the format class but explicitly not declared `const`, so it can be updated, even based on runtime decisions. See the left side of Snippet 8.12 for an example. In SeqAn2, the same can only

---

[17] Many components in pipelines or even hardware like sequencers output custom extensions. If these cannot be reconfigured it, the files need to be renamed before being passed to SeqAn2.

be achieved by defining a custom format that specialises FASTA and overwrites multiple metafunctions.[18]

While record-based reading and writing is the mechanism used by almost all libraries and modules evaluated, in some situations the data is not stored record-wise but field-wise, e.g. all sequences are in one container, all IDs are in another container etc. SeqAn2 offered extra functions to read/write data in this format, but in SeqAn3 it is very simple to adapt this data using a view. This is shown on the right side of Snippet 8.12: `seqan3::views::zip` creates a view over the three existing ranges (no copy involved) that appears like a single range with tuples as elements. This range can then simply be piped into the output file which perceives it as a range of records and writes them to disk.

These are just two examples of how the new I/O design is very versatile, even if usage scenarios differ from the "intended" ones.

8.5.5 Compactness

Since SeqAn3's *I/O* module still implements considerably fewer formats than SeqAn2 (see also Table 8.3), it does not make sense to compare the size of the implementations (i.e. lines of code). However, the structure is already a lot cleaner, with a single *I/O* module and four submodules for formatted files (three implemented + one planned).

Concerning the size of the codebase, the *I/O* module has revealed an expected conflict between the design goals of *performance* and *compactness*. In general, SeqAn3 relies more strongly on the standard library (favouring compactness), but for performance reasons facilities like the stream iterators were reimplemented. This is a minor violation of *compactness*, but the benchmarks have shown that it is necessary to reach a performance comparable to and even better than SeqAn2 (see Sect. 8.5.1). Unlike SeqAn2, SeqAn3 needs no further low-level utilities (custom stream types, operating system-specific I/O) to reach this performance.

However, the conflict also became visible when deciding on the style of implementation **within SeqAn3's format code**.[19] Snippet 8.13 shows how using views can dramatically reduce the complexity and size of the code. But as the benchmarks in Sect. 8.5.1 show, using this style throughout the format code incurs a total slowdown of 2–10%.

The code in Snippet 8.13 is the actual code found in the library; currently, it is possible to switch between the two implementations via a macro. This can be used to track performance changes with respect to compiler versions and different flags. But in the long term, the future maintainers of SeqAn need to decide on

---

[18] This is non-trivial, see e.g. https://github.com/seqan/seqan/issues/2054.

[19] All of this code is implementation-detail and none of the issues discussed here affect the user-visible design of formatted files.

```
   /* implementation without views */
 2 auto it = stream_view.begin();
   auto e = stream_view.end();
 4 for (; (it != e) && (is_id || is_blank)(*it); ++it)                    // skip leading "> "
   {}
 6
   bool at_delimiter = false;
 8 for (; it != e; ++it)                                                  // read line
   {
10     if (is_char<'\n'>(*it))
       {
12         at_delimiter = true;
           break;
14     }
       id.push_back(assign_char_to(*it, value_type_t<id_type>{}));        // convert alphabet
16 }

18 if (!at_delimiter)
       throw unexpected_end_of_input{"FASTA ID line did not end in newline."};
```

```
   /* implementation with views */
 2 std::ranges::copy(stream_view | views::take_line_or_throw               // read line
                                 | std::views::drop_while(is_id || is_blank)  // skip leading "> "
 4                               | views::char_to<value_type_t<id_type>>,     // convert alphabet
                     std::ranges::back_inserter(id));
```

Code snippet 8.13: Using views inside parsing code. The snippets show code for parsing the ID-line within the FASTA-format. The top snippet works without views, and the second snippet uses views. Namespace `seqan3::` is assumed

an implementation. This should follow a general discussion on how to weigh the different design goals. Ideally, future compilers will improve so that there is no overhead associated with using views in the described way.

Chapter 9
The Search Module

The search module offers data structures and algorithms for efficiently finding exact
and approximate matches between the so-called *query* sequences and a *text* (also
called *subject* sequence(s) or the *reference*). Query and subject may each be a single
sequence or a collection of sequences, and typically the total subject size is much
larger than the total query size. Bioinformatical examples of the model are DNA
reads searched in a genome or translated transcriptome reads searched in a protein
database.

Algorithms that perform the search in an unprocessed text are called *online
search* algorithms. They require no pre-processing, but complexity of the search
is at best in $O(n)$, where n is the total size of the text. For most use-cases, this
is not feasible, so demand for such methods has dwindled and SeqAn3 offers no
such algorithms at the moment. Should demand arise again, in the light of new
biotechnological developments or other algorithmic constraints, they can easily be
readded.

The alternative to online search is *indexed search*. See Table 9.1 for an overview
of search module indexed search requires pre-processing the input (usually the text)
to create an index data structure. This is associated with a high computational cost
initially but allows for fast searches whose complexity is no longer dominated by
the length of the text.[1] The additional data structures also demand notably more
memory.

Indexing methods have evolved rapidly within the last 30 years and still differ
greatly in which kind of performance they offer and how much extra space they
need. SeqAn1 already offered k-mer-indexes (also called q-gram indexes) and
various suffix tree-like indexes. k-mer indexes will be a part of SeqAn3, see
Sect. 9.2. In the past, the most common suffix tree-like indexes were suffix arrays
(SAs) and enhanced suffix arrays (ESAs)—which already improve substantially
over the traditional suffix tree. But for all practical purposes in sequence analysis,

---

[1] The length of the text may for example be present as factor of $O(log(n))$ or be entirely absent.

© The Author(s), under exclusive license to Springer Nature Switzerland AG 2022
H. Hauswedell, *Sequence Analysis and Modern C++*, Computational Biology 33,
https://doi.org/10.1007/978-3-030-90990-1_9

Table 9.1 Search module overview

Search module	
Submodules	Algorithm, configuration, FM-index, k-mer-index

these have been superseded by FM-indexes (also called compressed suffix arrays) which require even less memory; see Sect. 9.1 for SeqAn3's implementations. After SeqAn3's indexes are introduced, I will explain the common search interface (Sect. 9.4) and discuss the *Search* module's impact on SeqAn3's design goals (Sect. 9.6). For more historical and theoretical background on the different index types and how they relate to sequence analysis and next-generation sequencing, I recommend Reinert et al. (2015) and Pockrandt (2019).

One assumption shared by most indexed search approaches is that the indexed input (usually the text/database) is fairly constant over a longer period of time so that its construction time amortises. With the event of metagenomics and pangenomics, this is not strictly true any longer. Dadi et al. (2018) elaborate on this problem and propose the DREAM framework to solve it. The DREAM framework is not yet part of SeqAn3 but is a high priority for many involved parties, so I expect first prototypes based on SeqAn3 to appear soon.

Almost all approaches in indexed search pre-process the text and perform little to no pre-processing of the query. Exceptions are the Masai application (Siragusa et al., 2013) and the first version of LAMBDA (Hauswedell et al., 2014). These perform so-called *double-indexing* where both inputs to the search are indexed. Since both applications have given up on this design (Masai was superseded by Yara which does not perform double-indexing; for LAMBDA, see Part III), SeqAn3 does not currently offer this functionality. However, it would be trivial to extend the existing interfaces later on, if desired.

While I have shaped the design decisions regarding the *Search* module, I am not the main implementer of the features presented in this chapter. As such I will focus on discussing the design (both high-level and concrete API examples) and give less space to performance benchmarks and quality of implementation. Many parts of the search API are currently still in flux, so the designs introduced here are based on my vision of the interfaces and do not fully reflect the current state of SeqAn3's master branch.[2] However, the set of *features* introduced in this chapter is available as described (Table 9.2).

9.1   The FM-Index Submodule

The FM-index, also called compressed suffix array (CSA), is a data structure devised by Ferragina & Manzini (2000). It is based on the Burrows–Wheeler

---

[2] This affects mostly naming and structuring of configuration elements.

Table 9.2 FM-index submodule overview

Search: FM-index submodule	
Index types	`seqan3::fm_index, seqan3::bi_fm_index`
Cursor types	`seqan3::fm_index_cursor,` `seqan3::bi_fm_index_cursor`
Enumeration types	`seqan3::text_layout`
Type aliases	`seqan3::sdsl_default_index_type,` `seqan3::sdsl_wt_index_type`

Transform (BWT; Burrows & Wheeler (1994)) and allows for searches similar to those possible on suffix arrays—but with lower memory requirements. Compressed suffix arrays have replaced suffix arrays in most bioinformatics and computer science applications. A comprehensive introduction is given by Pockrandt (2019). See Table 9.2 for an overview of the submodule.

SeqAn3 uses the compressed suffix array data structures of the SDSL (Gog et al., 2014); see also Sect. 4.4.1. Since the interfaces provided by the SDSL are not very user-friendly and it would add even more complexity to redirect SeqAn's users to the API documentation of the SDSL, SeqAn3 provides its own data structures as wrappers around the SDSL data structures. These provide a more object-oriented "feel" and give SeqAn3 developers more control over the implementation. The latter is necessary because certain core features (indexing collections of texts and bidirectional indexes) are implemented "on top" of the data structures of the SDSL.

While SeqAn3's FM-indexes allow direct access, it is recommended to use the search interface (see Sect. 9.4) for most use-cases. This is much simpler and automatically chooses highly efficient *search schemes*.

9.1.1 Unidirectional FM-Index

`seqan3::fm_index` is a class template that takes three template parameters:

`alphabet_t` The text's alphabet type and must be a `seqan3::semialphabet`.[3]

`text_layout_mode` Either `seqan3::text_layout::single` or `::collection`. It specifies whether a single sequence (range) or multiple sequences (range of ranges) are indexed.

`sdsl_index_t` The type of the underlying SDSL index; `seqan3::sdsl_default_index_type` by default.

The FM-index class is constructed with the text which can help auto-deduce the template parameters and immediately creates the internal data structures (this

---

[3] Indexing and search are data structures/algorithms that operate purely on the rank interface of the elements.

involves suffix array construction, sampling, BWT-generation etc.). Details of the concrete algorithmic steps involved in this creation are currently not exposed. Especially, for suffix array construction, a second (optional) parameter might be added to the constructor that either indicates the choice of a different algorithm via an enum value, or the parameter may itself be a function (object) that performs the creation. This decision depends on whether the SDSL picks up such features (that SeqAn3 would simply delegate to) or whether SeqAn3 will want to provide such features on top of the SDSL interfaces. Currently, the step most computationally expensive, the suffix array construction, is performed by a local fork of the *libdivsufsort* library[4] inside the SDSL.

It was an intentional design decision to not store the text within the index (as SeqAn2 did), because no SeqAn3 index requires the text to operate. And, since the text is often also required independently, unsuspecting users of SeqAn2 often ended up with the text stored in memory twice. This means that the type of the text is also not a template parameter, only its alphabet and whether a sequence or a collection of sequences is indexed (the latter implies that certain transformations need to happen on the positions returned by the search).[5]

```
   /* Create data */                          /* Create empty data */
2  std::vector<seqan3::dna4> genome =      2  std::vector<seqan3::dna4> genome;
       "ATCGATCGAAGGCTAGCTAGCTAAGGGA"_dna4;     seqan3::fm_index<seqan3::dna4,
4  seqan3::fm_index index{genome};          4      seqan3::text_layout::single> index;

6  /* Create output archive */             6  /* Create input archive */
   std::ofstream os{"example.foo",             std::ifstream is{"example.foo",
8                  std::ios::binary};       8                  std::ios::binary};
   cereal::BinaryOutputArchive archive{os};    cereal::BinaryInputArchive archive{is};
10                                          10
   /* Store data */                            /* Load data */
12 archive(genome);                         12 archive(genome);
   archive(index);                             archive(index);
```

Code snippet 9.1: FM-index creation and serialisation. The left shows the snippet for an "indexer" application that only needs to be run once; it creates the index and stores it on disk. The right shows the respective code in the "mapper" application that loads the index and text stored previously

Since this process of creating the index is slow, it is recommended to perform it in a separate application (or application mode) and use serialisation (see also Sect. 8.2) to store/load it. An example is shown in Snippet 9.1. Note that when loading the index from disk, the full type needs to be specified. As the text is not copied into the index, it is stored together with it in this example.

The third template parameter of the index is always optional. It is the fully instantiated SDSL compressed suffix array (CSA) type. Since the type

---

[4] https://github.com/y-256/libdivsufsort.

[5] It is not strictly required that the alphabet be part of the type, but not including it reduces type safety and complicates error handling in other interfaces like the search.

is a nested template with many parameters, SeqAn3 provides an alias called
`seqan3::sdsl_wt_index_type` which sets template parameters to good
defaults (see Snippet A.9 in the appendix for the full definition). Application
developers knowledgeable of the SDSL can tune the behaviour of the index
by providing their own custom template argument. It is likely that SeqAn3
will provide more aliases for specialised use-cases in the future. The type
`seqan3::sdsl_default_index_type` is an "alias alias", and it always refers
to the `seqan3::sdsl_X_index_type` that is currently recommended (and is the
default template argument for the third parameter).

```
   seqan3::fm_index_cursor cur = index.cursor();          // create a cursor
2  cur.extend_right("AAGG"_dna4);                          // search the pattern "AAGG"

4  std::cout << "Number of hits: " << cur.count() << '\n'; // outputs: 2

6  seqan3::debug_stream << "Positions in the genome: ";
   for (size_t const pos : cur.locate())                  // outputs: 8 22
8      std::cout << pos << ' ';
```

Code snippet 9.2: The FM-index cursor. The `.cursor()` member of the index
returns a `seqan3::fm_index_cursor` that represents a single specific "traversal"
of the conceptional suffix tree of the index. Outputs given reflect the index created
in Snippet 9.1

Other than inducing the creation of the required data structures upon construction,
the main task for `seqan3::fm_index` is to expose an "index cursor" via the
`.cursor()` member function. An example is shown in Snippet 9.2. When creating
the `seqan3::fm_index_cursor`, it points to the root of the conceptional suffix tree.
`.extend_right()` can be used to traverse the tree by one or multiple characters,
narrowing the search/moving to a lower node in the tree. Further important members
are `.count()` which counts the number of leaves below the current node and
`.locate()` which resolves them to text positions. If the index is created over a
collection of sequences, the text positions would be pairs of integers indicating the
index of the sequence together with the offset inside that sequence. `.cycle_back()`
can be used to change the last character in the current pattern to the next letter in the
alphabet (this corresponds to moving to a sibling node in the tree).

The FM-index cursor plays a similar role as the *virtual suffix tree iterator* in
SeqAn2. SeqAn3 avoids calling it an *iterator* because it does not satisfy the standard
library's requirements for an iterator and SeqAn3's indexes are not designed as
ranges.

9.1.2   Bidirectional FM-Index

Bidirectional FM-indexes expand on unidirectional FM-indexes by allowing extension of the searched sequence to the left and right. This allows for more efficient
search schemes (see Sect. 9.4.1) at the cost of almost doubling the space requirements. See the literature for an in-depth description of the theoretical background
(Pockrandt, 2019).

SeqAn3 offers `seqan3::bi_fm_index` which is designed very similar to
`seqan3::fm_index` but returns a bidirectional cursor (`seqan3::bi_fm_index_`
`cursor`) when its `.cursor()` member is invoked. Notably, this cursor also
provides `.extend_left()` in addition to `.extend_right()` as well as
`.cycle_front()` to change the first character in the query. The SDSL does not
(yet) support bidirectional indexes, so SeqAn3 implements bidirectional semantics
by creating two unidirectional indexes internally and keeping them in-sync.

EPR-dictionaries (Pockrandt et al., 2017) are an important improvement (especially) for bidirectional indexes. They reduce asymptotic complexity of a single
character extension in the bidirectional index to $O(1)$. Adding support for them to
SeqAn3 requires certain changes to the SDSL. These are currently in the process of
being added by former SeqAn contributor C. Pockrandt and current SeqAn team
member Enrico Seiler. Subsequently, SeqAn3 will be updated to contain index
aliases that allow easy use of EPR-dictionaries.

9.2   The k-Mer-Index Submodule

The set of all k-mers of a sequence is the set of all (overlapping) substrings of length
k. In domains other than bioinformatics, they are often called q-grams or n-grams.
A k-mer-index of a text is a map of every k-mer to a list of that k-mer's occurrences
in the text. See Table 9.3 for an overview of the submodule.

There are a great variety of possible implementations with different implications
for performance and storage requirements. Most involve hash-tables and typically
the lookup of a single k-mer is assumed to happen in amortised $O(1)$, i.e. entirely
independent of the text size. This is an advantage over FM-indexes where—even
including the aforementioned optimisations—lookup of a sequence (especially
longer sequences) is on average more expensive. The main disadvantage of k-mer-

Table 9.3 k-mer-index submodule overview

Search: k-mer-index submodule	
Class types	`seqan3::shape`
Auxiliary types	`seqan3::bin_literal, seqan3::ungapped`
Function objects	`seqan3::views::kmer_hash`

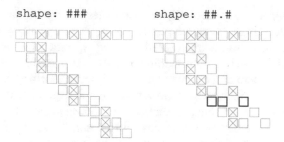

Fig. 9.1 Gapped shape used in filtering. This example illustrates the superiority of gapped shapes: three errors may prevent every ungapped 3-mer of a query of length 11 from being found while at least one gapped 3-mer is guaranteed to be found. ⊠ represents sequence mismatches. Image kindly provided by Knut Reinert

indexes is that only sequences of length k can be searched directly and that even for exact matches sequences longer than k, search strategies need to be used. These often include searching all the query's k-mers and performing various filters or counting algorithms to identify possible matching regions (Burkhardt et al., 1999).

Another important restriction of k-mer-indexes is that it is not possible to use the flexible search strategies available for FM-indexes (see Sect. 9.4.1), because one cannot backtrack in a k-mer-index. A popular alternative that allows performing inexact searches[6] is using k-mers that do not represent contiguous substrings but that "skip" certain text positions. Since k-mers overlap, all text positions are still searched and if desired full sensitivity can be reached (at a higher specificity). These k-mers are called "discontiguous", "spaced" or "gapped" in the literature (Burkhardt & Kärkkäinen, 2001). A specific "gap"-configuration of a k-mer is called a *shape*; an example of ungapped versus gapped shapes is shown in Fig. 9.1.

SeqAn3's *k-mer-index* submodule is in a very early development stage, and currently no index or search algorithm has been implemented. However, shapes are already available and introduced in the next subsection.

9.2.1   Shapes in SeqAn3

Excursus: Strong Types

Parameter names in C and C++ carry no significance for the developer invoking a function, i.e. given the following interface...

```
void foobar(int alpha, int beta) { /* ... */ }
```

---

[6] See Sect. 9.4.1 for a definition of "searching with errors" (inexact searches).

...the developer always sets the first parameter via the first argument and the second parameter via the second argument to the function call. `alpha` and `beta` only have descriptive value, and there is no way specified by the language how the names can be retrieved from within the code.[7] This means users can confuse the order of parameters of the same type and will receive neither compile-time nor runtime diagnostics which can lead to severe bugs that are difficult to track down. Beside the problem of user error, the absence of *named arguments* also means that the developer cannot create two overloads with the same (single) parameter type and different semantics. The latter would be desirable when creating e.g. a `Circle` type with one constructor for `double radius` and one for `double area`.

```
   struct alpha_t { int _v; };                      struct radius_t { double _v; };
 2 struct beta_t { int _v; };                      2 struct area_t { double _v; };

 4 void foobar(alpha_t alpha, beta_t beta)          4 struct Circle
   {                                                  {
 6    /* ... */                                     6   double diam = 0;
   }                                                    Circle(radius_t r) { diam = r._v / 2;            }
                                                    8   Circle(area_t a)   { diam = std::sqrt(a._v / M_PI); }
   // Invoke:                                           /* rest of class ...*/
10 foobar(alpha_t{3}, beta_t{7});                  10 };

12 // Errors:                                      12 // Invoke:
   // foobar(3, 7);                                   Circle c1{radius_t{2.5}};
14 // foobar(beta_t{3}, alpha_t{7});               14 Circle c2{area_t{23.7}};
```

Code snippet 9.3: Strong types. In the left example, strong types prevent errors in using the interface; in the right example, they allow multiple single-parameter overloads

The solution to both of these problems is the so-called *strong types* (Boccara, 2016). These types are created from an existing type (typically a built-in numeric type like `int` or `double`) and behave more or less exactly like the original type without being (implicitly) constructible from/convertible to that type. The value of the wrapped type can be exposed as a member variable or via a dedicated function like `.get()`. Examples are shown in Snippet 9.3.

The simplest kinds of strong types can be defined as just a `struct` that wraps a value of the desired type. But SeqAn3 also provides infrastructure for easily defining new strong types based on a given type and potentially emulating certain operators of that underlying type. This is useful in multiple scenarios, but the specifics are not important here.

---

[7] This is an important difference to the members of a class type, for which *designated initialisers* were introduced in C99 and C++ 17, see Sect. 9.5.1 on p. 257.

```
  seqan3::shape s0{seqan3::ungapped{5}};                    // represents "11111", i.e. ungapped 5-mer
2 seqan3::debug_stream << s0 << '\n';                        // prints "[1,1,1,1,1]"
  seqan3::debug_stream << std::ranges::size(s0) << '\n';    // prints "5"
4
  seqan3::shape s1{seqan3::bin_literal{0b101}};             // represents "101", i.e. gapped 3-mer
6 seqan3::debug_stream << s1 << '\n';                        // prints "[1,0,1]"
  seqan3::debug_stream << std::ranges::size(s1) << '\n';    // prints "3"
8
  seqan3::shape s2{0b101_shape};                            // same as previous (by shape-literal)
10 seqan3::debug_stream << s2 << '\n';                        // prints "[1,0,1]"
  seqan3::debug_stream << std::ranges::size(s2) << '\n';    // prints "3"
```

Code snippet 9.4: Constructing a shape in SeqAn3. `seqan3::shape` can be constructed from two auxiliary strong types: one that creates ungapped shapes and one that interprets the argument as a bit pattern. There is also a user-defined literal type that takes a binary literal and returns an object of type `seqan3::shape`

Shapes

Shapes in SeqAn3 are implemented as very simple extensions of `seqan3::dynamic_bitset`; in fact, `seqan3::shape` inherits from `seqan3::dynamic_bitset` and only adds two constructors and the constraint that the represented pattern must begin and end with a 1. This means that it is `constexpr`-safe, very compact (stored as one `uint64_t`) and comes with many convenient member functions for manipulation.

The two constructors serve to differentiate between two different semantics of "constructing by number". The first case represents constructing an ungapped shape of width *n*, and the other case is using `n`'s binary representation to encode the shape. The latter is especially useful since numbers can be given as binary literals since C++ 14.[8] To prevent mistaking one interface for the other, both require the usage of an auxiliary strong type. Snippet 9.4 shows how this looks in practice.[9] A user-defined literal (see Sect. 6.1.1) `N shape` is also defined which allows creating shape object as literal expressions (it is implemented in terms of the `seqan3::bin_literal`-constructor).

Hashing

Computing the hash of a *k*-mer is an essential operation and in many cases the hashes are computed of consecutive, overlapping *k*-mers. This is the case both during indexing and during the search of a sequence that is longer than *k*. Since this operation is a transformation of a sequence of letters to a sequence of hash-values, SeqAn3's general design suggests implementing this as a view. A trivial

---

[8] Integer literal examples: `5` is decimal; `05` is octal; `0x5` is hexadecimal; `0b101` is binary.

[9] Note that it is not the literal notation that is the differentiating factor here but the strong type. `0b101` is identical to `5`.

```
    std::vector<seqan3::dna4> text{"ACGTAGC"_dna4};
2
    /* trivial hashing view */
4   auto v0 = text | ranges::views::sliding(3) | std::views::transform(/*hash func.*/); // [6,27,44,50,9]

6   /* rolling hash with dedicated view */
    auto v1 = text | seqan3::views::kmer_hash(seqan3::shape{seqan3::ungapped{3}});    // [6,27,44,50,9]
8   auto v2 = text | seqan3::views::kmer_hash(0b101_shape);                           // [2,7,8,14,1]
```

Code snippet 9.5: A view for *k*-mer-hashing. Note that `views::sliding` is not yet part of the standard library or SeqAn3 but shipped with *range-v3*

implementation for ungapped shapes could look like l. 4 in Snippet 9.5. This does, however, not work for gapped shapes and it also does not take into account that subsequent *k*-mers overlap and that a full re-computation of the hash is not necessary. Instead, a so-called *rolling hash* computation can be used that only recomputes the difference between current and last hash. This is especially useful for large *k*-mers. SeqAn3 offers such a view called `seqan3::views::kmer_hash` that is also displayed in Snippet 9.5.

9.3   General Algorithm Design

```
    //                      ↓ Input data ↓           ↓ Single config object
2      auto        algo(data1,   data2,          config)
    // ↑ A view over the results (or void)
```

Code snippet 9.6: The general design of algorithm interfaces. *This is schematic and not valid SeqAn3 code*

Important algorithm functions like search and alignment follow a common design whose goal is to provide integration with ranges and to avoid large overload sets and complicated parameter-handling. Algorithms are restricted to having as few data parameters as possible: the input to the computation, e.g. query sequences and index. They additionally take a configuration object as parameter which contains algorithm-specific options as well as generic parameters, e.g. for parallel execution. See Snippet 9.6 for a schematic example. The config object is a *configuration* that consists of multiple *config elements*. These config elements can be runtime parameters (e.g. the number of allowed errors) or compile-time parameters (e.g. which information to include in the result type). Some options can even be set either way enabling a trade-off decision between flexibility and compile-time/binary size. Configuration elements are combined with `operator|` (like binary flags) and the result is an object of type `seqan3::configuration`.[10] This configuration is a template with deduced type, which means it cannot be assembled step by step; it needs to be defined as one expression (see Sects. 9.5 and 10.4 for examples). The common configuration infrastructure is provided by the *Core* module, but the algorithm-specific configuration elements are defined in the respective (sub)modules and typically also in a custom namespace.

Every algorithm defines a result type that contains the result of a single computation, e.g. one alignment or one hit in the database. Often the available members of the result type depend on the configuration. There are two ways of returning and processing results:

1. By default, a single-pass view over the results is returned immediately by the algorithm. This view buffers a certain number of results and invokes the algorithm again to produce new results whenever the buffer is consumed. This may happen on-demand in the main thread or (depending on configuration and algorithm) may also happen asynchronously in a background thread. The results are always processed by the main application thread when iterating over the view (even if they were produced by different background threads).
2. Alternatively, a delegate/callback function can be set via a config parameter.[11] This activates a more functional usage of the algorithm: every result is immediately passed to the callback which then processes that result in the same thread it is being called from. The return type of the algorithm becomes `void` and the algorithm does not return until all results are computed and processed.

These two designs allow very different usage patterns. While the first is much simpler and interacts well with views, the second may provide a slightly better performance for certain use-cases. However, in the second design, the user is responsible for ensuring that the callback function does not trigger data races when accessing shared state. The following section illustrates the design with examples of the search algorithm.

9.4 The (Search) Algorithm Submodule

The main search algorithm is defined together with its result type in the *Algorithm* submodule of the *Search* module. See Table 9.4 for an overview. Configuration elements for the algorithm are provided by the *Configuration* submodule (Sect. 9.5).

Table 9.4 Algorithm submodule overview

Search: Algorithm submodule	
Functions	`seqan3::search()`
Class types	`seqan3::search_result`

---

[10] This form of combination is commutative and entirely unrelated to the mechanism used by views.

[11] This mode is not yet available in the SeqAn3 master branch, but proof-of-concepts exist and the method is less novel than the first.

Since *k*-mer-indexes have not been implemented yet, the search algorithm only operates on FM-indexes, but this will change in the future. Certain configuration elements and members of the search result type are specific to the index type. This is documented in the API documentation and `static_assert()`s notify the developer of conflicts at compile-time.

```
    std::vector           queries{"ACGT"_dna5, "AGGA"_dna5, "TAGC"_dna5};
2   std::vector           text{"ACGACACACACACACACT"_dna5, /*...*/};
    seqan3::fm_index      index{text}; // normally index is instead loaded from disk
4   seqan3::configuration cfg{/* see Section 8.5 */};

6   auto results = seqan3::search(queries, text, index, cfg);

8   for (seqan3::search_result & result : results)  // results are computed lazily during iteration
    {
10      // print various details of the result
        std::cout << result.query_id()                 << ' '
12                << result.subject_id()                << ' '
                  << result.subject_begin_position()    << ' '
14                << result.substitutions()             << ' ';
    }
```

Code snippet 9.7: Short search example

A short example of using the search algorithm is shown in Snippet 9.7. The search function is designed as described in Sect. 9.3, and the arguments are as follows:

`queries`   The sequence(s) to search for. Either a range (one sequence) or range of ranges (multiple sequences). The alphabet type must be convertible to the alphabet type of the index/text.

`text`   The text that the index was created from. This argument is optional, and however certain optimisations are only performed when it is provided and certain members of the result type depend on it. Either a range (one sequence) or range of ranges (multiple sequences).

`index`   The index to search in. `seqan3::fm_index`, `seqan3::bi_fm_index` or a *k*-mer-index (not yet supported).

`cfg`   The algorithm configuration; see Sect. 9.5.

In Snippet 9.7, the input data is created directly before invoking the algorithm, but typically the index would be pre-computed and stored together with the text on-disk, and the query sequence(s) would be user-provided. Although providing the text is optional, most applications need it for pre- or post-processing steps anyway. It enables faster searches (see Sect. 9.4.1) and allows accessing `.subject_subrange()` on the results.

The class template `seqan3::search_result` has the following members:

`.query_id()`   The index of the query sequence in the set of inputs.

`.subject_id()`   The index of the subject sequence that the hit was found in.

`.subject_begin_position()`   The position in the subject sequence where the hit begins.

`.subject_end_position()` The position one after the end of the hit.

`.subject_subrange()` A subrange of the hit region in the subject sequence.

`.substitutions()` The difference between subject subrange and query sequence (the number of substitutions).

`.insertions()` The difference between subject subrange and query sequence (the number of insertions).

`.deletions()` The difference between subject subrange and query sequence (the number of deletions).

`.index_cursor()` The `seqan3::fm_index_cursor` / `seqan3::bi_fm_index_cursor` at the end of the search and can be used to continue more fine-grained "manual" search.

Not all of these are available in all configurations, see Sect. 9.5. The example in Snippet 9.7 returns the results as part of a view. The other method of returning results is also shown in Sect. 9.5.

9.4.1 Search Strategies

Searching for an exact match is fairly simple with the cursor: simply call `.extend_right()` with the query sequence as argument. However, in sequence analysis, it is very important to also find *approximate matches*, i.e. matches with up to k "errors", where k depends on the application and settings. Meaningful matches that include errors can be the result of:

Technical/statistical error The sequence was "correct" *in vivo*, but the sequencing method or post-processing introduced an error that prevents an exact match with the query.

Biological change Small individual mutations or long-term evolution leads to sequences that differ to varying degrees. Finding these related sequences ("homology search") is an important part of biological research.

Errors can be either *substitutions* (the letter at the given position in query and subject differs) or *insertions/deletions* (either query or subject has a letter that is not aligned to a letter of the other sequence).[12]

The simplest approach to approximate string matching (other than searching all permutations of the string with k errors) is *backtracking*. Backtracking traverses the conceptional suffix tree visiting all branches that would have been visited by searching all permutations with k errors but without searching common prefixes multiple times (Ukkonen, 1993). It is still comparatively expensive, especially when the conceptional suffix tree is dense, but it works on unidirectional indexes compared to other methods introduced below.

---

[12] Whether such a difference is interpreted as a deletion in one sequence or an insertion in the other is primarily a matter of perspective.

Faster (but lossless) approaches are based on the observation that it is beneficial to split the query into non-overlapping pieces which are searched in different combinations and where some are always searched exactly. These approaches include using the pigeon-hole principle (Herstein, 1964), *01*0*-seeds (Vroland et al., 2016) and optimum search schemes (Kucherov et al., 2016; Kianfar et al., 2017). The latter have been shown to provide the best performance, but all approaches require a bidirectional (FM) index, because matches of pieces may need to be extended in either direction.

An optimisation that can be used with almost all search strategies is *in-text verification*. The idea behind this is that, when the remaining depth of the to-be-searched conceptional suffix tree is still high but the number of represented text positions is already low, one can switch to directly resolving the text positions and performing comparisons (possibly a banded alignment) in the text instead of backtracking in the tree. Pockrandt (2019) provides theoretical background and in-depth comparisons of the search strategies and in-text verification.

SeqAn3 implements optimum search schemes (these are not part of the SDSL). But the search strategies are not directly exposed to the user, instead the `search()` algorithm simply picks the best strategy based on the provided error configuration and index. This may change in the future for advanced use-cases like selecting subsets of search schemes for heuristic purposes. Support for in-text verification is planned but not yet implemented. A significant amount of engineering time went into implementing search schemes, but this is not covered here as it is not my work and the focus of this book is discussing the publicly visible aspects of the library.

9.5   The Configuration Submodule

An overview of the configuration submodule is shown in Table 9.5 and the detailed list of available config elements is shown in Table 9.6. They are either types/type templates or objects. Those that are types need to be given initialising arguments to be useful; the type of the arguments is also given in Table 9.6 and syntax highlighting should further help distinguish these. Ultimately, the difference for users is only that some config elements have no further arguments and some do. Multiple config elements are combined into a `seqan3::configuration` via logical OR (see Sect. 9.3). The following excursus explains how those config elements that take arguments can be easily initialised.

Table 9.5 Configuration submodule overview

Search: Configuration submodule	
Namespaces	`seqan3::search_cfg::`
Config elements	See Table 9.6

9.5.1   Excursus: Aggregate Initialisation and Designated Initialisers

```
 2  struct Foo
    {
 4      int a = 0; // member initialisers
        int b = 0;
 6      int c = 0;
    };

 8  /* aggregate initialisation */
    Foo f0{};                    // all are 0
10  Foo f1{42, 23};              // c is 0

12  /* designated initialisers */
    Foo f2{.a = 42, .c = 23};    // b is 0
```

```
 2  struct Foo
    {
 4      int a = 0;
        Foo set_a(int _a)
        {
 6          a = _a; return *this;  // return copy
        }

 8      /* same for b and c ...*/
10  };

12  /* chained member function call */
    Foo f2 = Foo{}.set_a(42).set_c(23); // b is 0
```

Code snippet 9.8: Designated initialisers. Member initialisers and aggregate initialisation are available since C++ 11. Designated initialisers are a C99/C++ 20 feature. It
is possible to emulate this style in pre-C++ 20 with chained function calls; `f2` has
the same value in both snippets.

The attentive reader might have realised that I have used curly braces ({ })
instead of parentheses (()) to initialise variables in all Modern C++ code in this
book. Many rules have changed regarding initialisation, and I do not want to
cover them extensively here, but an important aspect of initialising with braces is
that so-called *aggregate types* (types without user-provided constructors, `virtual`
functions or `private:`/`protected:` members)[13] can be initialised as if they had a
constructor accepting values for their data members.

As Snippet 9.8 shows, this allows easily constructing objects of such types and
avoids having to write lots of "boilerplate code" for constructors. It does, however,
suffer from similar problems as function interfaces with many parameters of the
same type, i.e. that it is possible to confuse the order (see Sect. 9.2.1 on p. 249).
Furthermore, it is only possible to omit member variables "from the end", e.g. it is
not possible to provide only a custom value for the last data member.

To solve this problem, the so-called *designated initialisers* were added to C++ 20
(ISO/IEC 14882:draft, 9.3). These are originally a C99 feature and allow explicitly
setting the value of certain members by name as shown in Snippet 9.8. It is also
possible to omit any members in this fashion.

This feature is already available in GCC8 and GCC9, so only GCC7 as a target
platform cannot make use of it. Of course, it is possible to simply use aggregate
initialisation instead, but a more flexible workaround is also available: member
functions are provided that set the individual data members and return a copy of the

---

[13] For a formal definition of *aggregate types*, see ISO/IEC 14882:2017 (11.6.1).

object itself. These can be chained to create a single expression that has some data members "initialised" to custom values.[14] This style is also shown in Snippet 9.8 (on the right).

Note that an alternative solution to this problem would be to not use aggregate types but to instead define a class type that takes strong types for the individual parameters. But to achieve the same flexibility, constructors would have to be added for accepting all combinations of these strong types. This would not only be burdensome for the developer but would also complicate the invocation syntax for the user since these strong types would likely need to be put into an extra namespace or have very verbose names (e.g. `Foo f2{FooArgA{42}, FooArgC{23}};`).

In the following, I will use designated initialisers, but the same data structure can be used with the workaround mechanics on GCC7.

9.5.2 Search Config Elements

An overview of the config elements is given in Table 9.6. Elements are grouped according to whether they influence the computation of the actual matches, whether

Table 9.6 Configuration elements for the search. Namespace `seqan3::search_cfg::` is assumed for the left column. The user can provide arguments of the following types: `u` unsigned integer; `f` floating point; `m` a `search_cfg::mode` other than `dynamic`; `fun` a function (object). `x` is the number of CPU cores detected

Affect match computation	
`max_error{u,u,u,u}`	Max. abs. error counts (total, subst., insert., delet.).
`max_error_rate{f,f,f,f}`	Max. rel. error percentages (total, ...).
`mode::all`	All hits within the specified error bounds.
`mode::all_best`	All hits with minimal error count.
`mode::best`	Any one single hit with minimal error count.
`mode::strata{u}`	`all_best` but also hits with `u-1` errors more than best.
`mode::dynamic{m}`	Can be set to a `search_cfg::mode` at runtime.
Affect result type	
`without_subject_info`	Hits will not be expanded to subject positions.
Affect execution	
`parallel{u}`	Number of threads.
`on_result{fun}`	Callback-mode, takes a callback/delegate.
Aliases	
`default_configuration`	`mode::all | max_error{0} | parallel{x}`

---

[14] Although the compiler will likely optimise away the multiple copies, this style is only recommended for non-critical paths of the program or objects that are light-weight. Both are true for creating algorithm configurations.

they affect the characteristics of the result type or whether they influence the execution of the algorithm. An alias for a default configuration is also provided.

There are two ways to define which kind of errors are permitted when performing a search. By default, the search looks for exact matches of the query sequence in the index and allows no errors. But error counts can be specified as absolute values (`seqan3::search_cfg::max_error`) or percentage of the query length (`seqan3::search_cfg::max_error_rate`). Both config objects can be constructed with up to four parameters for the (1) total number of errors, (2) number of substitutions, (3) number of insertions and (4) number of deletions. In all cases, the counts are upper bounds and hits with fewer errors will also be returned. If both a total number and individual numbers for substitutions, insertions and deletions are given, the strictest bound applies. Designated initialisers (see the excursus above) can be used to easily initialise config objects without having to memorise the order of members or having to set all of them. This approach is shown in Snippet 9.9. The fallback solution for GCC7 is shown in Snippet 9.10 (the effects of the statements are identical).

```
  seqan3::configuration cfg = seqan3::search_cfg::max_error{.total = 1, .substitutions = 1} |
2                             seqan3::search_cfg::mode::all_best                            |
                              seqan3::search_cfg::parallel{4};
4
  auto search_results = seqan3::search(queries, text, index, cfg);
```

Code snippet 9.9: Search configuration examples I. One error is allowed, but only as substitution. If there are hits without errors for a given query, only those are returned. Four queries are searched in parallel with four CPU threads

The impact of errors on the search results differs depending on the `seqan3::search_cfg::mode`. It can be one of `::all`, `::all_best`, `::best` and `::strata{n}`, where n is an integer ≥ 1. These modes are frequently used in read mapping software and are defined as follows:

`::all` All hits within the specified error bounds are returned, and this is the default.

`::all_best` Only hits with minimal error count are returned, e.g. if up to one error is allowed but a single hit with zero errors is found, only hits with zero errors are returned.

`::best` Only a single hit of minimal error count is returned per query sequence (i.e. search can stop if a hit with 0 errors is found).

`::strata{n}` Like `::all_best` except that n "levels" are returned, e.g. if the best hit has one error and n $==$ 2, all hits with one or two errors are returned.

The search mode is typically set at compile-time (applications have a fixed mode); this is shown in Snippet 9.9. But the search mode can also be set as a runtime parameter (so it can be exposed via the argument parser to users); this is shown in Snippet 9.10. `seqan3::search_cfg::mode::dynamic` can be assigned any of the other values and stores that internally inside a `std::variant` (see Sect. 3.10).

Setting the parameter to a fixed value at compile-time has no benefit other than reduced build-time and binary size.[15]

```
     /* errors */
 2   auto e = seqan3::search_cfg::max_error{}.set_total(1).set_substitutions(1);

 4   /* mode */
     seqan3::search_cfg::mode::dynamic m;
 6   if (only_best_option == true)   m = seqan3::search_cfg::mode::best;
     else                            m = seqan3::search_cfg::mode::all;

 8
     /* result processing */
10   auto delegate = [] (auto & res) { std::cout << res.query_id() << '\n';};
     auto d = seqan3::search_cfg::on_result{delegate};

12
     /* config and search */
14   seqan3::configuration cfg = e | m | d;
     seqan3::search(queries, text, index, cfg); // calls delegate(r) for every search result r
```

Code snippet 9.10: Search configuration examples II. The config elements are predefined. Element `e` has the same effect as in Snippet 9.9. Element `m` depends on a user-provided runtime parameter. The algorithm does not return a range of results, instead a delegate is invoked for every result

The `seqan3::search_cfg::parallel` and `seqan3::search_cfg::on_result` config elements are based on common base classes in the *Core* module and are shared with other algorithms. The first activates parallelisation of the query sequences, i.e. up to `n` queries are searched in parallel. It can be seen in Snippet 9.9. This also means that no parallelism takes place if only a single query is supplied.

`seqan3::search_cfg::on_result` changes the way results are processed as described in Sect. 9.3. Its use can be seen in Snippet 9.10: the `search()` function passes every result to the `delegate` function object for direct processing; the function itself returns `void`. If this option is combined with `seqan3::search_cfg::parallel`, the callback function will be called from different threads, so care has to be taken to synchronise access to shared data or even output streams.

For some applications, it may not be desirable to process the search hits immediately at all—and to instead continue a fine-grained search within the index, e.g. by extending the initial seed further. This can be achieved through the `.index_cursor()` of a `seqan3::search_result`. However, a single index cursor often belongs to multiple hits (see Sect. 9.1.1) and resolving the subject information like ID and positions is associated with notable overhead. Setting the config element `seqan3::search_cfg::without_subject_info` will result in all the `.subject_*` members of the returned `seqan3::search_result` to become "disabled" and only one result per cursor being returned. The application developer

---

[15] The impact may however be quite substantial, because runtime parameters in combination with a lot of static typing lead to the expansion of many codepaths.

can then further fine-tune the search via the means described in Sect. 9.1 and resolve the subject information later via `.locate()` on the cursor.

Configuration elements that change the composition of the result type do so at compile-time. The result type will still have all member functions, but trying to call disabled members will result in a failed `static_assert()` during build-time that explains why this member is not usable.

9.6 Discussion

The search and index module in SeqAn3 is not yet feature-complete. Certain optimisations are missing from FM-indexes, and k-mer-indexes still lack the bulk of the implementation. They are a requirement for the promising DREAM index that is planned for the future. However, the current design already covers most user-visible aspects of this module and outlines well what can be expected in the following SeqAn3 releases.

None of the well-known bioinformatics suites written in Python offer comparable functionality, so the *Search* module is only compared against SeqAn2 in this section.

9.6.1 Performance

Search

The performance of searching depends on several factors and there are many options for fine-tuning the behaviour of the index data structures. Most options like the sampling rate of the suffix array (or using a bidirectional versus a unidirectional index) constitute a trade-off between execution speed and memory requirements.

The results of the search micro-benchmarks can be seen in Table 9.7. They are based on the current default FM-index types of SeqAn2 and SeqAn3 which use wavelet-trees. The alphabet is `seqan::Dna`/`seqan3::dna4`. When searching without errors on a unidirectional index, SeqAn2 has a strong and surprising lead over SeqAn3.[16] For most other benchmarks, the results are within ±20% of each other. In general, SeqAn2's search times are better than SeqAn3's for lower error rates and unidirectional searches, whereas SeqAn3's are sometimes better for bidirectional searches and/or higher error rates. However, throughout all benchmarks, SeqAn2's numbers improve more strongly when machine-specific optimisations are turned on. The most important effect of this is single-instruction `popcnt()`, which is used heavily in indexed search.

---

[16] Note in this context also that the runtimes for 0 errors should be identical between unidirectional and bidirectional searches—which they are not for SeqAn2.

Table 9.7 Search micro-benchmarks. Very small indexes used, either unidirectional or bidirectional wavelet-trees. Unidirectional is standard backtracking, and bidirectional search uses optimum search schemes. Errors are hamming distance. Suffix array sampled at 10%. Hits are counted and located. "Native" times are produced by binaries with per-machine optimisations including the `popcnt` instruction

	Errors	Wavelet-trees		Bi. wavelet-trees	
		SeqAn2	SeqAn3	SeqAn2	SeqAn3
Runtime	0	7 ms	22 ms	23 ms	21 ms
Runtime [native]	0	4 ms	17 ms	13 ms	17 ms
Runtime	1	124 ms	164 ms	85 ms	75 ms
Runtime [native]	1	74 ms	127 ms	54 ms	61 ms
Runtime	2	1146 ms	1058 ms	124 ms	115 ms
Runtime [native]	2	678 ms	823 ms	79 ms	95 ms
Runtime	3	6868 ms	6165 ms	113 ms	106 ms
Runtime [native]	3	4153 ms	4937 ms	78 ms	85 ms

Table 9.8 Search application benchmark. 1000 Illumina reads of length 200 are searched without errors in human chromosome 13. Bidirectional indexes used in all cases, either based on wavelet-trees or EPR-dictionaries. Suffix array sampled at 10%. Hits are only counted (not located). Time for searching is given (not start-up and I/O). Native time includes machine-optimised `popcnt` instructions

	Bi. wavelet-trees			Bi. EPR-dictionaries		
	SeqAn2	SeqAn3	SDSL	SeqAn2	SeqAn3	SDSL
Runtime	48 ms	55 ms	56 ms	26 ms	36 ms	35 ms
Runtime [native]	44 ms	50 ms	50 ms	24 ms	33 ms	33 ms
Memory	420 MB	114 MB	114 MB	468 MB	240 MB	240 MB

Independent of library, the speed-up through using bidirectional indexes is very notable for inexact searches. For three errors, the difference is 50x–60x!

A drawback of the micro-benchmarks is that the index is very small (not much larger than the sum of the query sequences). This leads to caching-effects that might over-pronounce certain differences that would otherwise be diminished. To attain numbers more indicative of typical use-cases like read mapping, I created a benchmark based on the test suite by Pockrandt using human chromosome 13 as the index file; it is ~95 MB big. I have also included an experimental branch of the SDSL that has EPR-dictionaries; SeqAn3 also makes use of this experimental SDSL branch in this benchmark. The benchmark focuses on the "pure" index speed as no locate operations are performed and no errors are allowed (search schemes are not used).[17] Runtimes and memory usages are shown in Table 9.8.

---

[17] Note that lambda3 which is discussed in Part III provides a more integrated "real-world" test case.

The most obvious result is that SeqAn2 is 10–30% faster than SeqAn3 (depending on settings). On the other hand, it requires 2–3 times more memory. For SeqAn3 and the SDSL, EPR-dictionaries require twice the memory of wavelet-trees which is expected.[18] It is unclear why SeqAn2 uses so much memory in the WT benchmark.

Speed and memory usage of the "pure" SDSL implementation and SeqAn3 are almost identical, showing that no notable overhead can be attributed to SeqAn3's intermediate layer. The impact of machine-specific optimisations is closer to 10% over the 20–30% observed in the micro-benchmark, but no locate operations happen here (these likely profit from single-instruction `popcnt`). For SeqAn2, the relative speed-up of using EPR-dictionaries over wavelet-trees is higher than for SeqAn3/SDSL, which suggests that more optimisation needs to happen in the latter.

I would summarise the performance analysis by concluding that SeqAn3 and SeqAn2 are in a similar overall performance range with SeqAn3 currently favouring a lower memory footprint but not reaching SeqAn2's speed. The SDSL (and thereby SeqAn3) definitely needs to receive a fully functioning implementation of the EPR-dictionaries, and other optimisations available in SeqAn2 should also be ported. This includes implicit sentinel characters (currently the alphabet size in the SDSL is always increased by 1) and the handling of sequence collections (this is currently implemented "on-top" in SeqAn3 and also leads to an increase of the alphabet size by 1, not benchmarked here). Based on assessments by developers deeply knowledgeable of both SeqAn2 and the SDSL, better speed–memory combinations should be possible with the latter. Other optimisations like in-text verifications can provide even stronger speed-ups. Once these features have been implemented, the SeqAn team should re-evaluate the impact of different parameters on speed and memory and predefine simple-to-use aliases for its users, e.g. `seqan3::sdsl_index_type_fast` (favouring speed) or `seqan3::sdsl_index_type_low_memory` (favouring space).

k-Mers and Shapes

Table 9.9 Performance of k-mer hashing. Throughput is shown for hashing the given shape on a DNA4 sequence of length 50k. Naïve implementation is as in l. 4 of Snippet 9.5 on p. 252

Shape	Naïve	SeqAn2	SeqAn3
Ungapped 8-mer	144.9 MB/s	145.7 MB/s[a]	137.0 MB/s[a]
Ungapped 30-mer	29.7 MB/s	144.9 MB/s[a]	137.0 MB/s[a]
Gapped 8-mer	–	144.2 MB/s	122.2 MB/s
Gapped 30-mer	–	58.8 MB/s	46.6 MB/s

[a] These cells are computed by a rolling-hash algorithm

[18] The total length of bitvectors is $\log_2(\sigma) * n = 2.3n$ for WTs and $\sigma * n = 5n$ for EPR (but of course other factors also contribute to memory usage).

Table 9.9 shows throughput of the k-mer hashing implementations. The through-put seems to have an upper bound on my system at approximately 145 MB/s. For the short 8-mer, this bound appears to be reachable independent of whether the shape is gapped or ungapped and independent of rolling-hash versus full-rehash.

The advantage of the rolling-hash scheme can be seen in the difference for hashing the ungapped 30-mer in the naïve approach versus the default implementations in SeqAn2 and SeqAn3. A speed-up of almost 5x is very significant.

SeqAn3's numbers are similar to SeqAn2's, although consistently lower by a noticeable margin. As initially stated, I am not the implementer of this module, but this gap seems bridgeable.[19] Since the naïve implementation reaches the upper bound with the 8-mer and is also based on a view (see l. 4 of Snippet 9.5 on p. 252), the view design itself is definitely not responsible for a lower speed here.

9.6.2   Simplicity

Indexes

SeqAn2's documentation spends a lot of time explaining the different indexes and their interfaces. Indexes in SeqAn2 have the so-called *fibres* which include the text, a hash-table or one of the various tables used by FM-indexes and enhanced suffix arrays (LCP table, occurrence table, BWT etc.). Algorithms in SeqAn2 can then access these tables directly which is very generic. But it requires a very high algorithmic understanding from the user and provides no benefit to most use-cases that simply involve searching. Since existence and usability of fibres also depend on template parameters (that are often manipulated by global metafunctions), even experienced SeqAn2 users struggled with configuring their indexes correctly. Crucial settings like the integer width used to represent positions in the index have to be set at compile-time, for some indexes via template parameters and for others via "metafunction overloading" (see Sect. 2.3.4). Index iterators (comparable to SeqAn3's cursors) behave differently when used on tree-like abstractions (e.g. `seqan::IndexSa`) than when used on trie-like abstractions (e.g. `seqan::FMIndex`). Furthermore, some indexes are suffix-based and others are prefix-based making it difficult to switch between them in generic contexts, because the user has to reverse their input and/or adjust the computed positions.

SeqAn3 shares certain details with SeqAn2 but sets very different priorities. The following is suggested for application developers:

---

[19] When looking at this particular code, the numbers for SeqAn3/gapped were in fact much lower (53 MB/s and 17 MB/s). Through a short debugging and optimisation session, I was able to bring them to the current performance level, so I am confident that some more time spent on this code can improve it further.

Most use-cases   The index is constructed from the text (or stored/loaded from disk) and is then passed to `seqan3::search()`. Developers do not need to understand anything about the internals of the index or its interface, they only need to decide on an index type (`seqan3::fm_index`, `seqan3::bi_fm_index` or a future *k*-mer-index) and the search does the rest.

Custom search   Advanced users interested in performing a custom search can do so via the index cursor. The cursor's interface is not named and designed after the tree structure (`goDown()`, `goRight()`) but on usage patterns (`extend_right()`, `cycle_back()`). This makes it more agnostic of the actual structure, e.g. `seqan3::bi_fm_index_cursor`'s interface is a strict superset of `seqan3::fm_index_cursor`'s.

Index type manipulation   For performance fine-tuning of FM-indexes, the SDSL type specifications can be changed. By default, SeqAn3 will offer predefined use-case-based aliases (instead of cryptic names based-algorithm or paper author). But expert users may define their own template specialisations.

This multi-step recommendation follows the principle that the most frequent usage-patterns should be simple, but that power-users are still given the opportunity to fine-tune. In some cases (e.g. separating the text from the index, not exposing fibres), it was decided to reduce the complexity overall. SeqAn3's FM-indexes (by ways of the SDSL) also choose the smallest possible integer width dynamically (based on the text input) and do not need to be configured to behave optimally in this regard.

```
/* SeqAn2 */
2  typedef Index<CharString, FMIndex<>> TIndex;
   CharString text =
4      "How much wood would a woodchuck chuck?";
   TIndex index(text);
6
   Iterator<TIndex, TopDown<>>::Type it(index);
8
   if (goDown(it, "doow")) // <- query reversed!!!
10     for (auto occ : getOccurrences(it))
          std::cout << occ << std::endl;
```

```
/* SeqAn3 */
2
   std::string text =
4      "How much wood would a woodchuck chuck?";
   fm_index index{text};
6
   auto it = index.cursor();
8
   if (it.extend_right("wood"))
10     debug_stream << it.locate() << '\n';
```

Code snippet 9.11: Manual search in SeqAn2 (left) and SeqAn3 (right). Prints '9' and '22'. The respective namespaces are assumed. Note how verbose SeqAn2 is regarding templates and that the user has to know that they need to reverse the query sequence. Respective namespaces are assumed

Snippet 9.11 illustrates some similarities and differences. As explained above, this is not even the simplest interface recommended. And due to said auto-detection, SeqAn3's index is based on 8-bit numbers, while SeqAn2' uses `uint64_t` by default.

Search

```
    typedef StringSet<String<Dna5>> TText;
2   typedef Index<TText, FMIndex<>> TIndex;

4   TText           text;        appendValue(text, "ACGACACACACACACACT"); /*...*/
    TText           queries;     appendValue(queries, "ACG"); appendValue(queries, "CACT"); /*...*/
6   TIndex          index{text};
    Finder<TIndex>  finder{index};

8
    for (auto & query : queries)
10  {
        clear(finder);
12      std::cout << query << '\n';
        Pattern<String<Dna5>> p{query};
14      while (find(finder, p))
        {
16          std::cout << beginPosition(finder) << ' '
                      << endPosition(finder)   << '\n';
18      }
    }
```

Code snippet 9.12: Finder-interface of SeqAn2. This is the SeqAn2 version of
Snippet 9.7 on p. 254. Namespace `seqan::` is assumed

Similar to `seqan3::search()`, SeqAn2 provided the `seqan::find()` interface
for iterative searching. Instead of modelling the results as a range or providing an
iterator, the so-called `Finder` objects were used. An implementation of a simple
search in SeqAn2 similar to SeqAn3's Snippet 9.7 on p. 254 is displayed in
Snippet 9.12.

The main disadvantages of SeqAn2's interfaces in regard to simplicity are:

- Multiple intermediate objects (`seqan::Finder`, `seqan::Pattern`) with individual syntaxes are needed. Omitting to for example `clear()` the finder will not
  result in a compile-time or runtime error but in wrong results.
- There is a huge number of top-level individual `find()` and `_find()` overloads,
  at least 20 in the index module and another 22 in the find module.
- Configuration of the search (and selection of overload) sometimes happens via
  specialisations of `Finder`, sometimes via specialisations of `Pattern`, sometimes via additional parameters and often via a combination of these. This is
  confusing and creates many opportunities to create incompatible combinations
  inadvertently.
- Certain overloads only take a single subject sequence (i.e. `Finder` over a single
  subject sequence) or a single query sequence (i.e. `Pattern` over a single query
  sequence), while others take a collection of either or both.
- Due to gradual extension of the modules over time, many potentially valid
  combinations were not implemented.

Table 9.10 Shapes in SeqAn2. Template specialisations (that in part need to be specialised further) are shown for `seqan::Shape`. In SeqAn3, there is only `seqan3::shape` (not even a template)—which encompasses all functionality

Specialisation	Modifiable	Number of gaps
UngappedShape	Compile-time	0
GappedShape	Compile-time	Any
SimpleShape	Runtime	0
OneGappedShape	Runtime	0 or 1
GenericShape	Runtime	Any

SeqAn3 solves all of these problems. There is a single search interface with very few overloads[20] and all configuration happens via a single configuration object. Most importantly, this is very simple to document: users find everything in one place. The documentation contains a single table that displays which configuration parameters are compatible with each other. When incompatible options are selected, a verbose static assertion is printed at compile-time.

Parallelisation happens generically and independently of other algorithmic details or specialised behaviour. Both query and subject can be a single sequence or a collection. This means no nested searches or nested loops are required. By default, the results are returned as a range which allows simple iteration via a `for`-loop; no repeated calls to the algorithm are necessary. Storing all results in a vector or performing range-based transformations can be achieved with a single line of code.

k-Mers and Shapes

The shape is prime example of reduced complexity in SeqAn3. It is implemented as a single data structure (`seqan3::shape`) that can be set at runtime and compile-time. In the latter case, the properties of such types (gapped or ungapped, specific pattern etc.) can be used to select optimised codepaths without incurring a runtime overhead. SeqAn2 needs multiple type specialisations to cover these use-cases (see Table 9.10). Furthermore, storage of `seqan3::shape` is a single `uint64_t` (stack storage), while some of SeqAn2's shapes contain dynamically allocated vectors of integers or bools.

Especially, the syntax for creating static shapes in SeqAn2 is very unintuitive as can be seen in Snippet 9.13. SeqAn2 requires the text alphabet to be given as a template parameter, because it incorporates simple hashing into the shape itself. However, to achieve rolling hashes iteratively, one needs to resort to the `seqan::hashNext()` function. It requires initialisation and is orthogonal to other hashing mechanisms.

---

[20] Mainly for handling the special case of `char *` input which needs to be converted to a proper range before processing to not mistakenly also search for the null-terminator.

```
   /* Dynamic gapped shape */                       /* Dynamic gapped shape */
2  Shape<Dna5, GappedShape<GenericShape>> s1{"1101011"};   2  shape s1 = 0b1101011_shape;

4  /* Static gapped shape */                         4  /* Static gapped shape */
   Shape<Dna5, GappedShape<HardwiredShape<1,2,2,1>>> s2;     constexpr shape s2 = 0b1101011_shape;
```

Code snippet 9.13: Constructing shapes in SeqAn2 (left) and SeqAn3 (right). The respective namespaces are assumed. The static notation in SeqAn2 encodes the distance to the next `1` in the shape (the encoded shape is the same for all examples).

In SeqAn3, hashing happens via `seqan3::views::kmer_hash`. When incrementing the view's iterators, they automatically use rolling-hash mechanism to produce the next hash-value. But, conveniently, the view also allows random access to the `i`-th hash in constant time (this will lead to a full hash computation).

9.6.3 Integration and Adaptability

The primary way the *Search* module interacts with user-provided types and third party libraries is through the generic interfaces of index construction and search. These are based on the range and alphabet concepts as introduced in Sects. 3.6.2 and 6.1.3. This also means they work very well with standard library types. The indexes are themselves usable in third party algorithms, and they are generic in the sense that the interface of `seqan3::fm_index` and its cursor is a strict subset of `seqan3::bi_fm_index` and its cursor. This allows writing algorithms that accept either—optionally taking advantage of bidirectional features if desired.

Indexes of the SDSL cannot be used as is, because SeqAn3's indexes wrap around them and the search expects this particular interface.[21] However, specific specialisations of the SDSL index type (choice of wavelet-tree type, suffix array sampling rate etc.) that are known to provide good results for particular use-cases can be passed as is to SeqAn3's FM-indexes resulting in the same internal type.

As discussed in Sect. 9.6.2, the module provides a multi-layered approach to searching that offers many ways to adapt the search without compromising simplicity. Returning results as a range ensures good integration with the other SeqAn3 modules and makes extending the search, e.g. by a post-processing filter, very easy. This is displayed in Snippet 9.14. Note that this particular example only makes sense if the region parameters are dynamic as it would otherwise be much more efficient to simply create an index directly over the desired region instead of the full text.

---

[21] One could of course write a custom wrapper (that would be accepted by `seqan3::search()`!), but this would likely result in something very similar to `seqan3::fm_index`.

```
   /* some example region in the text */
2  size_t region_begin = 13908;
   size_t region_end   = 17432;
4
   auto results = seqan3::search(queries, text, index, seqan3::search_cfg::default_configuration)
6                | std::views::filter([&] (auto & r)
                 {
8                        return (r.subject_begin_position() >= region_begin) &&
                               (r.subject_end_position()    <= region_end);
10               });
12 for (seqan3::search_result & r : results)  // process filtered results
   {
14     /*...*/
   }
```

Code snippet 9.14: Applying a view onto a search. Only the results that appear in a certain region of the text are considered by the loop

9.6.4 Compactness

The main difference in compactness between SeqAn2 and SeqAn3 with regard to searching is that SeqAn3's scope is different. It is narrower in the sense that SeqAn3 does not attempt to replicate all of SeqAn2's features in this area. As mentioned in the introduction to this chapter, there has been virtually no demand for SeqAn2's online search capabilities—and these are very extensive. The implementation of the different online search algorithms in SeqAn2 alone is more than twice the size of the entire search module of SeqAn3 (5,539loc versus 2,459loc).

On the other hand, SeqAn3 will support new indexing techniques like the DREAM framework in the future. This and the addition of smaller features and optimisations will definitely make the *Search* module grow in size. It may even pick up an online search algorithm again, but I do not expect that it will amount to the 35,000loc that SeqAn2 currently contains in the *find* and *index* modules (combined).

One important reason beyond a different focus is delegating a lot of the algorithmic "heave lifting" to the SDSL. The SDSL provides many FM-index implementations (balanced wavelet-trees, Huffman-encoding wavelet-trees etc.) which are available to SeqAn3 "for free" (in regard to compactness).

Even if overall numbers are difficult to compare based on the different feature sets, I think that it is safe to assume that the Modern C++ techniques used contribute considerably to compactness. Comparing a single feature like the shape implementations underlines this strongly. SeqAn2 requires 1,117loc to implement its shapes (excluding code for hashing), while SeqAn3's implementation is 36loc. This grows to 650loc if including the full definition of `seqan3::dynamic_bitset`, but the dynamic bitset is a very generic and useful type in and of itself (and even the sum is almost half the size of SeqAn2's code).

A different aspect of compactness is the sheer number of types and functions that are necessary to perform basic tasks and the amount of "glue code" that is required to keep these parts together. The examples given in this chapter highlighted that both have been strongly reduced compared to SeqAn2.

Chapter 10
The Alignment Module

Sequence alignment is an arrangement of two or more sequences that visualises which regions are conserved between the set of sequences and which regions differ. An overview of the alignment module is given in Table 10.1. Typically, one assumes that the compared sequences are of common evolutionary descent and that mutation events have introduced changes between them, but differences may also be the result of errors of the sequencing technology or in subsequent processing steps (e.g. normalisation). For the individual positions in every alignment, one differentiates between *matches* (a symbol is preserved between sequences), *mismatches* (a substitution event or error occurred) and insertions/deletions (*indels*; a biological event or error resulted in removal or addition of symbol(s) in one sequence). In the latter case the position marking a deletion (or an insertion in the other sequence) is denoted by a *gap* symbol (typically `'-'`). These symbols are placeholders that allow shifting certain regions of the sequences against each other more freely to yield an improved overall alignment. Section 10.1 introduces the data structures used in SeqAn3 to represent such aligned sequences.

Many alignments are possible between a pair of sequences, but one is usually interested in the alignment(s) that minimise the number of mutation events necessary to explain relatedness of the two sequences (because the events themselves are unlikely). To measure this relatedness a *score* is computed for the alignment and the goal of alignment algorithms is to compute the alignment with optimal or near-optimal score.[1] The *Scoring* submodule is covered in Sect. 10.2.

Alignments can be computed between two sequences ("pairwise alignment", Fig. 10.1) or between many sequences ("multiple sequence alignment", MSA, Fig. 10.2). Pairwise sequence alignment is more common; the respective interfaces and a brief introduction to the theoretical background are given in Sect. 10.3.

---

[1] Sometimes *distance* measures are used instead of scores, in that case the goal would be to minimise the distance.

© The Author(s), under exclusive license to Springer Nature Switzerland AG 2022
H. Hauswedell, *Sequence Analysis and Modern C++*, Computational Biology 33,
https://doi.org/10.1007/978-3-030-90990-1_10

Table 10.1 Alignment module overview

Alignment module	
Submodules	Aligned range, Configuration, Pairwise, Scoring

```
GGTGGTTTAGAACGATCTGGTCTTACCCTGCTACCAACTGTTCATCGTTATTGTTGGAG
|||||  ||||||| ||  |||||||||||||   ||||| ||  ||||||||||  |   |||
GGTGGGGTAGAAC-ATTTGGTCTTACCCTGAAACCAATTGCTCATCGTTA--G-GGGAC
```

Fig. 10.1 Pairwise sequence alignment (DNA). The notation is a common output of many applications. Matches are highlighted with pipe symbols

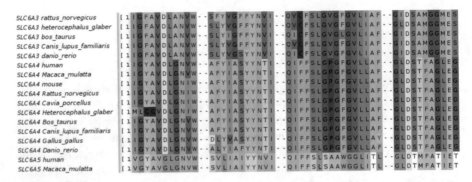

Fig. 10.2 Multiple sequence alignment (protein). This graphic shows a part of an MSA over several trans-membrane proteins (also from various species). Colour is used to group similar amino acids. Screenshot taken by myself; content produced with UGENE (Okonechnikov et al., 2012)

MSAs are not yet implemented in SeqAn3 and not covered in this chapter. They will receive their own submodule in the future and build on the existing pairwise alignment code and graph algorithms provided by the Lemon library (see Sect. 4.4.1).

A separate submodule is dedicated to the configuration of the alignment algorithm, because most of these are expected to be shared between pairwise alignments and MSA. It is discussed in Sect. 10.4.

10.1   The Aligned Range Submodule

An alignment always consists of multiple sequences: two in the case of a pairwise alignment or more in the case of an MSA. While it would have been possible to model an alignment as a single data structure, this appeared to only reduce genericity and flexibility.

For SeqAn3 it was decided to focus on a single *aligned range*, i.e. one row of the alignment. An overview of the aligned ranged submodule is shown in Table 10.2. Full alignments can then be designed as tuples/pairs of such aligned ranges or as a range of ranges. The first approach allows different underlying data types (e.g. regular vector versus bitcompressed vector) and/or different but comparable

Table 10.2 Aligned range submodule overview

Alignment: Aligned range submodule	
Concepts	seqan3::aligned_range, seqan3::writable_aligned_range, seqan3::resettable_aligned_range
Function objects	seqan3::insert_gaps, seqan3::is_gap, seqan3::remove_gaps
Class types	seqan3::gap_decorator

alphabet types (e.g. `seqan3::dna4` and `seqan3::rna4`). This is especially useful for pairwise alignments. The second approach requires that all rows of the alignment have the same type but allows a dynamic number of rows in the alignment; it is particularly useful for MSA.

The concepts and customisation point objects for aligned ranges are described in Sect. 10.1.1. Specific implementations and their theoretical background are explained in Sect. 10.1.2.

10.1.1 Concepts and Function Objects

Two main groups of "aligned ranges" exist: containers that own all elements and decorators that only own the gap information and hold a reference/pointer to an underlying "ungapped" range (see Sect. 10.1.2). The design for the aligned range concepts needs to cover both equally well. It is based on the existing range concepts (Sect. 3.6.2) and also split into multiple read/write concepts (Sect. 6.1.3). A sketched-out version of these concepts has been agreed upon by the SeqAn team but is not yet fully implemented in the current master branch.

The following three concepts and respective CPOs are the solution that I propose. I will give examples for how `std::vector` over `seqan3::gapped<seqan3::dna4>` would satisfy these as well as `seqan3::gap_decorator<std::vector<seqan3::dna4>>`, but I will spare the reader the full specification (see Sect. 6.1.3 for how to define and specialise customisation points).

`seqan3::aligned_range<T>` is the "read-only" concept, it requires that `T` be a range and that the CPO `seqan3::is_gap(r, it)` be valid (where `r` is the range and `it` an iterator of that range). The CPO returns `true` or `false`. It is defined by default for all ranges over alphabet types comparable with `seqan3::gap`, so it works automatically for containers over e.g. `seqan3::gapped<seqan3::dna4>` and gap decorators. But it also allows adapting ranges where the gap information cannot be queried directly or that use an entirely different alphabet to indicate gap symbols (it will look for member functions or free functions that provide the functionality, see Sect. 6.1.3).

`seqan3::writable_aligned_range<T>` is the "writable" concept, it requires that an aligned range also be callable with the two CPOs `seqan3::insert_gaps`

(r, it, n) and seqan3::remove_gaps(r, it, sen). The former inserts
n gaps before it into r and returns an iterator behind the last inserted
element. The latter removes all gaps between it and sen in r and returns
the number of removed gaps. Default insertion-implementations are provided
for container interfaces (standard .insert(it, n, seqan3::gap{})) and
the .insert_gaps() member is selected for gap decorators (regular CPO
behaviour). To remove gaps from containers the standard library algorithm
std::ranges::remove_if(it, sen,
  seqan3::is_gap) is called while the .remove_gaps() member function
is used for gap decorators. If only one gap is inserted/removed from the end of a
container, .push_back(seqan3::gap{})/.pop_back() are called instead.
seqan3::resettable_aligned_range<aligned_t, unaligned_t> is a
third concept that takes two parameters. It checks whether an aligned range
can be "reset" to the value of a second (unaligned) range. This is performed via
the CPO seqan3::reset_aligned_range(r, src) where r is the aligned
range and src an unaligned range. The exact semantics depend on the types,
but it is assumed the statement std::ranges::equal(r, src) be valid and
return true after the reset-CPO is invoked. For containers an implementation is
provided that clears the aligned range and then copies all elements from src to
r. For gap decorators r.reset(src) is called which clears the gap information
and changes the internal pointer to src.

The last concept can be used to constrain interfaces that take two parameters
to make sure that they match. But it is also important if a data structure internally
already holds a fixed unaligned range type (e.g. std::vector<seqan3::dna4>) to
make sure only aligned range types are accepted that can be "rebased" off of that
existing type.

10.1.2   Gap Decorators

As the default implementations for the CPOs have already indicated, the simplest way to make an aligned range from a non-aligned range is to create a
std::vector<seqan3::gapped<T>> where T is the alphabet of the original range
and copy all elements into this vector interspersing gap symbols where needed. This
kind of aligned range has the best possible read performance and especially for small
ranges, the cost of random inserts is still tolerable. Since the entire original range
needs to be copied, the space overhead, however, is very noticeable—assuming that
the original range needs to be kept in memory, too.[2]

To reduce space and (potentially) increase random write performance, various
adaptors can be devised that only hold a pointer to the original range and store

---

[2] For small ranges this might still be worth it!

the gap information in separate data structures. Multiple approaches and their algorithmic complexities are shown in Table 10.3. Here are brief descriptions:

Vector of gapped The full-fledged vector over a gapped alphabet. Fast, but huge size overhead.

Sparse bitvector An adaptor that holds an `sdsl::sd_vector` of the aligned sequence's size (n+g) with gap positions indicated by a 1. Useful if g/k is small and no modifications happen after construction.[3]

Width-vector An adaptor that holds a vector encoding the lengths of consecutive sequence and gap intervals, e.g. `AC-G--T` would be `2,1,1,2,1`. This was present in SeqAn2 as the `seqan::ArrayGaps` specialisation of `seqan::Gaps`. Random reads have bad asymptotic complexity but are cache-efficient for small ks.

Set of anchors An adaptor that stores an ordered set of `(orig_pos, cu_size)` where every pair represents one gap interval, `orig_pos` is the position in the original sequence where the interval is "anchored" and `cu_size` is the size of the interval plus all preceding gap characters. It allows performing faster random reads and has no asymptotic disadvantages over the previous method.[4] It is implemented in SeqAn2 as the `seqan::AnchorGaps` specialisation.

Blocked An adaptor that works like a nested version of "Set of anchors". It creates blocks of fixed width on the original sequence and inside each a set is stored similar to that of "Set of anchors" except that the values are relative to the beginning of the block. Additionally, a vector is created that contains a pair for each block (with values relative to the beginning of the range). Access then happens in two steps performing first a binary search on the blocks and then inside the target block. This should reduce the average random read/insert times for large k. Especially inserts should profit, because only trailing block counters need to be updated, not the elements inside.

Currently, only "set of anchors" is implemented in SeqAn3 as simply `seqan3::gap_decorator`. First implementations of the other data structures exist, but—as per the design goals—extensive benchmarks will have to show which data structures actually provide a benefit in practice (and where the "sweet spots" for using each lie).

As the name suggests, these data structures are "decorators", i.e. they are adaptors on existing ranges but also provide their own data that might grow proportionally to the size of the underlying or resulting range (see also Sect. 7.1). Simple usage of `seqan3::gap_decorator` can be seen on the left of Snippet 10.1.

The gap decorators satisfy all the syntactic requirements of a `std::ranges::random_access_range` (e.g. `operator[]`) if the underlying range does so. However, they do not satisfy the semantic requirement that random access is in $O(1)$ (affects all gap decorators, see Table 10.3). Because the low impact

---

[3] All modifications result in full reconstruction of the bit-vector, because SDSL data structures are not designed for dynamic modification.

[4] Inserts are still linear, because after an insert, the tail needs to be updated.

Table 10.3 Gap decorator possibilities. Complexity of removal is the same as for insertion and is each for one element. Worst-case and average case complexity are identical for all but "blocked". **n**: orig. seq. length; **g**: no. of gap symbols; **k**: no. of contiguous gap intervals; **b**: block size

Range	++it	[]	Insert	Insert back	Space
Vector of gapped	$O(1)$	$O(1)$	$O(n + g)$	$O(1)$	$O(n + g)$
Sparse bitvector	$O(1)$	$O(\log \frac{n}{g})$	$O(n + g)$	$O(n + g)$	$O(g + g \log \frac{n}{g})$
Width-vector	$O(1)$	$O(k)$	$O(k)$	$O(1)$	$O(k)$
Set of anchors	$O(1)$	$O(\log(k))$	$O(k)$	$O(1)$	$O(k)$
Blocked (worst)	$O(1)$	$O(\log \frac{n}{b} + \log(k))$	$O(\frac{n}{b} + k)$	$O(1)$	$O(\frac{n}{b} + k)$
Blocked (avg.)	$O(1)$	$O(\log \frac{n}{b} + \log \frac{bk}{n})$	$O(\frac{n}{b} + \frac{bk}{n})$	$O(1)$	$O(\frac{n}{b} + k)$

```
 2   /* construct original and decorator */     void print4(std::ranges::random_access_range auto && r)
     std::vector orig = "ACGTAC"_dna4;        12  {
     seqan3::gap_decorator alig{orig};
 4                                                  seqan3::debug_stream << r[4];
     /* Insert two gaps before position 2 */   14  }
 6   alig.insert_gaps(alig.begin() + 2, 2);
                                                    /* Does not model random_access_range */
 8   /* Print alignment */                         // print4(alig);
     seqan3::debug_stream << alig;             18
10   // "AC--GTAC"                                  /* "Fakes" a random access range signature */
                                               20   print4(alig | seqan3::views::enforce_random_access);
```

Code snippet 10.1: Using gap decorators. Simple usage is shown on the left. The snippet on the right side shows how to make a gap decorator "pretend" to have $O(1)$ random access—it is the only effect of that respective view and it requires that the input is a `seqan3::pseudo_random_access_range`

of random access is crucial to maintaining the guarantees that algorithms give on their complexity and because the notion of what "random access range" entails is very widely known and excludes such ranges, SeqAn3 does not treat them as `std::ranges::random_access_range`s by default, only as `std::ranges::bidirectional_range`s. An additional concept called `seqan3::pseudo_random_access_range` denotes this special in-between position. It further allows users to make such ranges pretend to have constant-time random access via `seqan3::views::enforce_random_access`.

An example can be seen on the right side of Snippet 10.1. This may seem overly cautious and one might think that SeqAn3 should declare by default that every `seqan3::pseudo_random_access_range` be a `std::ranges::random_access_range` (like SeqAn2 which makes no difference between them), but a small technical excursion should demonstrate why this is not the case (and why such design questions are important to consider!):

The realm of ranges has many cases where a specialisation is chosen based on the advertised strength of the input range, and falsely assuming that `operator[]` is in $O(1)$, when it is not, can actually decrease performance. For example, for all ranges, it is assumed that calling `operator++` on the iterator is in $O(1)$ and usually views that adapt another range implement their iterator's operators in terms of the underlying range's iterator's operators. So if a bidirectional range is passed to a view, the new iterator's `operator++` is defined in terms of the underlying iterator's `operator++`. But if random access ranges are passed to the view, it is convenient to implement most operations in terms of the underlying range's `operator[]`. This is a valid design choice, because that is guaranteed to have the same complexity as `operator++`. However, if the random access range passed to the view *lied* about its complexity, suddenly the `operator++` becomes much slower; had the range correctly been identified as only bidirectional, the derived view would have chosen a different implementation with complexity of $O(1)$.

Thus, a more conservative approach is better and `seqan3::pseudo_random_access_range`s by default result in the behaviour designed for `std::ranges::bidirectional_range`s. Users that are aware of the implications may opt-in to the "true" random access label via `seqan3::views::enforce_random_access`. This may be particularly useful for algorithms that *only* work on `std::ranges::random_access_range`s.

10.2   The Scoring Submodule

Scoring is an essential part of sequence alignment, although it can also be used stand-alone to score two sequences (aligned or not) or to just score two individual characters. This submodule makes the latter possible (Sect. 10.2.1) and also provides facilities for interpreting gap characters and gap intervals (Sect. 10.2.2). (Re-)Scoring an existing alignment can be easily implemented on top of these features. Computing an optimal alignment (including the score) between two unaligned sequences is described in Sect. 10.3. This submodule is one of the (few) parts of the *Alignment* module that I did not only help design but also implemented myself. See Table 10.4 for an overview.

Table 10.4 Scoring submodule overview

Alignment: Scoring submodule	
Concepts	`seqan3::scoring_scheme`
Scheme types	`seqan3::aminoacid_scoring_scheme`, `seqan3::gap_scheme`, `seqan3::nucleotide_scoring_scheme`, `seqan3::scoring_scheme_base`
Auxiliary types	`seqan3::gap_open_score`, `seqan3::gap_score`, `seqan3::match_score`, `seqan3::mismatch_score`
Enumerator types	`seqan3::aminoacid_similarity_matrix`

10.2.1   Alphabet Scoring Schemes

The common denominator of all alphabet scoring schemes is that they be able to
return a score for two objects of the alphabet(s) they support. This is reflected
in the `seqan3::scoring_scheme` concept. It is a three-argument concept
(`<T, TAlph1, TAlph2 = TAlph1>`) that states the single requirement that an
object of type `T` provide a member function `.score(a1, a2)` where `a1` and
`a2` are objects of type `TAlph1` and `TAlph2` respectively and that this function
return some arithmetic type. In the context of previously discussed concepts this
represents the "read-only"-aspects modelled by scoring schemes. While SeqAn3's
scoring schemes also share certain interfaces for configuring the scoring scheme
(see below), these are not formalised as a concept at this point (because there was
no use-case for such a concept).

To reduce complexity when scoring SeqAn3's alphabets, only two scoring
schemes are provided by default:

`seqan3::nucleotide_scoring_scheme`   Supports scoring any combination of
   types that are explicitly convertible to `seqan3::dna15`. This is true for all
   alphabets shipped with SeqAn3 that model `seqan3::nucleotide_alphabet`.

`seqan3::aminoacid_scoring_scheme`   Supports scoring any combination of
   types that are explicitly convertible to `seqan3::aa27`. This is true for all
   alphabets shipped with SeqAn3 that model `seqan3::aminoacid_alphabet`.

Both scoring schemes derive from `seqan3::scoring_scheme_base` which
avoids code-duplication but is not required to model the concept. The schemes
contain an $n \times n$ matrix internally with dimensions of the largest respective alphabet
(15 for nucleotides, 27 for amino acids). All input alphabets are then converted
to `seqan3::dna15`/`seqan3::aa27` and the score is computed by looking up the
ranks in the matrix. This ensures a high degree of flexibility with regard to different
inputs of one domain (e.g. nucleotides) without supporting nonsensical comparisons
between different domains (e.g. nucleotides and amino acids).

The scoring schemes offer a number of member functions to manipulate the
scoring behaviour:

`.set_custom_matrix(new_matrix)`   Directly overwrite the internal matrix.
   This enables maximum control over the scoring behaviour, but is very verbose.

`.set_simple_scheme(m, mm)`   Set the diagonal of the matrix to `m` and
   the rest to `mm`. The two parameter types are the arithmetic strong types
   `seqan3::match_score` and `seqan3::mismatch_score` (this prevents
   accidentally confusing the order).[5]

`.set_hamming_distance()`   Set a simple scheme of $(0, -1)$.

---

[5] See Sect. 9.2.1 for details on strong types.

```
   seqan3::dna4 d = 'A'_dna4;                    seqan3::aa27 a0 = 'L'_aa27;    // Leucine
2  seqan3::rna5 r = 'C'_rna5;                  2 seqan3::aa27 a1 = 'I'_aa27;    // Isoleucine

4  /* Hamming distance is default */           4 seqan3::aminoacid_scoring_scheme ss{};
   seqan3::nucleotide_scoring_scheme ss{};       ss.set_similarity_matrix(
6                                               6     seqan3::aminoacid_similarity_matrix::BLOSUM62);
   int s = ss.score(d, r);    // s == -1
8                                               8 int s = ss.score(a0, a1);   // s == 2
   /* U and T are a match */
10 d = 'T'_dna4;                               10 /* overwrite value in matrix */
   r = 'U'_rna5;                                  ss.score(a0, a1) = 3;
12 s = ss.score(d, r);        // s == 0        12 s = ss.score(a0, a1);       // s == 3

14 ss.set_simple_score(seqan3::match_score{2}, 14 /* reset to BLOSUM62 */
                  seqan3::mismatch_score{0});     ss.set_similarity_matrix(
16 s = ss.score(d, r);        // s == 2        16     seqan3::aminoacid_similarity_matrix::BLOSUM62);
```

Code snippet 10.2: Using scoring schemes. On the left simple usage is shown as well as the ability to compare values of different types. On the right amino acids are scored by the BLOSUM62 matrix and manipulation of single values in a matrix is shown

`.set_similarity_matrix(matrix_id)`   Select a matrix from a predefined internal set of matrices. The argument is a value of the enumeration type `seqan3::aminoacid_similarity_matrix`, e.g. `::BLOSUM62` (Henikoff & Henikoff, 1992). [Only available for `seqan3::aminoacid_scoring_scheme`].

Some of these are demonstrated in Snippet 10.2. The snippet on the right side also shows how simple an individual value in a matrix can be changed: by assigning to the return value of the score function (it returns a reference to the position in the matrix).

While the presented scoring schemes cover all use-cases that SeqAn3 developers encountered, it is not required to use them and custom schemes can easily be created (see Snippet A.8).

10.2.2   The Gap (Scoring) Scheme

The scoring of gaps is handled separately in SeqAn3 from the scoring of sequence characters, because gaps are usually not scored on a per-character basis. The reason for this is that occurrences of consecutive gap characters (the result of insertions and deletions) are not independent biological events; adjacent gap characters are typically the result of a single event. Thus, it makes little sense to penalise two consecutive gap character twice as strongly as a single gap character (*linear* gap costs).

Instead, the so-called *affine* gap scoring is very common. This assigns a single (large) cost to the existence of a gap interval of arbitrary length and adds a smaller cost per gap character. Other schemes like *convex* or *dynamic* (Urgese et al., 2014) are also possible but less widely used.

SeqAn3 offers the `seqan3::gap_scheme` that currently supports the linear and affine models but could be extended to support other approaches. The data structure provides a `.score()` member function that takes the length of the gap interval as the only parameter. Depending on the chosen model and score values, it will compute the score for the entire interval.[6] The gap model can be selected with:

`.set_affine(g, go)`   Takes parameters of type `seqan3::gap_score` and `seqan3::gap_open_score` respectively.

`.set_linear(g)`   Takes one parameter of type `seqan3::gap_score`.

Note that `seqan3::gap_scheme` does not model the `seqan3::scoring_scheme` concept, because one cannot use it to score two alphabet values. Since the gap model strongly influences the alignment algorithm, there is currently also no other concept or abstraction. The reason is that the dynamic programming algorithm does not "see" gap intervals as a whole so it cannot use the specified interface. Instead, it checks which gap model and scores are set in the scheme and adapts the algorithm accordingly. This behaviour depends on the intrinsics of the type `seqan3::gap_scheme`. Providing an arbitrary gap scheme type with custom semantics is incompatible with this design, so more thought will have to be put into the matter of whether this can be made more generic in other ways. In any case, adding more gap models to `seqan3::gap_scheme` in the future is possible if the alignment algorithm is adapted accordingly as well.

10.3   The Pairwise (Alignment) Submodule

This submodule provides the algorithm and a dedicated return type. See Table 10.5 for an overview.

10.3.1   Algorithm Interface

The pairwise alignment algorithm follows the general algorithm design presented in Sect. 9.3. As shown in Snippet 10.3, it only takes two parameters:

Table 10.5 Pairwise (alignment) submodule overview

Alignment: Pairwise submodule	
Functions	`seqan3::align_pairwise()`
Class types	`seqan3::align_result`

---

[6] Note that SeqAn3 always uses "scores" and never uses "penalties". The latter are simply implemented as negative scores which avoids ambiguity.

```
     std::vector seq_pairs{std::pair{"AGTGGCTACG"_dna4, "AGTGCCTACG"_dna4},
2                          std::pair{"AGGACTACG"_dna4, "AGTAGACTACGG"_dna4}};

4    seqan3::configuration cfg = /* see Section 9.4 */;

6    auto results = seqan3::align_pairwise(seq_pairs, cfg);

8    for (seqan3::align_result & result : results)  // results are computed lazily during iteration
     {
10       // print various details of the result
         std::cout << result.score()   << ' '
12                 << result.seq1_id() << ' '
                   << result.seq2_id() << '\n';
14   }
```

Code snippet 10.3: Pairwise alignment interface. The pairwise alignment algorithm takes sequence-pairs and the configuration object

`seq_pairs`   A range of tuples/pairs. Each tuple contains either two sequences (seq1, seq2) or additionally two integer identifiers (seq1, seq2, id1, id2). The IDs are user provided or generated by views (see below).

`cfg`   The configuration object.

In general the interface is meant to be called with many to-be-computed alignments at once—and not repeatedly with a single sequence pair. One reason is that there are certain initial setup costs (e.g. parsing and transforming the configuration, creation of buffers) that will impact performance negatively if repeated frequently.[7] Another important reason is that parallelisation and vectorisation happen to a large degree between multiple sequences/sequence-pairs. So the greater the number of sequence-pairs given, the better the work can be distributed and the higher the speed-up.

```
     std::vector seqs1{"AGTGGCTACG"_dna4, "AGGACTACG"_dna4};
2    std::vector seqs2{"AGTGCCTACG"_dna4, "AGTAGACTACGG"_dna4};

4    seqan3::configuration cfg = /* see Section 9.4 */;

6    /* I:   1-to-1 pairing of input -> two pairs                            */
     auto results1 = seqan3::align_pairwise(seqan3::views::zip(seqs1, seqs2),      cfg);
8
     /* II:  n-to-m pairing of input -> four pairs                           */
10   auto results2 = seqan3::align_pairwise(seqan3::views::all_pairs(seqs1, seqs2), cfg);

12   /* III: n-to-n pairing within one input collection -> one pair          */
     auto results3 = seqan3::align_pairwise(seqan3::views::all_pairs(seqs1),       cfg);
```

Code snippet 10.4: Providing input to pairwise alignment

Different designs are possible for accepting sequence data as input to the algorithm, e.g. two parameters for two collections that are then "paired". But it was

---

[7] Making e.g. the buffers "externalisable" via the configuration is being discussed but not yet implemented.

decided to instead take a single range of tuples, because this is much more flexible, especially when combined with views. There are many situations where a single sequence shall take part in multiple alignments, e.g. when aligning all sequences of one collection against all sequences of another collection. On the one hand, puzzling together the correct collections without copying the sequences would have been non-trivial for users; on the other hand providing multiple distinct interfaces (single-parameter, double-parameter) with different semantics also increases the chance for user error.

Snippet 10.4 shows how a single interface solves all of these problems with one data parameter and the help of two views: `seqan3::views::zip` and `seqan3::views::all_pairs`.[8] The former combines every i-th element in the left collection with every i-th element in the right, creating a new collection of n sequence-pairs (the two input collections need to be the same size).[9] The second view creates all possible pairing between two collections (when created with two arguments). This constitutes the many-against-many approach: $n * m$ pairs are returned. If the view is only applied to a single collection of n sequences, it creates the $\binom{n}{2}$ unique pairs within that collection. This covers all widely used patterns for pairwise sequence alignments. And users can of course provide their own range of sequence-pairs if they have different requirements.

Another important advantage of using views here is that the views "know" where the sequences come from and can thus add matching IDs to the tuples that are passed to the algorithm. This has the effect that e.g. in case II of Snippet 10.4 four pairs are generated and the `seqan3::align_result` objects will have the ID pairs (0,0), (0,1), (1,0) and (1,1) respectively—instead of (0,0), (1,1), (2,2) and (3,3). Users very likely expect this behaviour but it would be difficult to achieve with a different approach.

10.3.2   Alignment Result Type

The pairwise alignment algorithm is similar to the search algorithm in that it returns a view of dynamically created result objects.[10] These are of type `seqan3::align_result` which is also quite similar to `seqan3::search_result`. It has the following members:[11]

`.score()`   Integral or floating pointing value indicating the quality of the alignment.

---

[8] See also Sect. 7.3.3.

[9] Technically the pairs only hold references to the original sequences—no data is being copied.

[10] Unless it is configured to execute the callback function—which in that case is given the result object.

[11] The naming of this `struct`'s members is not yet final.

`.alignment()` Pair of `seqan3::gap_decorator`s that represent the alignment (trace).

`.seq1_id()` Index of the first (of the pair) input sequence.

`.seq1_begin_position()` Position on the first input sequence where alignment begins.

`.seq1_end_position()` Position on the first input sequence where alignment ends +1.

For all `seq1_*` members there is a corresponding `seq2_*` member for the second sequence of the pair. The `score()` and `seqX_id()` members are always usable; validity of the remaining members depends on the configuration (see Sect. 10.4).

10.3.3 Theoretical Background and Implementation Details

Sequence alignment has long been a core part of algorithms research in bioinformatics and also computer science in general. Almost all generic algorithms used today go back to the dynamic programming design of Needleman and Wunsch (1970) and Smith and Waterman (1981). The former laid the foundation for *global* sequence alignment, i.e. finding the best alignment between two sequences as a whole. The latter extended this approach to *local* alignments, i.e. finding the best alignment between any subsequences of two sequences. Subsequent optimisations followed for *affine gap costs* (Gotoh, 1981), reduced space consumption (Myers & Miller, 1988) and restricting the alignment to a band around the diagonal (Chao et al., 1992).

In combination with different alphabet and range types, a huge diversity of sequence alignment algorithms is possible. Implementing these generically while delivering the best possible performance is exceptionally difficult. Members of the SeqAn team achieved this in cooperation with Intel in the late stages of SeqAn2 (Rahn et al., 2018). This implementation of the generic alignment algorithm can handle (almost) all configuration parameters and supports inter-sequence and intra-sequence vectorisation and parallelisation (see Fig. 10.3). Additionally, an even faster version based on Myer's bitvector algorithm (Myers, 1999) is also available in SeqAn2 for a subset of configurations.

The degree to which these two major implementations have been ported to SeqAn3 varies. The examples given in Sect. 10.4 are supported, but many combinations theoretically possible are not yet available. Especially vectorisation support is missing for most combinations and the entire submodule needs more performance optimisation. But, in contrast to SeqAn2, the details of the algorithm are not exposed to the user. Even selecting the generic algorithm versus selecting the bitvector implementation happens automatically based on the chosen options and provided input data; there is just one function in SeqAn3. As such the implementation status impacts the user-interfaces only marginally and since this area of research has not been the focus of my studies, I will not cover the details here further.

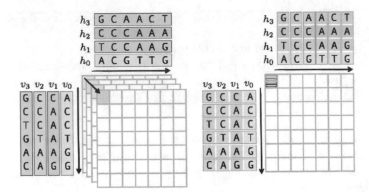

Fig. 10.3 Inter-Sequence alignment vectorisation. The vectorised dynamic programming matrix is shown. Image kindly provided by René Rahn

10.4   The Configuration Submodule

The *Alignment* module's *Configuration* submodule is similar to the *Search* module's (Table 10.6). It currently provides configuration elements for the pairwise alignment algorithm (Sect. 10.3) but will also contain the configuration elements for MSA in the future (many are shared). At the time of writing the naming and usage patterns for configurations were still in flux. Presented here is my proposal, but all config elements introduced here (except `seqan3::align_cfg::on_result`) are available in SeqAn3 in one way or another.

An overview of the available configuration elements is shown in Table 10.7. The config elements in the first section directly affect the progression of the dynamic programming and reflect the aforementioned algorithms. One `seqan3::align_cfg::mode::` has to be set, it determines whether the Needleman & Wunsch or Smith & Waterman cell computation is used. Similarly, the `seqan3::align_cfg::band` option activates band computation. Whether Gotoh's affine gap optimisation is used depends solely on the selected `seqan3::gap_scheme`. It is intentional that the underlying algorithms are not exposed individually and by their author's names, because users are not expected to know these.

The alignment's score is always computed, but most remaining members of `seqan3::align_result` (including the alignment itself!) are optional. That is because computing them is associated with (significant) overhead and there are valid use-cases where only the score is necessary (e.g. counting alignments that score above a certain threshold). To explicitly enable the computation of features like the alignment, the user can add the option `seqan3::align_cfg::with_alignment`. However, options that are implied by another option (i.e. they add no computational overhead) are always added, e.g. begin and end positions are automatically added when `::with_alignment` is specified. The alignment is currently always given as

Table 10.6 Alignment configuration submodule overview

Alignment: Configuration submodule	
Namespaces	`seqan3::align_cfg::`
Config elements	*see Table 10.7*

Table 10.7 Configuration elements for the alignment. Namespace `seqan3::align_cfg::` assumed for the left column. The user can provide arguments of the following types: `b` bool, `i` integer, `u` unsigned integer; `f` floating point; `m` an `align_cfg::mode` other than `dynamic`; `fun` a function (object)

Affect (score) computation	Description
`mode::global`	Global alignment
`mode::local`	Local alignment
`mode::dynamic{m}`	Can be set to `::global` or `::local` at runtime
`free_gaps{b,b,b,b}`	Configure forms of semi-global alignment
`band{i,i}`	Only compute band around diagonal of DP matrix
`max_error{u}`	A form of "dynamic band"
`scoring{s}`	Scoring scheme (see Sect. 10.2.1)
`gap{g}`	Gap scheme (see Sect. 10.2.2)
`using_score_type<intX_t>`	Type to use for computing scores
Affect traceback	
`with_alignment`	Compute alignment (trace)
`with_begin_positions`	Compute front-coordinate of DP matrix
`with_end_positions`	Compute back-coordinate of DP matrix
Affect execution	
`parallel{u}`	Number of threads
`vectorised`	Enable use of SIMD extensions
`on_result{fun}`	Callback-mode, takes a function (object)
Aliases	
`edit_distance`	Global, edit-distance scoring schemes

two `seqan3::gap_decorator`s but as soon as more gap decorator types become available, this will become configurable, too.

The config elements that affect execution are similar to the respective search options. The alignment additionally has a config element for vectorisation that enables the use of SIMD if available on the platform.

Snippet 10.5 displays an example of a valid configuration for performing alignments. The alignment algorithm is "semi-global", i.e. the gaps at the beginning and ending are nor penalised. This is indicated by `::global`-mode and `::free_gaps` with the respective parameters. Semi-global alignment is typically used when the sequences in one input collection are much shorter than in the other one—but one is still interested in matching them completely. Mapping reads to a genome is one such use-case. The other configuration elements in the example are fairly standard: the alignment will be computed and affine gap scores are implied by the given scheme.

```
     seqan3::nucleotide_scoring_scheme ss{seqan3::match_score{5}, seqan3::mismatch_score{-4}};
2    seqan3::gap_scheme                gs{seqan3::gap_score{-1}, seqan3::gap_open_score{-10}};

4    seqan3::configuration cfg = seqan3::align_cfg::mode::global                              |
                                 seqan3::align_cfg::free_gaps{.seq1_front = true, .seq1_back = true} |
6                                seqan3::align_cfg::scoring{ss}                               |
                                 seqan3::align_cfg::gap{gs}                                   |
8                                seqan3::align_cfg::with_alignment;

10   auto results = seqan3::align_pairwise(seq_pairs, cfg);
```

Code snippet 10.5: Semi-global alignment configuration. The example shows multiple config elements in use. `::global` combined with `::free_gaps` results in semi-global alignment computation. Definition of `seq_pairs` not shown

```
     seqan3::configuration cfg = seqan3::align_cfg::edit_distance | seqan3::align_cfg::max_error{4};
2
     auto results = seqan3::align_pairwise(seq_pairs, cfg);
4
     for (seqan3::align_result & result : results)  // results are computed lazily during iteration
6    {
         seqan3::debug_stream << result.score()              << ' '
8    //                       << result.alignment()          << ' ' // would lead to static_assert
                              << result.seq1_id()            << ' '
10                            << result.seq2_id()            << ' ';
     }
```

Code snippet 10.6: Fast alignment configuration. `::edit_distance` implies global alignment and respective scoring schemes. No alignment (traceback) is computed. Definition of `seq_pairs` not shown

The example in Snippet 10.6 shows a configuration that will lead to the selection of Myers bitvector algorithm internally. This is because the `seqan3::align_cfg::edit_distance` config element is selected and no other specified config elements conflicts with using the bitvector algorithm.[12] `::edit_distance` implies a global alignment, the hamming distance scoring scheme (match score 0, mismatch score -1) and a gap scheme with a gap score of -1 and linear gap model. `seqan3::align_cfg::max_error` activates an additional (heuristic) optimisation where cells of the DP matrix column are no longer computed if they fall beneath the given threshold (Ukkonen, 1985). This example also illustrates that if the computation of the alignment itself is not explicitly requested, it will not be available on the result objects. Calling the respective member function would lead to a `static_assert()` failing at compile-time and a readable error message being printed.

---

[12] A table in the API documentation states which config elements are compatible with each other and which enable the faster algorithm internally.

10.5 Discussion

Usability-wise the *Alignment* module offers most features that are planned and the interfaces are in an almost final design-state (even if not all the latest decisions have been realised, yet). This subsection discusses these interfaces in detail and compares them with SeqAn2 and in some cases also with BioPython.

However, with regard to a high performance, many crucial parts of implementation are still missing. Since the implementation and optimisation of these algorithms are a whole field of research by themselves, I will only briefly discuss the performance here.

10.5.1 Performance

As previously discussed, the optimisation of the *Alignment* module is an ongoing process and part of the work of other SeqAn team members. An early micro-benchmark does, however, already show promising results (Table 10.8). It is based on changes not yet merged into SeqAn3 and the numbers were kindly provided by René Rahn (see Sect. A.2 for the specifications of the system). The base performance (one thread, no vectorisation) of SeqAn3 is almost twice as slow as SeqAn2, but SeqAn3 scales better with number of threads than SeqAn2. SeqAn3 also scales much better with vectorisation, surpassing even the numbers of SeqAn2 in five of six tests.[13] Independent of the library, a variability in the sequence length decreases the performance of vectorised alignments, because sequence lengths in the same batch are "rounded up" to the longest length present.

The numbers shown in Table 10.8 with a peak performance of 11 GCUPS at four threads are in line with the results of SeqAn3 published by Rahn et al. (2018) who measured 106.2 GCUPS at 40 threads.[14] It should still be noted that this is a micro-benchmark and more application data will have to be collected to provide accurate predictions about the performance.

Table 10.8 Alignment micro-benchmarks (SeqAn2 vs SeqAn3). Speed given as (million) cell updates per second (CUPS). Global alignment with affine gap model selected; `seqan::Dna/seqan3::dna4` alphabets; only the score is computed (no traceback)

Seq. length	Threads	SIMD	SeqAn2	SeqAn3
150 ± 0	1	–	386 M/s	219 M/s
150 ± 32	1	–	378 M/s	220 M/s
150 ± 64	1	–	384 M/s	221 M/s
150 ± 0	4	–	1362 M/s	951 M/s
150 ± 32	4	–	1306 M/s	1005 M/s
150 ± 64	4	–	1333 M/s	1000 M/s
150 ± 0	1	SSE4	3606 M/s	3071 M/s
150 ± 32	1	SSE4	2007 M/s	2032 M/s
150 ± 64	1	SSE4	1545 M/s	1634 M/s
150 ± 0	4	SSE4	10,758 M/s	11,183 M/s
150 ± 32	4	SSE4	5442 M/s	8574 M/s
150 ± 64	4	SSE4	5036 M/s	6584 M/s

```
/* SeqAn2 */                                      /* SeqAn3 */
2   typedef Dna TChar;                        2
    typedef String<TChar> TSequence;
4   typedef Align<TSequence, ArrayGaps> TAlign;    4

6   TSequence seq1 = "ACAGAGCCT";             6   std::vector        seq1 = "ACAGAGCCT"_dna4;
    TSequence seq2 = "ACTAGACT";                  std::vector        seq2 = "ACTAGACT"_dna4;
8   TAlign align;                             8

10  resize(rows(align), 2);                   10  seqan3::gap_decorator ali1{seq1};
    assignSource(row(align, 0), seq1);            seqan3::gap_decorator ali2{seq2};
12  assignSource(row(align, 1), seq2);        12  std::tuple         align{std::tie(ali1, ali2)};

14  insertGap(row(align, 0), 2);              14  ali1.insert_gaps(ali1.begin() + 2);
    insertGaps(row(align, 1), 6, 2);              ali2.insert_gaps(ali2.begin() + 6, 2);
16                                            16
    std::cout << align << '\n';                   seqan3::debug_stream << align << '\n';
18  /*    AC-AGAGCCT                          18  /*    AC-AGAGCCT
         // /// //                                     // /// //
20       ACTAGA--CT      */                   20       ACTAGA--CT      */
```

Code snippet 10.7: Alignment data structure in SeqAn2 & SeqAn3. The example on the left is adapted from the official SeqAn2 tutorial. Namespace `seqan::` assumed for the left snippet. In SeqAn3 the `align` object is a tuple of references to the individual rows

10.5.2 Simplicity

Using the Gap Decorator

Snippet 10.7 shows how `seqan3::gap_decorator` is used in comparison to SeqAn2's code. In general the usage patterns have not changed much, but I would argue that the lack of visible templates and the availability of member functions does make the snippet simpler. It should be noted, though, that most use-cases do not require users to insert gaps themselves. Usually users just receive such objects from e.g. the alignment algorithm. The only thing possibly unexpected in SeqAn3's interface is that iterators are used instead of numbers to denote the insert positions in the alignment row. This has several reasons:

1. As explained in Sect. 10.1, random access is not in $O(1)$ (also not in SeqAn2). However, in most algorithmic contexts it is likely that an iterator exists to the position currently in focus, so using iterators for inserts is often faster.[15]
2. `seqan3::gap_decorator` is more generic than the likes in SeqAn2. It can also be created on e.g. `std::list` which means that the decorated container[16] does not have any random access at all and needs to insert by iterator.
3. The interface is strongly modelled after the standard containers' `.insert()`-member which also takes iterators.

   Position-based access is easy to achieve via `.begin() + i` as shown in the snippet. So there is no drawback to this approach.

The Algorithm Interface

Snippet 10.8 shows a side-by-side comparison of SeqAn2, BioPython and SeqAn3 invoking an alignment over potentially multiple pairs of input sequences (although only one pair is actually aligned in the example). SeqAn2 is very verbose in assembling the type of and constructing the `aligns` object which contains both the input sequences and later also the alignment generated by the function. This in itself is already a usage paradigm very specific to C++ (98) and uncommon in other languages. SeqAn2 also executes a traditional function that blocks until all alignments are computed.

---

[13] In the given configuration, the theoretical maximum factor is 4 threads × 16 vector units = 64. SeqAn3 reaches 51x, SeqAn2 reaches 28x (but starts out with a better base performance).

[14] Both results are computed with the same alignment configuration and using SSE4 on the Intel Sylake architecture (although one is measured on a Desktop/Mobile unit and the other on a server processor).

[15] In SeqAn3 the `.insert_gaps()` function returns an iterator that can be reused for subsequent inserts.

[16] Remember that—although not obvious in Snippet 10.7 because of CTAD (Sect. 3.1.2)—the decorator is a template that is specialised over the underlying range.

```
                                                # BioPython
                                              2 aligner = Align.PairwiseAligner()
                                                aligner.match_score = 2
                                              4 aligner.mismatch_score = -1
/* SeqAn2 (namespace seqan:: assumed) */        aligner.extend_gap_score = -1
2 typedef Dna TChar;                            6 aligner.open_gap_score = -10
  typedef String<TChar> TSequence;               aligner.mode = 'global'
4 typedef Align<TSequence, ArrayGaps> TAlign;   8
  typedef String<TAlign> TAligns;                 s1 = "TACCG"
6                                            10 s2 = "ACG"
  TSequence seq1 = "TACCG";
8 TSequence seq2 = "ACG";                     12 results = aligner.align(s1, s2)
                                                for result in results:
10 TAligns aligns;                            14   print(result.score)
   resize(aligns, 1);                              print(result)
12
   TAlign & align = aligns[0];
14 resize(rows(align), 2);                       /* SeqAn3 (namespace seqan3:: assumed) */
   assignSource(row(align, 0), seq1);         2 nucleotide_scoring_scheme ss{
16 assignSource(row(align, 1), seq2);             match_score{2}, mismatch_score{-1}};
                                              4 gap_scheme gs{
18 Score<int, Simple> scoreScheme;               gap_score{-1}, gap_open_score{-10}};
   setScoreMatch(scoreScheme, 2);             6
20 setScoreMismatch(scoreScheme, -1);            configuration cfg =
   setScoreGapExtend(scoreScheme, -1);        8   align_cfg::mode::global        |
22 setScoreGapOpen(scoreScheme, -10);             align_cfg::scoring{ss}         |
                                             10   align_cfg::gap{gs}             |
24 auto results =                                 align_cfg::with_alignment;
     globalAlignment(aligns, scoreScheme);   12
26 for (size_t i = 0; i < length(results); ++i)  auto s1 = "TACCG"_dna4;
   {                                          14 auto s2 = "ACG"_dna4;
28   // results only contains scores
     // alignment returned as in-out parameter 16 auto results =
30   std::cout << results[i] << '\n'              align_pairwise({std::tie(s1, s2)}, cfg);
             << aligns[i]   << '\n';          18 for (auto & result : results)
32 }                                               debug_stream << result.score    << '\n'
                                             20               << result.alignment << '\n';
```

Code snippet 10.8: Alignment interfaces compared with BioPython

On the other hand, BioPython and SeqAn3 are quite alike in that they first prepare the options for the alignment and then create a generator/view that dynamically produces results when being iterated over. In BioPython all options are set as members on the `aligner` object which follows the object-oriented programming paradigm and is arguably a little simpler than SeqAn3's configuration system. This is, however, only possible because all options in Python are runtime options. SeqAn3's config system also allows to seamlessly set compile-time options— the presence/absence of config elements changes the type of the configuration.[17] Considering this, the configuration is still very simple and quite similar, only the scoring schemes are predefined to improve readability of the snippet.

As shown in Snippet 10.9, SeqAn2 has a huge variety of function interfaces for computing alignments. Depending on whether only the score is to be computed or also the alignment (trace), either `seqan::globalAlignmentScore()` or `seqan::globalAlignment()` is invoked. However, both functions return only the

---

[17] This is not visible because the full type of the configuration is not shown anywhere, it is deduced by the compiler (see Sect. 3.1.2).

```
/* SeqAn2 */
  TScoreCollection globalAlignment([exec,] alignCollection, scoringScheme, [alignConfig,] [lowerDiag, upperDiag]);
  TScoreCollection globalAlignment([exec,] gapsHCollection, gapsVCollection, scoringScheme, [alignConfig,] [lowerDiag, upperDiag]);
4 TScoreVal globalAlignment(align, scoringScheme, [alignConfig,] [lowerDiag, upperDiag,] [algorithmTag]);
  TScoreVal globalAlignment(gapsH, gapsV, scoringScheme, [alignConfig,] [lowerDiag, upperDiag,] [algorithmTag]);
  TScoreVal globalAlignment(frags, strings, scoringScheme, [alignConfig,] [lowerDiag, upperDiag,] [algorithmTag]);
  TScoreVal globalAlignment(alignGraph, scoringScheme, [alignConfig,] [lowerDiag, upperDiag,] [algorithmTag]);
8
  TScoreCollection globalAlignmentScore([exec,] seqHCollection, seqVCollection, scoringScheme[, alignConfig][, lowerDiag, upperDiag]);
  TScoreCollection globalAlignmentScore([exec,] seqH, seqVCollection, scoringScheme[, alignConfig][, lowerDiag, upperDiag]);
  TScoreVal globalAlignmentScore(strings, scoringScheme[, alignConfig][, lowerDiag, upperDiag][, algorithmTag]);
12 TScoreVal globalAlignmentScore(seqH, seqV, {MyersBitVector | MyersHirschberg});
  TScoreVal globalAlignmentScore(strings, {MyersBitVector | MyersHirschberg});


  /* SeqAn3 */
2 auto align_pairwise(seq_pairs, cfg);
```

Code snippet 10.9: Global alignment functions in SeqAn2. The top shows the global alignment function signatures according to SeqAn2's API documentation. The local alignment has again as many

score and the alignment is written into the second function's first parameter(s).[18] These can either be an alignment object or two separate parameters for the first and second "alignment row"—or a collection of either. The other parameters are similar to some configuration elements introduced in Sect. 10.4, in particular alignConfig is comparable to seqan3::align_cfg::free_gaps and the *Diag parameters are the equivalent of seqan3::align_cfg::band.

algorithmTag is a curious option as not even the author is entirely sure what its effects are. Some tags can clearly affect a switch between the generic and the bitvector implementation, but others are ambiguous. The documentation states that one of the following tags can be specified: AffineGaps, DynamicGaps, Gotoh, Hirschberg, LinearGaps, MyersBitVector, MyersHirschberg and NeedlemanWunsch (the last is an alias of Hirschberg). I would have expected at least AffineGaps and Gotoh to be aliases of each other, but this is not the case. More confusingly, tags for local alignment algorithms exist also (SmithWaterman and WatermanEggert), however they cannot be passed to the globalAlignment() functions and their localAlignment() counterparts do not accept any algorithmTag parameter. The entire option can also not be provided when aligning collection of sequences.

As previously noted, part of shifting focus from academic questions to high-quality software is that whenever a solution or optimisation is clearly superior, it should always be performed.[19] Furthermore, an option that can be deduced from the input data, should be; e.g. affine versus linear gaps should depend on the gap costs set in the scheme, not an additional tag parameter. If options are explicitly given to the user, their names should be intuitive or at least reflect properties of

---

[18] This is the out-parameter style, see Sect. 4.2.6.

[19] Why would a user want affine gap costs but not the optimisation by Gotoh? Since NeedlemanWunsch is an alias for Hirschberg in SeqAn2, this principle seems to have been followed at least sometimes. Although it is unclear why these options are provided at all under the circumstances.

the algorithm. Naming them after their inventor(s) is not helpful for most users—although good API documentation should of course include such references.

Another drawback of configuring the algorithm via multiple parameters is that their order matters and is hard to memorise when the list is long. Additionally, defaulted parameters can only be omitted if all optional parameters *after them* are omitted, too. So it is not possible to e.g. provide upper and lower diagonal without also specifying the `alignConfig` (configuration parameter for semi-global alignment). Finally, it is unclear and always a little arbitrary which configurations warrant their own function name. In SeqAn2 local and global are differentiated by function name as well as score-only and alignment (trace), but arguments could well be made for making e.g. the interface for collections an extra function (`globalAlignments()` ?).

All of these problems are solved in SeqAn3. There is a single interface with two parameters: the data and the configuration. Like most functions/function objects in SeqAn3, the name contains a verb which expresses that an action is performed. Results of the algorithm are returned as return values. The configuration object can be assembled from different config elements whose order is irrelevant and where optional elements can always be omitted. Most config elements are independent of each other and a single table in the documentation states which are not. Ultimately, the interface provided by SeqAn3 allows more configuration options while being much simpler.

10.5.3   *Integration*

The *Alignment* module's primary interface to external data types and libraries is through the handling of ranges and alphabets. It builds upon the concepts defined in Sects. 3.6.2 and 6.1.3, but it also defines its own range concepts and it gives special meaning to alphabets that represent gaps. Important in this respect is that these concepts are equally friendly to the integration of third party types. For the aligned range concepts, this happens through customisation point objects (CPOs; see Sects. 3.7, 4.2.3). These work out-of-the-box for standard library containers in combination with the `seqan3::gap` alphabet but can be extended to work on any user provided ranges and alphabets.

Scoring schemes are also used through a generic interface regulated by a concept, however, this concept does not require a CPO, it simply looks for a member function. The reason for this is that SeqAn3's notion of a scoring scheme is very specific to SeqAn3's algorithms and adapting a third party type "as-is" for this role is not a likely use-case. If a new scoring scheme with custom semantics is required, it can quickly be written (an example in 11 lines of code is given in Snippet A.8). This method is different from how raw sequence data or pre-aligned sequences are adapted, because these appear more often in custom user code and would be expensive to convert at runtime. So in that case the extra layer of genericity through CPOs helps to use the data "as is".

10.5.4 *Adaptability*

```
1  /* SeqAn2 */
2  struct MySpec{};
3  template <typename TValue> struct Score<TValue, MySpec> : Score<TValue, Simple>{};
4
5  template <typename TValue, typename TSeqHVal, typename TSeqVVal>
6  inline TValue score(Score<TValue, MySpec> const & me, TSeqHVal valH, TSeqVVal valV)
7  {
8      if (valH == 'N' && valV == 'N')      return scoreMismatch(me);
9      else if (valH == valV)               return scoreMatch(me);
10     else                                 return scoreMismatch(me);
11 }
12
13 Score<int, MySpec> ss;    setScoreMatch(ss, 3);    setScoreMismatch(ss, 0);
14 int s = score(ss, Dna5('N'), Dna5('N'));                              // s == 0
```

```
1  /* SeqAn3 */
2  nucleotide_scoring_scheme ss{match_score{3}, mismatch_score{0}};
3  ss.score('N'_dna15, 'N'_dna15) = 0;
4  int s = ss.score('N'_dna15, 'N'_dna15);                              // s == 0
```

Code snippet 10.10: Extending scoring schemes. A scoring scheme is created with a match score of 3 and a mismatch score of 0 but modified so that the letter N shall not match against itself. Top snippet is SeqAn2 (requires template subclassing and function overloading), the bottom is SeqAn3 with a more object-oriented interface. Respective namespaces assumed

Aligned ranges have improved genericity and extensibility over their predecessors in SeqAn2, because they are based on concepts and CPOs instead of template subclassing. Currently, two models of the aligned range concepts are provided: any standard container over a gapped alphabet or `seqan3::gap_decorator` over an existing range. But I have also given an overview of other possible implementations and their benefits.

The scoring schemes are also more generic by having a concept abstraction. By splitting scoring schemes and gap schemes, unnecessary *coupling* between components has been reduced and it has become simpler to add custom scoring schemes. While the provided scoring schemes are specific for groups of alphabets (nucleotides, amino acids), it is also possible to provide an alphabet-independent scoring scheme that simply does comparisons in a few lines of code (see Snippet A.8). But since SeqAn3's scoring schemes are based on matrices, it is often not necessary to define a custom type; existing schemes can easily be adapted. This can be seen in Snippet 10.10, a use-case that required a custom type in SeqAn2 but no longer does in SeqAn3.

The alignment algorithm is generic insofar as it accepts data input based on concepts, but it is not an extension point for developers, i.e. the algorithm cannot be manipulated other than through the configuration system. This, on the other hand, is very versatile, and in contrast to configuration via multiple function parameters (as in SeqAn2), it is also very simple for SeqAn developers to extend the configuration by new config elements in future releases. Such an addition would only require changes inside the algorithm and not in the algorithm interface.

```
   auto seq_pairs = /*...*/;
 2 seqan3::configuration cfg = /*...*/;

 4 int min_score = 42;
   auto results = seqan3::align_pairwise(seq_pairs, cfg)
 6             | std::views::filter([&] (auto & r) { return r.score >= min_score; })
               | std::views::take(20);

 8
   for (seqan3::align_result & result : results)
10 {
       /* iterate over first 20 results with score of at least 42; then stop */
12 }
```

Code snippet 10.11: Combining the alignment with views. Alignments are computed on-demand during the loop and low-scoring alignment are discarded. When 20 non-discarded alignments have been generated, computation stops

Since the alignment algorithm returns a view, it can be combined with other views to perform powerful and expressive post-processing. Snippet 10.11 shows an example of this. Because views are lazy-evaluated, this form of "post-processing" can even influence the algorithm progression. In this case the computation is terminated when the combined view is parsed fully (`std::views::take` drops elements after the first n).[20]

10.5.5   Compactness

Many important parts of the *Alignment* module regarding parallelisation and vectorisation are still missing and more implementations of aligned ranges will likely be added in the future, as well. This makes it difficult to compare the size of the respective codebases at this point. However, many examples as in the previous

---

[20] Strictly speaking the algorithm might have computed *very few* alignments more than necessary since the underlying algorithm computes alignments in batches for performance reasons (e.g. vectorised computation). But this is always a "constant" amount and usually very small compared to the total input size.

section (Snippet 10.10) suggest that the programming techniques and new designs will strongly reduce the size of the codebase.

What can clearly be assessed is that the size of the API has been reduced significantly, i.e. the number of public types and function interfaces that users need to learn about is much smaller than before. Based on the current designs I am fairly confident that this difference will remain.

Part III
Lambda

The third part of this book introduces *lambda3*, a new version of the LAMBDA local alignment application. Background information on homology search, local alignment computation and prior research in this area is given. This includes the author's contributions to this domain as well as an analysis of the current applications of other authors. But the main purpose of this part is to present an application design based on SeqAn3's library design and to document the process of porting an application from SeqAn2 to SeqAn3. *lambda3* is the first application to be built on SeqAn3 and thus serves as a showcase for the features and techniques introduced in the previous chapters.

Chapter 11
Lambda: An Application Built with SeqAn

11.1 Introduction

A wide variety of use-cases exist in sequence analysis that involve searching for the so-called *query* sequences in existing, annotated databases also called reference or *subject* sequence(s). These can widely be classified into searches that try to find exact or close-to-exact matches of the query, and such that are also interested in finding partial or fuzzy matches.

The first class of problems includes identifying the query sequences (via their genomic origin) but also appears in contexts such as genome-assembly or the identification of structural variants. It typically entails *read mapping*, a form of search that maps the full length of the query sequence to a position in the reference and allows only very few errors. This means the form of alignment is *semi-global*: the query is expected to match fully against a similarly sized subsequence of the reference/database. Ideally all query sequences have a single unique hit in this kind of search (although this is not usually the case in practice). The domain of these searches is almost always nucleotide-space and not protein-space. A schematic is shown in Fig. 11.1.

The second class of problems is very diverse, although most are rooted in the search for *homologues*. Homologues are sequences of common evolutionary descent—either within a species (*Paralogues*, result of a duplication event), or between species (*Orthologues*, result of speciation). Homology search plays an important role in determining the species content in a mixed species sample or inferring the relatedness of such species through a full taxonomic classification. Such samples are common in *metagenomics* and *metatranscriptomics*, emerging research areas with applications in fields so diverse as ecology (Mackelprang et al., 2011) and cancer research (Schwabe & Jobin, 2013). Implementing this form of search usually means looking for a *local alignment*, i.e. a well-scoring match with possibly many errors that may span just a part of the query sequence (often only subsequences are well-preserved by evolution). Most of the time one is also

© The Author(s), under exclusive license to Springer Nature Switzerland AG 2022
H. Hauswedell, *Sequence Analysis and Modern C++*, Computational Biology 33,
https://doi.org/10.1007/978-3-030-90990-1_11

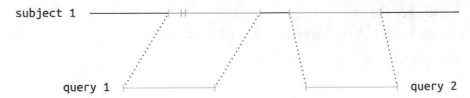

Fig. 11.1 Read mapping (schematic). Many short query sequences are mapped completely (and with few errors) against substrings of comparatively few long subject sequences. The sequence type is typically DNA/RNA

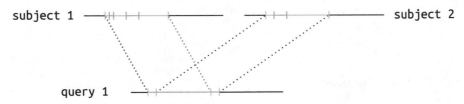

Fig. 11.2 Homology search (schematic). Potentially different substrings of each query sequence are mapped fuzzily against substrings of the subject sequences. Many hits per query are expected (and desired). There are typically more subject sequences than in read mapping, and they may each be shorter (but still longer than the query). The sequence type is often protein

interested in finding many hits per query sequence, because this demonstrates how rare a sequence is and can help infer the taxonomy. A schematic is shown in Fig. 11.2.

The complexity of the second class of problems is usually considered greater than the first, because inexact search is computationally more expensive than exact search and the data sets are often much larger. This is true for, both, the databases (which can contain sequences of many species) and the query sequences (which can contain millions of reads from environmental samples). As such, *heuristic* algorithms are employed more often to solve these problems, i.e. programs do not *guarantee* to find all hits with the given properties but work well in practice. Giving up on this promise of exactness and completeness allows programs to be faster than exact solutions by many orders of magnitude—often a necessary condition to perform such analysis at all.

In the context of this class of problems, protein-searches also play a much more important role. One reason is that protein sequences (and the genes they are built from) are highly functional compared to most untranslated genomic sequences. This means they are more strongly conserved which in turn increases the chance to infer homology. On the other hand, the redundancy within the genetic code[1] and the functional similarity of certain amino acids[2] allow certain mutations within genes to

---

[1] Different triplets of RNA bases code for the same amino acid, see also Fig. 6.3 on p. 167.

[2] Some amino acids have very similar biochemical and physical properties while other are very different.

have little or no impact on the function of the protein and thus be more probable. As a result, comparing the translated sequences (protein-space) may reveal homology that would hardly be visible in the DNA sequences alone. Another reason for using protein-searches is that protein databases are on average much better annotated and results are thus more likely to be useful (e.g. for taxonomic analysis). A third reason is sheer size: cross-species protein databases are still much smaller than their respective full-genome counterparts, so some searches are only feasible on the former and not the latter.

This chapter discusses Lambda, a tool for performing local alignment searches with many optimisations for searches in protein-space and large input data.

11.1.1 Previous Work

BLAST (Altschul et al., 1990, 1997; Camacho et al., 2009) is by far the most popular tool for local alignment search, and it is one of the most well-known and most highly cited bioinformatics applications to date. The authors not only produced the BLAST program(s), but they also developed the statistics that are used for assessing the significance of local alignment matches. These are widely accepted and most local alignment search tools today implement them in one form or another. But also the applications themselves are still very popular. In regard to sensitivity they are still the reference, but compared with newer software they are very slow. This is especially problematic considering the grown amounts of data that are usually processed nowadays (see Chap. 1).

With respect to sequence classification, I already summarised several years ago:

Bazinet and Cummings (2012) give an overview of the various programs that have been developed to address this problem. Of the approaches they compare, 11 of 14 use BLAST in their pipeline. Hence, BLAST (Altschul et al., 1997) can be seen as the de facto standard used for trying to solve this problem. Bazinet and Cummings (2012) also note in their study that '[the] BLAST step completely dominates the runtime for alignment-based methods'. For the two programs with the highest precision in their comparison, CARMA (Gerlach & Stoye, 2011; Krause et al., 2008) and MEGAN (Huson et al., 2007), the BLAST step actually made up 96.40 and 99.97% of the runtime. Another metagenomic study (Mackelprang et al., 2011) states that 800,000 CPU hours at a supercomputer center were required to conduct the study. Hence, since some time there is an effort to replace the BLAST suite by algorithms and tools that are much faster while not sacrificing too much accuracy. That means the tools aim at finding the same alignment locations as BLAST and possibly an alignment of similar quality (expressed by bit score) (Hauswedell et al., 2014).

Several BLAST alternatives have been developed; noteworthy applications prior to the first LAMBDA release are: BLAT (Kent, 2002), UBlast (Edgar, 2010), RAPSearch2 (Zhao et al., 2012) and PAUDA (Huson & Xie, 2014). I have discussed all of these in detail and including their algorithmic choices (Hauswedell, 2013).

The different BLAST program modes are shown in Table 11.1. Except BlastN, all modes effectively perform alignment and search in protein-space but some modes

Table 11.1 BLAST program
modes and input alphabets

BLAST mode	Query alphabet	Subject alphabet
BlastN	Nucleotide	Nucleotide
BlastP	Amino acid	Amino acid
BlastX	Translated nucl.	Amino acid
TBlastN	Amino acid	Translated nucl.
TBlastX	Translated nucl.	Translated nucl.

Table 11.2 Noteworthy LAMBDA releases

Version	Date	Notable features
0.4.0	2014-11-10	SeqAn1, published version, double-indexing
1.0.0	2016-08-18	SeqAn2, single-indexing default, SAM/BAM
2.0.0	2019-01-11[†]	variable-length seeding, EPR-dict., SIMD, taxonomy
3.0.0	tba	SeqAn3, bisulfite-mode

[†] 2.0.0 is a rebranded 1.9.5 which was releases on 2018-05-30

translate input data beforehand. Most tools focus on and only support these protein
modes of BLAST, others try to cover all the functionality.

Since the release of LAMBDA (Hauswedell et al., 2014), several new tools have
been published, including Diamond (Buchfink et al., 2015). It performs BLAST
protein modes and has become very popular. MALT (Herbig et al., 2016) on the
other hand also performs (untranslated) nucleotide searches. Many more tools have
been published that claim better performance for very closely related sequences,
e.g. PALADIN (Westbrook et al., 2017), but the lack of e-value statistics makes
it difficult to compare these fairly. Furthermore, all previous tools can also be
configured to perform better at the cost of losing (lower-scoring) results.

11.1.2   History of LAMBDA

LAMBDA was published in 2014 (Hauswedell et al., 2014); a brief history is
shown in Table 11.2. It was—to my knowledge—the first local aligner to use
double-indexing.[3] Double-indexing refers to the non-trivial pre-processing of not
only the database but also the query sequences. In the case of LAMBDA, this meant
constructing a radix-trie of the query sequences and searching this in the database
index, a suffix array. Please see the literature for an in-depth discussion of the data
structures and algorithms used and how they compare to other tools at the time
(Hauswedell, 2013; Hauswedell et al., 2014).

This approach was given up in subsequent versions of LAMBDA, because the
size of the suffix array becomes prohibitively large and the performance benefits

---

[3] DIAMOND (Buchfink et al., 2015) later also included a form of double-indexing although the
involved index data structures are quite different from LAMBDA.

of double-indexing were a lot less pronounced when an FM-index was used for the database. The 1.x-series (including pre-releases named 0.9.x) gained many optimisations and clean-up. It also added support for writing SAM and BAM files and was based on SeqAn2.

The next major iteration was called *lambda2*, its first pre-release (1.9.0) was released in parallel with 1.0.0. Important changes include switching the FM-index implementation from using wavelet-trees to EPR-dictionaries (Pockrandt et al., 2017). Inspired by LAST (Kiełbasa et al., 2011), the search strategy was adapted to perform *variable-length seeding*. Bidirectional indexes and search were supported but did not provide any benefits in combination with the search strategy and default error configuration. The alignment step was changed to utilise vectorisation based on the work by Rahn et al. (2018). Due to technical limitations of SeqAn2 at the time and the very complex x-drop extension mechanism used by lambda2 for long sequences, this was only available for the alignment of short reads. Lambda2 also introduced support for adding taxonomic identifiers to search results and performing simple lowest-common-ancestor (LCA) computation.

Development of lambda3 began in March 2019 with the intent to port the entire application from SeqAn2 to SeqAn3. It is the first proper application built with SeqAn3 and as such serves as a test case for the designs discussed in Part II. Applying and evaluating the innovations of SeqAn3 through lambda3 is the main focus of this chapter. However, in the end, lambda3 should of course also become a viable general-purpose local aligner.

A colleague of mine is continuing the work on lambda3 and already helped notably with the porting effort. She observed that the mechanism of alphabet reduction, as used by LAMBDA for proteins during search, can also be used to simulate the alphabet reduction resulting from bisulfite sequencing. This will hopefully lead to a new program mode of lambda3 and proves how generic and extensible SeqAn and derived tools are.

In the following "LAMBDA" refers to the program as such (any version) and "lambda3" refers to the version currently in development.

11.2 Implementation

Lambda3 is implemented as a single application with multiple program modes, similar to `git pull`/`git push`. The program modes are:

`lambda3 mkindexn`	Creates an index file for nucleotide searches.
`lambda3 mkindexp`	Creates an index file for protein-searches.
`lambda3 searchn`	Performs a nucleotide search (BLASTN, MEGABLAST).
`lambda3 searchp`	Performs a protein search (any other BLAST mode).

Creation of index files is performed separately from the search because the step is computationally expensive and the index files can be reused (see also Sect. 9.1.1). The distinction between nucleotide and protein modes is made primarily to improve the command line interface.[4] In the discussion of the individual program steps below, this distinction is not made.

11.2.1 Index Creation

An overview of the `mkindex`* program modes is given in Fig. 11.3. Program execution starts with **parsing the command line arguments** via SeqAn3's *Argument parser* module (which will become stand-alone in the future, see Sect. 5.2.1). Certain runtime options are then transformed to compile-time options/types through a series of `switch` statements and nested function calls; these effectively instantiate all possible branches and select the correct one at runtime.

SeqAn3's `seqan3::sequence_file_input` from *Input/Output* module then facilitates the **reading of the database** (see Sect. 8.4). The result is a set of sequences in the original alphabet (`seqan3::dna5` or `seqan3::aa27`) and a set of `std::string`s with the respective IDs. To save memory, the IDs are truncated

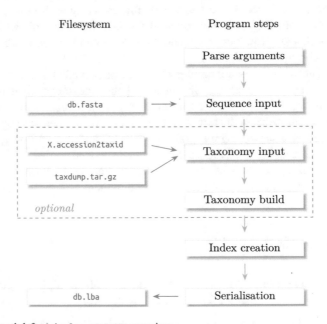

Fig. 11.3 Lambda3 `mkindex` program overview

---

[4] Certain options are only available in one mode or the other, and default values for other options differ. Separate program modes allow better documentation and validation.

at first white space.[5] Depending on program mode, the translated sequences are created from the original sequences and the reduced sequences are created from the translated sequences. This always happens through the use of views. For translation `seqan3::views::translate_join` is used (see Sect. 7.3.2 and Fig. 7.1 on p. 202), and for reduction `seqan3::views::convert` is used. In nucleotide mode, a special kind of reduction happens that converts the alphabet from `seqan3::dna5` to `seqan3::dna4` but that replaces `'N'_dna5` with a random character of `seqan3::dna4`.[6] A smaller alphabet improves performance of the index, and—since later steps in the search access the original sequences again—this does not result in false positives.[7] Since views are used to represent the transformed sequence sets, no space overhead is incurred for any of these steps.

Optionally, the indexer program mode can read a mapping file that associates one or more **taxonomic IDs** with an accession number.[8] Lambda provides parsers for the NCBI format of the mapping files (`*.accession2taxid`) and also UniProt's format (`idmapping.dat`). If such a file is provided and read, all accession numbers are extracted from each sequence ID (before possible truncation) and the respective taxonomic ID(s) are extracted from the mapping and stored; a single subject sequence can have multiple taxonomic IDs. The taxonomic IDs can later be used during search to annotate results.

In addition to merely associating the correct IDs with each other, Lambda can also parse the **full taxonomy** if a "taxdump" is provided. This is a large archive provided by the NCBI[9] with the taxonomic tree of all known taxonomic IDs. Lambda will process this and create a small binary representation of a tree with those taxonomic IDs that are present in the database being indexed. During search this allows the computation of the lowest-common ancestor (LCA) of all database results found for one query.

The next step is the actual **creation of the index**. This entails creating an FM-index (unidirectional or bidirectional depending on chosen options) with SeqAn3's *FM-index* submodule (see Sect. 9.1) of the *Search* module. The index is always created from the reduced sequences.

Finally, a description of the relevant parameters, the original sequences, the IDs, the index (and optionally taxonomic IDs and tree) are written to disk. This is performed using SeqAn3's **serialisation** support (see Sect. 8.2). A single, binary (but platform-independent) file is created that contains all the data structures. The

---

[5] This would be performed by all output formats other than "BLAST report" anyway. The option can be turned off if this output format is used primarily and users require the full length ID.

[6] This view is provided by lambda3 and is not part of SeqAn.

[7] Alignments of an `'N'_dna5` in the query against an `'N'_dna5` of the subject are typically not considered meaningful and scored nagatively, see e.g.: http://ftp.ncbi.nih.gov/blast/matrices/ NUC.4.4.

[8] "The accession number is a unique identifier assigned to a record in sequence databases such as GenBank" (Tatusova et al., 2013).

[9] ftp://ftp.ncbi.nlm.nih.gov/pub/taxonomy/taxdump.tar.gz.

extension for this is `.lba` (lambda **b**inary **a**rchive). For debugging purposes, a text archive can be created instead (`.lta`). This is in JSON format (Crockford, 2002) and fully functional (but very large). Note that the translated sequences and the reduced sequences are never written to disk.

11.2.2   Search

The search program modes start similarly to the indexer modes; an overview is given in Fig. 11.4. First the **command line arguments** are parsed (Sect. 5.2.1) and certain user-provided options are transformed into types. This step includes reading specifications from the index file (e.g. unidirectional versus bidirectional)[10] and auto-detecting the alphabet of the query sequence file.

Next, the index file is fully **deserialised** (Sect. 8.2). This deserialised object contains the original subject sequences and again a translated view and a reduced view are created from them (depending on program mode/selected options).

In contrast to the subject sequences (during index construction), the **query sequences** are not read en-bloc. The file is only opened (also via `seqan3::sequence_file_input`; Sect. 8.4), and then an asynchronous-reader view is created on top of the file (`seqan3::views::async_input_buffer`; Sect. 8.4.4). This buffers a certain number of records and refills the buffer dynamically from the file in a background thread. More threads[11] are started which loop over the records in the buffer and perform search, alignment and output until no records are left.

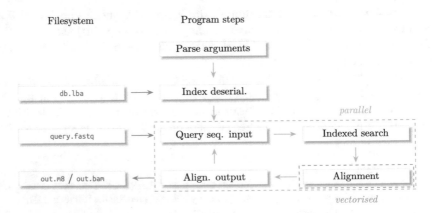

Fig. 11.4 Lambda3 `search` program overview

---

[10] Although the archive is binary, parts of it can be read without loading the entire file.

[11] The number can be configured, defaults to one thread per CPU(-core) available.

Before **the search**, the query sequence in the current record is translated and/or reduced via a view. The reduced query sequence is split into seeds that are then searched via SeqAn3's *Search* module in the index (Sect. 9.4). The seed length, number of allowed mismatches and degree of overlap depend on program mode and command line options. If the number of found hits per seed exceeds a certain threshold, the seed is elongated (Kiełbasa et al., 2011). This reduces the number of hits and thereby the complexity of subsequent steps. It is also based on the assumption that matches that appear ubiquitously are less interesting. Before a hit is stored for further processing, a very fast ungapped local alignment in the region around the hit determines its viability. Like all subsequent steps, this is performed on the potentially translated but unreduced sequences (search happened on reduced sequences).

All hits stored by the previous step are sorted and organised into batches by sequence length. The full query sequence is paired with a sufficiently large subsequence of the respective subject sequence to accommodate for insertions and deletions. These batches are then processed by SeqAn2's vectorised *Alignment* module. First, only the **local alignment** scores are computed and transformed to bit-cores/e-values. Hits that do not pass the e-value threshold are discarded before full local alignments (with tracebacks) are computed for the rest.

Finally, all matches are sorted, duplicates removed and a record of all the matches belonging to one query sequence is **written to disk**. This is currently still performed by SeqAn2's *Blast I/O* module for BLAST output formats and SeqAn2's *BAM I/O* module for SAM/BAM output. The output format depends on the given extension, and the exact content (including optional fields) depends on the command line arguments provided. If taxonomic data was incorporated into the index file and the respective field is requested as part of the output, the LCA will be computed for all matches that belong to a query sequence.

As long as input remains, the thread returns to taking a new query sequence from the file buffer and repeats the last steps.[12]

11.3 Results

The most important and obvious result of the porting effort is a functioning version 3 of LAMBDA. It builds on the well-established lambda2 code base and contains many advancements and simplifications of SeqAn3, although not all SeqAn2 code has been replaced, yet. Some notable features are discussed below (Sect. 11.3.1) and the performance of the current development state is shown in Sect. 11.3.2.

---

[12] This loop conceptionally iterates over individual query sequences, but in practice a batch of query sequences is always loaded and processed, because this improves the performance of the alignment step.

11.3.1   Notable Features

Beside the significant performance gains over BLAST, LAMBDA offers many popular features that other comparable tools lack. This includes support for a wide variety of input and output formats as can be seen in Table 11.3. These formats are also highly configurable, e.g. the composition and order of the output columns in BLAST Tabular formats can be changed and SAM/BAM files can be created with multiple optional fields. Among the local aligners previously evaluated (Hauswedell, 2013; Hauswedell et al., 2014) and including DIAMOND, MALT and PALADIN, lambda3 is the only aligner to natively support BAM output. In contrast to the SAM implementation by e.g. DIAMOND, lambda3's files conform to the specification.[13] And lambda3 will automatically receive support for formats that are added to SeqAn in the future.

Lambda3 is a general-purpose local aligner. Its options cover a huge range on the speed ↔ sensitivity scale, and it supports all modes offered by the original BLAST, not only BlastX like many other aligners. This means users only need to learn one application interface for many use-cases.

The built-in taxonomy related options that lambda2 introduced were quite novel at the time (for a local aligner) and still stand out, although DIAMOND and MALT have since adopted similar features.[14] When the bisulfite-mode is added, this will again provide a unique feature not found anywhere else.

LAMBDA has often been lauded for being very easy to use. It provides detailed help-pages as well as UNIX manual-pages that are installed with LAMBDA. The online-wiki[15] contains more detailed documentation and according to GitHub-statistics is read quite frequently.[16] Other convenience features include a progress-bar that indicates current progress of the search and a memory-check that verifies

Table 11.3 Input and output formats of lambda3's search. Additionally, all formats can be combined with GZip, BZip2 or BGZF compression

Input formats	Output formats
FASTA (`.fasta,...`)	Blast Pairwise (`.m0`)
FASTQ (`.fastq`, `.fq`)	Blast Tabular (`.m8`)
EMBL (`.ebl`)	Blast Tabular with Comments (`.m9`)
Genbank (`.gbk`)	SAM (`.sam`)
	BAM (`.bam`)

---

[13] Cursory analysis showed that DIAMOND e.g. writes protein sequences into the SEQ field which causes errors when files are processed by SAMTOOLS or other third party applications.

[14] Even earlier than LAMBDA they supported creating special output files to plug into MEGAN (Huson et al., 2007) for taxonomic analysis.

[15] https://github.com/seqan/lambda/wiki.

[16] Apparently also for its general documentation of BLAST and SAM output formats. It is the second search result on Google for "BLAST formats".

whether a system meets the estimated memory requirements before a run. These are especially helpful when working with large datasets on remote machines as the remaining time can be estimated and out-of-memory crashes do not surprise users after already running for multiple days.

11.3.2 Performance

The performance measurements are conducted via a custom benchmarks suite[17] that operates similarly to what I described in my previous analyses (Hauswedell et al., 2014). Focus of the benchmarks is to determine viable default parameters and compare lambda3 with previous versions, but comparisons against current versions of DIAMOND and MALT are also provided. Only the protein mode (BlastX) is compared here, proper tuning of lambda3 to nucleotide searches will happen in the context of developing the bisulfite features.

Query sequences are taken from two recent microbiome studies (see Table 11.4). They represent different query lengths and domains (Metagenomics/DNA and Metatranscriptomics/RNA). All benchmarks are performed against UniProt Swiss-Prot (The UniProt Consortium, 2019), downloaded on 2020-01-31. The computer environment is explained in Sect. A.2.

It should be noted that all programs can be configured in various ways, and performance depends on many factors, among them: size of query and database, length of the query sequences, number of CPU threads. The results shown here are only a subset of the test data accumulated, but they are indicative of the general trends observed and do not contradict any findings not shown.

Lambda3's Parameter Space

Results of the first benchmark are shown in Table 11.5. Lambda2's custom seeding strategy is not available in lambda3 which completely relies on SeqAn3/the SDSL. This seeding strategy meant searching for seeds of length 10 with one error but only permitting this error in the second half of the seed. It could be considered "one half" of the pigeon-hole bidirectional seeding strategy. To "compensate" for

Table 11.4 Query datasets used in LAMBDA benchmarks. Benchmarks do not use the full datasets but samples of different sizes

No.	Query-set	Technology	Length	Domain	Author/date
I	SRR6043351	Illumina	125	Metatrancriptomics	Visnovska et al. (2019)
II	ERR187768*	Illumina	251	Metagenomics	Bahram et al. (2018)

---

[17] https://github.com/h-2/labench.

Table 11.5 Exploring Lambda3's options. Lambda2's default mode and possible future default, `--fast` and `--sensitive` modes of Lambda3 are highlighted. Benchmarks performed on 100MB sample of Query-set I. "len" refers to seed length, "err" to the allowed hamming distance of the seed to index, "off" to the offset (if smaller than read-length, seeds are overlapping)

LAMBDA	Index		Alph.	Seeding			Hits		Performance	
version	FM	Dir.	Red.	len	err	off	# query	Total	Time	Memory
0.4.7	WT	uni	Mu10	10	1	10	193,070	4,469,891	126s	1137MB
1.0.3	WT	uni	Mu10	10	1	10	192,983	3,824,099	146s	897MB
2.0.1	**EPR**	**uni**	**Mu10**	**10**	**1**[†]	**5**	**194,552**	**3,815,294**	**49s**	**2480MB**
2.0.1[††]	EPR	uni	Mu10	10	1	10	194,534	3,867,987	76s	2758MB
3.0.0[††]	EPR	uni	Mu10	10	1	10	194,567	3,936,657	126s	1806MB
3.0.0	EPR	uni	Li10	10	1	10	194,566	3,815,873	91s	2034MB
3.0.0	EPR	uni	Li10	15	2	15	181,881	3,144,857	521s	1065MB
3.0.0	EPR	bi	Mu10	10	1	10	194,567	3,936,657	104s	2154MB
3.0.0	**EPR**	**bi**	**Li10**	**10**	**1**	**10**	**194,566**	**3,815,873**	**72s**	**2364MB**
3.0.0	EPR	bi	Li10	15	2	15	181,881	3,144,857	134s	1583MB
3.0.0	EPR	bi	Li10	11	1	11	188,490	3,360,862	31s	1687MB
3.0.0	**EPR**	**bi**	**Li10**	**12**	**1**	**12**	**187,546**	**3,182,364**	**25s**	**1583MB**
3.0.0	**EPR**	**bi**	**Li10**	**10**	**1**	**9**	**196,583**	**4,061,000**	**116s**	**3162MB**

[†] Lambda2 uses "half-exact" seeding by default that does not allow the error in the first half of the seed

[††] These modes are algorithmically almost identical between lambda2 and lambda3

only allowing errors in one half of the seed, the seeds overlap each other by half their length instead of being non-overlapping. This is very heuristic, but works well in practice; importantly it also works on unidirectional indexes. A simpler and more comparable strategy based on allowing one error anywhere in the seed and producing non-overlapping seeds is also shown and marked by [††].

The first important observation is that lambda3 is notably slower when configured similar to lambda2 (the entries marked by [††]). This is not mitigated fully by using a bidirectional index in lambda3 (which implicitly leads to half-exact seeding even without sensitivity loss). It does, however, improve performance by ~20% to use a bidirectional index. Another very important speed-up of ~25% is achievable by using the reduced alphabet of Li et al. (2003) instead of the reduction by Murphy et al. (2000). Although previous research came to slightly different conclusions (Knorr, 2017), almost no loss of sensitivity is associated with this change in this benchmark. These changes combined bring lambda3 closer to lambda2's performance (72s vs 49s).

The availability of optimum search schemes in combination with bidirectional indexes suggested that longer seeds with more allowed errors might provide configurations that are both faster and more sensitive. This did not turn out to be true as exemplified by the configurations with seed length 15 and two allowed errors. They do, however, highlight very strongly the impact of optimum search schemes which are automatically used with bidirectional indexes (runtime 134s) versus backtracking on unidirectional indexes (runtime 521s).

Simply increasing seed length (and offset accordingly) results in viable configurations that lean stronger to speed on the speed ↔ sensitivity axis. A configuration that is more sensitive (without incurring the full cost of reducing the seed length below 10) can be achieved by increasing the overlap of the seeds. This results in two further recommended profiles for lambda3 (highlighted in bold in Table 11.5).

Since the benchmarks are run with 32 threads and the index is very small, memory usage is dominated by the per-thread caches and buffers for hits and alignment computation.[18] It should be noted that all results shown are based on the experimental EPR-dictionary branches of the SDSL, SeqAn3 and lambda3. Benchmarks for lambda3 based on wavelet-trees were also conducted and consistently resulted in runtimes that were 50% higher and memory usage that was 500–600MBs lower. In general the impact of the index on the required memory is lower than for lambda2, because the SDSL indexes are generally smaller (see Sect. 9.6) but also because the index in the given example is quite small and the (compressed) suffix array's position types are automatically adjusted to the size of the index.

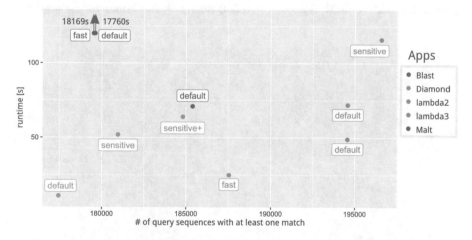

Fig. 11.5 Local aligner performance on 125bp RNA reads. The data table is available in the appendix (Table A.3). Blast data points would be well outside the graph; Blast's fast mode is indeed slower that the default mode

---

[18] The number of matches and the memory usage are not obviously correlated in the table, because the ratio of initially found seeds to valid matches varies greatly between the different configurations. For example, the configurations with seed length 15 have much fewer initial seeds (of which most are valid), so the memory usage is lower.

Speed and Sensitivity Compared to Other Applications

Figure 11.5 shows the results of searching 100MBs (~777,000 reads) of the first query dataset. As measure for sensitivity the number of sequences with at least one match was chosen, but the total number of matches is also provided in Table A.3.[19] BLAST is very strong on the latter and especially finds lower-scoring hits missed by other applications, but (perhaps surprisingly) it is not among the most sensitive tools when it comes to classifying the most reads; only DIAMOND's default mode is less sensitive here.

On the other hand, all modes of LAMBDA perform very well in this benchmark. Even the least sensitive lambda3 mode is more sensitive than all other applications while being faster than all others—except DIAMOND's default mode. In sensitivity, it beats the latter by more than 5%. Lambda3's more sensitive modes increase the difference to ~8% and ~10% respectively. Lambda2 has the same sensitivity as Lambda3's default mode but is almost 30% faster (see also Table 11.5).

BLAST's runtimes are expectedly very high; Lambda3's fast mode classifies more query sequences in 25 seconds than BLAST does in 5 hours.

The second shown benchmark (Fig. 11.6) is performed on the second query dataset but with a 10MB sample (~40,000 reads).[20] The trends of the previous

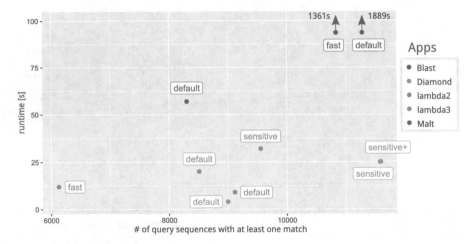

Fig. 11.6 Local aligner performance on 251bp DNA reads. The data table is available in the appendix (Table A.2). Blast data points would be well outside the graph. Diamond's sensitive modes perform almost identically

---

[19] For most use-cases the former is seen as the more important criterium, because missing a query entirely is more severe than finding $n - 1$ matches for a query instead of n. See Hauswedell (2013), Mackelprang et al. (2011), Huson and Xie (2014).

[20] Tests without BLAST were performed also on larger samples but produced similar relative results.

benchmark are now reversed with DIAMOND leading most speed ↔ sensitivity trade-offs. In particular, its sensitive modes surpass BLAST in sensitivity which is otherwise in the lead by almost 20% over all other program's default modes.

Lambda3's default mode is slower and less sensitive than Lambda2's which in turn is closer to DIAMOND's default mode in speed and sensitivity. Lambda3's fast mode costs a notable amount of sensitivity in this benchmark.

As in the previous benchmark, MALT provides neither speed nor sensitivity advantages over any other presented application. Moreover, its memory require-ments (shown in the appendix: Table A.2) are quite steep considering the size of the inputs.

The speed-ups over BLAST by all other programs are smaller compared to the previous benchmark, but the overall running times in this particular benchmark are quite low (due to the small query size), so one-time effects might skew the results slightly.

11.4 Discussion

Lambda3 is the first application that uses large parts of SeqAn3. Porting it was crucial for understanding how SeqAn3 designs actually play out in practice. It also helped discover many smaller issues that were since fixed.

However, not all SeqAn2 code has been replaced, yet. This was a conscious decision based on the lack of certain features and optimisations in SeqAn3, but it also allowed to analyse how SeqAn2 and SeqAn3 can be used simultaneously in a single application.

The performance of lambda3 is not on par with lambda2, yet. This is primarily due to SeqAn3's *Search* module lacking certain optimisations, but to a minor degree is also caused by different design decisions (async input, no custom seeding strategy). Nevertheless, the performance is still better than that of many other tools, and, on some tested datasets, lambda3 is at the same time faster and more sensitive than any other application, including BLAST. As previously shown for SeqAn3's *Search* module, the amount of memory used is lower for similar configurations.

Lambda3 will profit automatically from any performance gains in SeqAn3, but the algorithmic choices and intermediate filters need to be re-evaluated once all core components have been ported to SeqAn3 and received the necessary optimisations. The reason is that the interplay of the different heuristic steps is quite sophisticated, with changes in one part being the requirement for changes in another—or being counterbalanced by them. Finding the "sweet spot" of sensitivity and speed in this realm is not trivial. Furthermore, the effects of different heuristic parameters seem to vary between datasets as shown in Sect. 11.3.2.

The codebase of lambda3 has become much simpler than before and also 30% smaller. At 5.7K loc it is a fraction of DIAMOND's codebase (25K loc) which, similarly to lambda3, also depends on significant amounts of library code. I am confident that other developers can continue my work on LAMBDA. The bisulfite-mode will be a welcome addition, and the many planned features for SeqAn3 (distributed computing, device offloading, CRAM format and many more) promise to expand LAMBDA's status as the most general-purpose and feature-rich local aligner.

The following two subsections present a more detailed discussion of certain points raised in the last paragraphs.

11.4.1   From SeqAn2 to SeqAn3

Most parts of LAMBDA have been successfully ported from SeqAn2 to SeqAn3. This includes most notably all reading, handling and storage of sequence data. Lambda2 had a very complex case-handling system on when original sequences, translated sequences and/or reduced sequences were available in the program and how they were defined. Depending on program mode this would also be different for query and subject sequences, e.g. in certain cases the original subject sequences were not used (and therefore not stored), but the lengths were required and had to be stored separately. In lambda3 this is much simpler: the program now always stores only the original sequences and always creates translated and reduced sequences as views. If the selected program mode does not perform translation and/or reduction, the respective data structures are simply defined as an identity view. This allows writing the algorithm more abstractly, e.g. search is always performed on the "reduced sequences"—depending on the program mode these are actually reduced (lazily) or simply return the underlying translated/original sequences.

While reduction in lambda2 was also performed in a view-like way (see Sect. 7.4), translation always led to the creation of an extra set of sequences. Thus, the memory required by lambda2 was higher. This is even true for cases where the original sequences could be discarded after translation, because the cumulative size of the six protein frames (each one third the original length) is twice the size of the original sequences.[21]

Another area that profited strongly from being ported to SeqAn3 is the index (de-)serialisation. In lambda2 this resulted in more than 20 files being written to disk. To not clutter the working directory, these were put into a separate directory, but it was still something that confused users. The new single-file serialisation additionally enables the use of transparent compression which is very helpful considering the size of indexes. Furthermore, index files are now compatible across platforms, even different endiannesses. This does, however, come at the cost of slightly longer deserialisation times, because the on-disk format is no longer identical to the in-memory layout.

---

[21] This effect is even more pronounced when bit-compression is used since the nucleotide sequences can be compressed more strongly.

Since LAMBDA departed from the double-indexing approach very early, it was always the plan to not load the entire query file at program start. But since the processing of query sequences happens in parallel, there was no easily implementable solution based on SeqAn2 that would not result in very expensive locking. This changed when I implemented `seqan3::views::async_input_buffer` for SeqAn3. The view wraps a concurrent queue around the input range (in this case the file) and allows very simple parallel access to the underlying elements (see Sect. 8.4.4). Such an abstraction is not free; I measured a runtime overhead of about 10% over the version that pre-loads all query sequences. But it removes query size entirely from the space complexity of lambda3 which is an important advantage considering that the query sizes are ever growing.

The search has been ported to SeqAn3's *Search* module. Lambda2 previously used a custom backtracking implementation and did not use any of SeqAn2's `find()` interfaces. For performance reasons, it also provided its own suffix array construction algorithm. Removing both of these reduced the complexity of the codebase significantly. Lambda3 still provides variable-length seeding, but this is performed on the index cursor after it is returned from the search and works independently of whether the search used a backtracking algorithm (unidirectional index) or optimum search schemes (bidirectional index). This is a good example of how the SeqAn3 interfaces are both simpler and more powerful than their SeqAn2 counterparts. They allow extending/adapting existing solutions without forcing the developer to re-engineer basic behaviour like index backtracking.

Based on the benchmarks of SeqAn3's *Search* module (Sect. 9.6), it was clear that lambda3's search performance would not match lambda2's and that this would impact the overall runtime of lambda3.[22] However, the effect was a lot more notable than anticipated. This shows that application benchmarks are very important in addition to micro-benchmarks and that more resources need to be invested into improving the SDSL.

I decided to not port the alignment code to SeqAn3, yet, because the vectorisation support is not yet mature and the further impact on performance would have likely been very noticeable. The same decision was made for the writing of output files which is also still performed by the respective SeqAn2 modules, because SeqAn3 has no support for BLAST output files, yet. This did, however, present the opportunity to study the compatibility between SeqAn2 and SeqAn3. In the end, this worked out well, with lambda3 data types becoming consumable by SeqAn2 interfaces. Even creating a `seqan::Gaps` data structure (the equivalent of `seqan3::gap_decorator`) around nested sub strings of SeqAn3 translate views is possible; no sequences are ever copied at the interface between SeqAn2 and SeqAn3. However, due to SeqAn2's programming techniques and as I predicted in Sect. 2.4.4, a non-trivial amount of glue-code (function and metafunction overloads, template

---

[22] Depending on configuration, but especially for short query sequences, the search is the dominant factor in program runtime.

specialisations, etc.) is necessary. I still expect this to be very helpful for creating documentation on transitioning from SeqAn2 to SeqAn3.

11.4.2 Algorithmic Choices

RAPSearch (Ye et al., 2011) was the first protein search tool to prominently feature alphabet reduction. It has since been used by LAMBDA but also by many other tools. I previously discussed the algorithmic background of alphabet reductions in detail (Hauswedell, 2013; Hauswedell et al., 2014) and an undergraduate thesis that I oversaw empirically evaluated many reductions with LAMBDA (Knorr, 2017). This led to the reduction by Li et al. (2003) being added to SeqAn3 and now possibly becoming the default in lambda3. The reduction by Murphy et al. (2000) that was previously used, is still available. DIAMOND and MALT use a custom reduction derived from Murphy et al. (2000) that contains eleven symbols.

Lambda3 uses FM-indexes to perform fast searching, although many other comparable tools like DIAMOND (Buchfink et al., 2015) use k-mer-indexes. The advantages of FM-indexes over k-mer-indexes are allowing mismatches in arbitrary positions of the query and enabling adaptive-length seeding. Furthermore, many parameters that trade between speed and sensitivity (like seed length) can be chosen during the search and need not be fixed during index creation. k-mer-indexes, on the other hand, promise constant time look-ups and fast (re-)computation of the index data structure. Their use in other high-performance software shows that choosing one index type over the other is not always a clear call. Especially when the k-mer-based DREAM index becomes available in SeqAn3, it could make sense to re-evaluate the index choice in lambda3. Due to the modular and generic nature of the code, such a switch would not be very invasive.

Within the domain of suffix tree-like searches, various seeding strategies are possible. I explained the different strategies used in lambda2 and lambda3 in Sect. 11.3.2. Since SeqAn3, optimum search schemes are easily available and currently the bidirectional search with one error is used for performance reasons. However, since the overall performance of the *Search* module is not yet where it needs to be, it is difficult to predict which strategy will perform best in the end. As with all heuristic applications, this can only be deduced by rigorous testing on various data sets.

The alignment step currently performs vectorised but unbanded local alignments between the query and a subsequence of the subject. Adopting a banded approach[23] will certainly improve performance, especially for longer reads. Previous versions of LAMBDA used an x-drop approach modelled after BLAST (discussed extensively previously: Hauswedell, 2013; Hauswedell et al., 2014), but this proved to be

---

[23] Not yet available with vectorisation in either SeqAn2 or SeqAn3 for sequences of arbitrary lengths.

a performance bottle-neck, because it could not be vectorised. On the other hand, alignments for very long query sequences are more expensive to compute under the current model. Should even banded alignments appear too costly, other opportunistic strategies could be explored, e.g. computing the alignment in growing parallelograms or dividing the alignment into tiles along the diagonal that are conditionally computed expanding from the seed-tile. The basis for such implementations already exists in form of the wave-front model (Rahn et al., 2018).

Finally, I would like to underline the generic and modular nature of LAMBDA's codebase. Due to its integration of SeqAn, it is very simple to switch certain algorithmic components for others, compare them empirically and select the best. But it also means that lambda3 can only deliver a high performance if the underlying library components are properly optimised.

Part IV
Conclusion and Appendix

The last part of this book contains the conclusion with a summary of the previous discussion sections. It also provides references to other works. Finally, the appendix offers explanatory sections, details of the benchmarked software and hardware as well as further code snippets that did not fit into the regular body of the book.

Chapter 12
Conclusion

I began this book with introducing the reader to sequence analysis and explaining the requirement for highly efficient computing in this very important field of bioinformatics. Subsequently, the programming language C++ was established as one of the best technologies to produce such high-performing solutions. A central building-block in quickly developing these solutions has been the SeqAn library (version 1 and 2). The history of SeqAn, its motivation, as well as its strengths and weaknesses were discussed in detail. While the performance of SeqAn was found to be exceptional, I came to the conclusion that the library is prohibitively difficult to use and maintain and that it is less adaptable than desired. To find potential remedies for these problems, changes in the C++ programming language were explored. C++ has improved greatly in the last 10 years and many of the fundamental advances were illustrated in this book.

Based on this progress in C++ and rooted in the experience of working with SeqAn2, I devised a new library design. This includes revised design goals and the choice of very different C++ programming techniques than previously used—most importantly the use of C++ Concepts and dedicated customisation points as well as C++ Ranges and many more functional programming features of Modern C++. But beyond the immediate technical decisions, I also presented many thoughts on project management and administration, including guidelines for quality assurance, tooling and community involvement.

In successive chapters the implementation of this design, called *SeqAn3*, was explored. And finally a chapter was dedicated to covering the port of an application from SeqAn2 to SeqAn3. These chapters each had their own discussion sections, but I want to distil the results here.

The most profound difference of SeqAn3 compared to SeqAn2 is a library that is much **simpler** to use. An important part of this is a flexible approach regarding programming paradigms. While SeqAn3 is centred around generic programming, it incorporates aspects of functional programming and object-oriented programming whenever this improves usability (without compromising other design goals).

© The Author(s), under exclusive license to Springer Nature Switzerland AG 2022
H. Hauswedell, *Sequence Analysis and Modern C++*, Computational Biology 33,
https://doi.org/10.1007/978-3-030-90990-1_12

"Natural" function interfaces are a priority, i.e. functions use in-parameters and return values. I demonstrated repeatedly that SeqAn3's usage patterns are very comparable to Python libraries like BioPython, PySAM or PyBAM which in turn are widely regarded as the epitome of user-friendliness. Templates—while still at the heart of generic C++ code—have become invisible for almost all simple use-cases; in fact no angular brackets appear in the great majority of SeqAn3 snippets presented in this book. Concepts constrain all public templates so that mismatching types are easily diagnosed and compiler-errors are much more readable. It is important to note that there is a consistent design that encompasses the library as a whole and not a mismatch of different developer styles. A central aspect of this is that "everything is a range": data is modelled as ranges (e.g. containers of sequence data), transformations on data are modelled as ranges (e.g. nucleotide-to-amino acid translation view), files are modelled as ranges and even complex algorithms are modelled as ranges that lazily produce results (e.g. alignment computation). The API documentation of SeqAn3 is superb, counting as many lines as the library itself.

SeqAn2 is certainly among the **best-performing** libraries in sequence analysis and in many areas SeqAn3 performs equally well, with even some improvements. But in other areas it has also become clear that SeqAn3 has not yet reached the same performance level as SeqAn2, most-prominently in the indexed search and the alignment algorithms. Nevertheless, it is important to mention that core aspects of a high performance, like parallelisation and vectorisation, are part of SeqAn3's design and not applied a posteriori as in SeqAn2. Using them is thus much simpler and more consistent.

Due to the use of concepts and customisation point objects, SeqAn3 **integrates** third party code much better than SeqAn2. The standard library is used throughout SeqAn3, and the designs, usage patterns and naming conventions known from the standard library are applicable. Points of extension are clearly denoted as such and there are multiple ways to **adapt** user-provided types. These methods scale well to groups of types and thus allow combining complete libraries much more easily. Since requirements of generic code are formalised in concepts, it is simpler to define custom types that "fit" and to refine existing layers of abstraction.

The codebase of SeqAn3 is much more **compact** than SeqAn2's and it is structured more cleanly, i.e. there are fewer top-level modules with a manageable number of submodules each. While absolute numbers may be less meaningful in this regard, I repeatedly showed that the relative size of comparable components is strongly reduced. SeqAn3 has a clearly defined API, i.e. separation of stable interfaces and implementation detail. The size of the API (number of public types, functions, etc.) is even smaller, making those parts of SeqAn3 that users *need to understand* even more compact (compared to SeqAn2). These changes also benefit the maintainers of the library and contribute to quality-of-implementation.

SeqAn3 contains many crucial features for developing sequence analysis software and provides some exciting novelties that will make developing such software much easier in the future. Furthermore, the design is well suited to be applied to any future additions to the library. But I also understand now that I initially underestimated the magnitude of re-designing and reimplementing a library of this size. The parts of the codebase where I was not only responsible for the design, but also for most of the implementation (*Alphabet* module, *Range* module, generic I/O and *Sequence file* submodule, *STD* module, parts of the *Core* module) are now feature-complete (and perform very well!). They offer more functionality than the respective modules and submodules in SeqAn2, and they are completely in line with the design goals. Other parts, however, including the *Search* module and the *Alignment* module, have not yet reached this state. All evidence suggests that this is not the result of the designs and techniques presented here and that current deficiencies are merely a matter of not-yet-implemented features and (lacking) optimisation. The respective code in SeqAn2 was heavily optimised in multiple iterations over many years, and reproducing these properties in the new library is not trivial—especially since the SeqAn3 designs have been a "moving target" over the last years. Now that the design of SeqAn3 is complete and all core functionality is stable, I expect the SeqAn team to have a much easier job improving the performance and augmenting the library with new features. Several such desirable features have been mentioned throughout this book (some including design proposals), and I am aware of several others that are already being worked on.

I look forward to seeing the stable release of SeqAn-3.1 in the next months and to subsequently also see many new applications built with the library. I truly believe that SeqAn3 is a small revolution and that it not only benefits the bioinformatics research community but that the design also serves as a role-model for any new (C++) library developed in the years to come.

Correction to: Sequence Analysis and Modern C++

Hannes Hauswedell ⓘ

Correction to:
H. Hauswedell, *Sequence Analysis and Modern C++,*
Computational Biology 33,
https://doi.org/10.1007/978-3-030-90990-1

The original version of the book was inadvertently published without a volume number. This has now been amended in the book with volume number 33.

The updated online version of the book can be found at
https://doi.org/10.1007/978-3-030-90990-1

Appendix A

A.1 Notes on Reading This Book

A.1.1 References and Hyperlinks

Links to External Resources and Websites

All URLs for which no other date is specified were verified as working on 2020-04-20.

Cross-References Inside the Book

I frequently reference different entities within this book from inside the text. In digital format, these links are hyperlinks and take the reader to the respective places immediately. In printed form, the reader should be able to quickly find the respective places with the help of the table-of-contents. Unnumbered entities like subsubsections and paragraphs are always given as the section number with an additional page-reference to the exact place. Floating environments (figures, tables and code snippets) are usually found very close to the place that refers to them (i.e. the same page ± 1). When referring to floating environments in a more distant part of the book other than the appendix, an additional page-reference is provided.

A.1.2 How to Read Code Snippets

Since code snippets play an important role in this book, I have done my best to make them readable. This includes custom syntax highlighting and rules for line-numbering (both introduced below).

I have also tried to write snippets that appear like actual code would, although I often had to compromise due to size constraints and still wanting the snippets to be readable in the context where they are explained. This means that I usually do not declare the necessary includes and I also do not explicitly state the context of the snippet (e.g. whether the code is valid inside function scope or at namespace scope). In some cases snippets may even contain both (e.g. a function template declaration and a function invocation).

Unless otherwise specified, snippets do not assume a specific namespace and all names are given fully qualified. This follows my advice on not doing `using namespace seqan3;` and makes it easier for the reader to distinguish library names from local names. However, due to space constraints many snippets deviate from this default; they are marked as such. An exception to these rules is user-defined literals which are always assumed to be included (`namespace seqan3::literals`, see Sect. 6.1.1 on p. 148).

Function names are always given with parentheses to highlight that they are function names, e.g. `foobar()`. This does not imply that the function does not take arguments; the parameters/arguments may simply be omitted in-text for brevity. Function objects are typically given without parentheses (unless being invoked with arguments).

C-style comments (`/* foo */`) typically describe multiple following lines whereas C++-style comments (`// foo`) typically describe the line the comment is on or the single line after it. An ellipsis (`/*... */`) is used to denote sections of code that are omitted in a snippet but present in the actual implementation.

Line-Numbering in Code Snippets

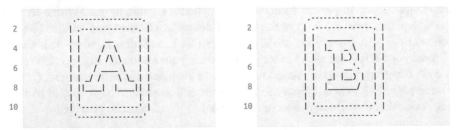

Code snippet A.1: Two independent snippets. Independent line-numbering indicates independent snippets

Many code snippet floating regions contain more than one snippet. This is done to enable line-by-line comparisons of different styles (e.g. C++ 98 vs C++ 20) or libraries (e.g. SeqAn2 vs SeqAn3). Such independent snippets have their own line-numbering (see Snippet A.1). In other cases a single snippet is split into two halves to use the space more efficiently. This can be seen in Snippet A.2 and is illustrated by contiguous line-numbering.

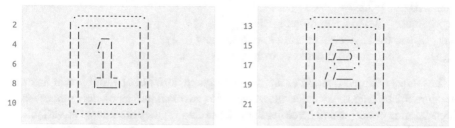

Code snippet A.2: Two consecutive snippets. Line-numbering indicates that the second snippet follows the first

Syntax Highlighting

I developed custom syntax highlighting and language parsing based on the Shiki project,[1] the Shiki-LATEX-frontend[2] and the Solarized theme.[3] Changes include parsing of Modern C++ constructs, importing specific SeqAn3 names and changes to improve overall consistency. Note that C++ is particularly difficult to parse and all parsing that is not done by an actual C++ compiler is bound to be imperfect.

The following colour codes are used (with examples each):

orange: Used for types (`int32_t`), type templates (`std::vector`), template type parameters (`T`), type placeholders (`auto`, also constrained: `std::integral auto i = /**/`), type aliases and the names of namespaces (`namespace seqan3`).

green: Used for keywords that open namespaces (`namespace`) or *introduce* type declarations (`struct`), templates (`template`) or type aliases (`typedef`).

blue: Used for function declarations (`int foo(int i);`), function calls (`foo(3);`) and invocations of function objects.

cyan: Used for...

- ...the const, volatile and reference qualification of types (`T const &`).
- ...various modifiers on function and variable declarations (`private:`, `static`).
- ...all operators (`+`, `->`).
- ...angular brackets on templates (`<>`).

---

[1] https://shiki.matsu.io/.

[2] https://github.com/leafac/shiki-latex/.

[3] https://ethanschoonover.com/solarized/.

purple: Used for include directives (`#include`), pragmas and macros
(`#define foo=bar`).

red: Used for literals: numeric (`3`), character (`'f'`), string (`''ACG''`) and user-
defined (`'C'_dna4`).

grey & italic: Used for comments (`/* foo */`).

bold: Used for keywords relating to control flow (`if`).

magenta: Used for all other keywords (`requires`).

The names of objects/variables are not highlighted, also not of variable templates.
Function objects appear as functions when being invoked and as objects otherwise.
When defining concepts and invoking them as predicates, they are treated as variable
templates (no highlighting). However, when they appear with `auto` they are treated
as part of the type (terse concepts syntax) and if they introduce a type name
(intermediate syntax) they are treated like `typename`.

Code highlighted within text has coloured background (`foobar()`) except in
those places where a background colour is already defined (e.g. tables). If such a
term appears as the label of a description/definition (used like enumerations), it is
instead highlighted by a framebox to distinguish it from the regular text and improve
readability:

$\boxed{\texttt{foobar()}}$ A very important function that calls `foo()` and `bar()`.

A.2   Software and Hardware Details

A.2.1   Benchmarking Environment

The different computer systems used in benchmarking are shown in Table A.1.
Benchmarks of the vectorised alignment code were conducted on "René Rahn's
system"; Lamba benchmarks were performed on the second system and all other
benchmarks were performed on the "Main system".

Table A.1 System specifications

	Main system	Lambda benchmarks	René Rahn's system
CPU model	AMD Ryzen 5 2600X	2x Intel Xeon E5-2667	Intel Core i7-6700HQ
CPU frequency	3.6 GHz	3.3 GHz	2.6 GHz
CPU cores/threads	6/12	16/32	4/8
RAM	32 GB	384 GB	16 GB
Storage	SSD (M.2)	HDD (spinning disk)	SSD
Operating system	FreeBSD 12.1	Debian GNU/Linux 10.3	macOS 10.14

All C++ code built for benchmarks was compiled with GCC7. All python code was run with Python-3.7. Unless otherwise noted, the following versions of other software packages were used: SeqAn2 (`-develop` on 2019-12-20), BioPython (1.73), LibGenomeTools (`-master` on 2020-02-09), DIAMOND (0.9.29), MALT (0.4.1).[4]

The source code for all "microbenchmarks" is available in the test suite of SeqAn3. All other library benchmark code (including other libraries and Python) is provided in the following repository:
https://git.fsfe.org/h2/thesis_macro_bench

The local aligner benchmark suite is available here:
https://github.com/h-2/labench

A.2.2 Helpful Software

Many Free and Open Source Software projects were instrumental in creating SeqAn3. The following is a non-exhaustive list:

CDash, CMake, Clang, Debian, FreeBSD, GCC, GDB, Git, GNU, CompilerExplorer, GoogleTest, GoogleBenchmark, KDE, Linux

Software that was vital in creating this book includes the following:

BibLaTeX/Biber, ggplot, KOMA-Script, LanguageTool, ŁTEX, LuaLaTeX, Minted, R, Shiki, TexStudio

I am grateful to all authors for providing such highly useful software and hope that my own contributions help others to be equally productive.

A.3 Copyright

All content taken directly or indirectly from other sources is marked as such.

A.3.1 SeqAn Copyright

Code snippets, images and text taken directly from SeqAn1/2 (source code or documentation), as well as code snippets taken from SeqAn3's source code are made available under the terms printed below ("three-clause BSD license"). Code snippets taken from the documentation of SeqAn3 are in the public domain/ⓒ⓪ . Text reproduced from the documentation of SeqAn3 is available under ⓒ①.

---

[4] MALT was not executable with current versions of JAVA anymore, but installing OpenJDK-10.0.2 manually worked.

A.4   Longer Code Snippets

```cpp
   #include <seqan3/alphabet/alphabet_base.hpp>
2  #include <seqan3/alphabet/concept.hpp>

4  namespace my_namespace
   {
6  // Derive from a CRTP base class, second template parameter is alphabet size
   struct my_alph : seqan3::alphabet_base<my_alph, 2>
8  {
   private:
10     // Map 0 -> 'A' and 1 -> 'B'
       static constexpr std::array<char_type, alphabet_size> rank_to_char{'A', 'B'};
12     // map every character to rank zero, except Bs
       static constexpr std::array<rank_type, 256> char_to_rank
14     {
           []() // initialise with an immediately evaluated lambda expression
16         {
               std::array<rank_type, 256> ret{};   // initialise all values with 0 for 'A'
18             ret['B'] = 1;                        // only 'B' results in rank 1
               return ret;
20         }()
       };
22     // make the base class a friend so it can access the tables
       friend alphabet_base<my_alph, 2>;
24 };

26 } // namespace my_namespace

28 // CPOs:
   static_assert(seqan3::alphabet_size<my_namespace::my_alph> == 2);
30 static_assert(seqan3::char_is_valid_for<my_namespace::my_alph>('B'));
   static_assert(!seqan3::char_is_valid_for<my_namespace::my_alph>('!'));
32 // Concept (seqan3::writable_alphabet subsumes the others)
   static_assert(seqan3::writable_alphabet<my_namespace::my_alph>);
```

Code snippet A.3: Example of a user-defined alphabet based on `seqan3::alphabet_base`. This snippet is a valid header file. The base class provides all the necessary members, one need only provide the two given tables

```
    #pragma once
2
    #include <seqan3/alphabet/concept.hpp>
4   #include <seqan3/core/detail/int_types.hpp>

6   namespace seqan3::detail
    {
8
    template <typename type>
10  constexpr bool is_char_adaptation_v = std::same_as<type, char>     ||
                                          std::same_as<type, char16_t> ||
12                                        std::same_as<type, char32_t> ||
                                          std::same_as<type, wchar_t>;
14  } // namespace seqan3::detail

16  namespace seqan3::custom
    {
18
    template <typename char_type>
20      requires detail::is_char_adaptation_v<char_type>
    struct alphabet<char_type>
22  {
        static constexpr auto alphabet_size = detail::size_in_values_v<char_type>;
24
        static constexpr char_type to_char(char_type const chr) noexcept
26      {
            return chr;
28      }

30      static constexpr auto to_rank(char_type const chr) noexcept
        {
32          return static_cast<detail::min_viable_uint_t<alphabet_size - 1>>(chr);
        }
34
        static constexpr char_type & assign_char_to(char_type const chr2, char_type & chr) noexcept
36      {
            return chr = chr2;
38      }

40      static constexpr char_type & assign_rank_to(decltype(alphabet::to_rank(char_type{})) const rank,
                                                    char_type & chr) noexcept
42      {
            return chr = rank;
44      }
    };
46
    } // namespace seqan3::custom
```

Code snippet A.4: Built-in character types adapted as alphabets

```cpp
class sam_dna16 : public nucleotide_base<sam_dna16, 16>
{
private:
    using base_t = nucleotide_base<sam_dna16, 16>;
    friend base_t;
    friend base_t::base_t;

public:
    constexpr sam_dna16()                           noexcept = default;
    constexpr sam_dna16(sam_dna16 const &)          noexcept = default;
    constexpr sam_dna16(sam_dna16 &&)               noexcept = default;
    constexpr sam_dna16 & operator=(sam_dna16 const &) noexcept = default;
    constexpr sam_dna16 & operator=(sam_dna16 &&)   noexcept = default;
    ~sam_dna16()                                    noexcept = default;

    using base_t::base_t;

protected:
    static constexpr char_type rank_to_char[alphabet_size]
    {
        '=', 'A', 'C', 'M', 'G', 'R', 'S', 'V', 'T', 'W', 'Y', 'H', 'K', 'D', 'B', 'N'
    };

    static constexpr std::array<rank_type, 256> char_to_rank
    {
        [] () constexpr
        {
            std::array<rank_type, 256> ret{};

            // initialize with UNKNOWN (std::array::fill unfortunately not constexpr)
            for (auto & c : ret)
                c = 15; // rank of 'N'

            // reverse mapping for characters and their lowercase
            for (size_t rnk = 0u; rnk < alphabet_size; ++rnk)
            {
                ret[         rank_to_char[rnk] ] = rnk;
                ret[to_lower(rank_to_char[rnk])] = rnk;
            }

            // set U equal to T
            ret['U'] = ret['T']; ret['u'] = ret['t'];

            return ret;
        }()
    };

    static const std::array<sam_dna16, alphabet_size> complement_table;
};

constexpr std::array<sam_dna16, sam_dna16::alphabet_size> sam_dna16::complement_table
{
    'N'_sam_dna16, 'T'_sam_dna16, 'G'_sam_dna16, 'K'_sam_dna16,
    'C'_sam_dna16, 'Y'_sam_dna16, 'S'_sam_dna16, 'B'_sam_dna16,
    'A'_sam_dna16, 'W'_sam_dna16, 'R'_sam_dna16, 'D'_sam_dna16,
    'M'_sam_dna16, 'H'_sam_dna16, 'V'_sam_dna16, 'N'_sam_dna16
};
```

Code snippet A.5: Full definition of `seqan3::sam_dna16`

```
   template <writable_alphabet sequence_alphabet_t>
2    requires std::regular<sequence_alphabet_t>
   class masked : public alphabet_tuple_base<masked<sequence_alphabet_t>, sequence_alphabet_t, mask>
4  {
   private:
6    using base_t = alphabet_tuple_base<masked<sequence_alphabet_t>, sequence_alphabet_t, mask>;

8  public:
     using sequence_alphabet_type = sequence_alphabet_t;
10   using char_type = alphabet_char_t<sequence_alphabet_type>;
     using base_t::alphabet_size;
12   using typename base_t::rank_type;

14   constexpr masked() = default;

16   using base_t::base_t;
     using base_t::operator=;
18
     constexpr masked & assign_char(char_type const c) noexcept
20   {
       using index_t = std::make_unsigned_t<char_type>;
22     base_t::assign_rank(char_to_rank[static_cast<index_t>(c)]);
       return *this;
24   }

26   constexpr char_type to_char() const noexcept { return rank_to_char[base_t::to_rank()]; }

28 protected:
     static constexpr std::array<char_type, alphabet_size> rank_to_char
30   {
       [] ()
32     {
         std::array<char_type, alphabet_size> ret{};
34       for (size_t i = 0; i < alphabet_size; ++i)
         {
36         ret[i] = (i < alphabet_size / 2)
                  ? seqan3::to_char(seqan3::assign_rank_to(i, sequence_alphabet_type{}))
38                : to_lower(seqan3::to_char(seqan3::assign_rank_to(i / 2, sequence_alphabet_type{})));
         }
40       return ret;
       } ()
42   };

44   static constexpr std::array<rank_type, detail::size_in_values_v<char_type>> char_to_rank
     {
46     [] ()
       {
48       std::array<rank_type, detail::size_in_values_v<char_type>> ret{};
         for (size_t i = 0; i < 256; ++i)
50       {
           char_type c = static_cast<char_type>(i);
52         ret[i] = is_lower(c)
                  ? seqan3::to_rank(seqan3::assign_char_to(c, sequence_alphabet_type{})) * 2
54                : seqan3::to_rank(seqan3::assign_char_to(c, sequence_alphabet_type{}));
         }
56       return ret;
       } ()
58   };
   };
```

Code snippet A.6: Full definition of `seqan3::masked`

```
 1  struct Dna5_ {};
 2  typedef SimpleType<unsigned char, Dna5_> Dna5;
 3
 4  template <> struct ValueSize<Dna5>
 5  {
 6    typedef uint8_t Type;
 7    static const Type VALUE = 5;
 8  };
 9
10  template <> struct BitsPerValue<Dna5>
11  {
12    typedef uint8_t Type;
13    static const Type VALUE = 3;
14  };
15
16  inline Dna5 unknownValueImpl(Dna5 *)
17  {
18    static const Dna5 _result = Dna5('N');
19    return _result;
20  }
21
22  inline void assign(char & c_target,
23                     Dna5 const & source)
24  {
25    c_target =
26      TranslateTableDna5ToChar_<>::VALUE[
27        source.value];
28  }
29
30  template <>
31  struct CompareTypeImpl<Dna5, uint8_t>
32  {
33    typedef Dna5 Type;
34  };
35
36  inline void assign(Dna5 & target,
37                     uint8_t c_source)
38  {
39    target.value =
40      TranslateTableByteToDna5_<>::VALUE[c_source];
41  }
42
43  template <>
44  struct CompareTypeImpl<Dna5, char>
45  {
46    typedef Dna5 Type;
47  };
48
49  inline void assign(Dna5 & target,
50                     char c_source)
51  {
52    target.value =
53      TranslateTableCharToDna5_<>::VALUE[
54        (unsigned char)c_source];
55  }
56
57  template <> struct CompareTypeImpl<Dna5, Iupac>
58  {
59    typedef Dna5 Type;
60  };
61
62
63
64  inline void assign(Dna5 & target,
65                            Iupac const & source)
66  {
67    target.value =
68      TranslateTableIupacToDna5_<>::VALUE[
69        source.value];
70  }
71
72  template <> struct CompareTypeImpl<Dna5, Dna>
73  {
74    typedef Dna Type;
75  };
76
77  inline void assign(Dna5 & target,
78                            Dna const & c_source)
79  {
80    target.value = c_source.value;
81  }
82
83  template <typename T = void>
84  struct TranslateTableDna5ToChar_
85  {
86    static char const VALUE[5];
87  };
88
89  template <typename T> char const
90  TranslateTableDna5ToChar_<T>::VALUE[5] =
91    {'A', 'C', 'G', 'T', 'N'};
92
93  template <typename T = void>
94  struct TranslateTableDna5ToIupac_
95  {
96    static char const VALUE[5];
97  };
98
99  template <typename T> char const
100 TranslateTableDna5ToIupac_<T>::VALUE[5] =
101   {0x01, 0x02, 0x04, 0x08, 0x0f};
102
103 template <typename T = void>
104 struct TranslateTableCharToDna5_
105 {
106   static char const VALUE[256];
107 };
108
109 template <typename T> char const
110 TranslateTableCharToDna5_<T>::VALUE[256] =
111   { /* 256 hard-coded values */ };
112
113 struct TranslateTableByteToDna5_
114 {
115   static char const VALUE[256];
116 };
117
118 template <typename T> char const
119 TranslateTableByteToDna5_<T>::VALUE[256] =
120   { /* 256 hard-coded values */ };
```

Code snippet A.7: SeqAn2's Dna5 type

```
struct my_simple_scheme
{
    int match{0}; int mismatch{-1};

    template <typename Tl, typename Tr>
        requires std::equality_comparable_with<Tl, Tr>
    constexpr int score(Tl && lhs, Tr && rhs) const
    {
        return lhs == rhs ? match : mismatch;
    }
};

/* Perform concept checks: */
static_assert(seqan3::scoring_scheme<my_simple_scheme, seqan3::dna5, seqan3::dna5>);
static_assert(seqan3::scoring_scheme<my_simple_scheme, seqan3::dna5, seqan3::rna5>);
static_assert(seqan3::scoring_scheme<my_simple_scheme, seqan3::aa27, seqan3::aa27>);
static_assert(!seqan3::scoring_scheme<my_simple_scheme, seqan3::dna5, seqan3::aa27>);
```

Code snippet A.8: Defining a custom scoring scheme

```
using sdsl_wt_index_type = sdsl::csa_wt<sdsl::wt_blcd<sdsl::bit_vector,
                                        sdsl::rank_support_v<>,
                                        sdsl::select_support_scan<>,
                                        sdsl::select_support_scan<0>>,
                             10,            // SA sampling rate
                             10'000'000,  // ISA sampling rate
                             sdsl::sa_order_sa_sampling<>,
                             sdsl::isa_sampling<>,
                             sdsl::plain_byte_alphabet>;

using sdsl_default_index_type = sdsl_wt_index_type;
```

Code snippet A.9: Definition of the default SDSL index type

A.5   Detailed Benchmark Results (Local Aligners)

See Tables A.2 and A.3.

Table A.2 Local aligner benchmarks; 10MB of Query-set II

Application		Bit-scores			Hits		Performance	
Name	Profile	Q_{25}	Q_{50}	Q_{75}	# query	Total	Runtime	Memory
lambda0	Default	60.8	79.3	103.2	8999	638,624	16s	826MB
lambda1	Default	61.2	79.7	104	8932	482,882	16s	645MB
lambda2	Default	60.8	79.0	103	9112	473,095	9s	1309MB
lambda3	Default	61.2	80.9	105	8510	431,698	20s	2111MB
lambda3	Fast	67.4	89.4	112	6129	273,877	12s	1694MB
lambda3	Sensitive	59.7	77.8	102	9553	522,611	32s	2639MB
Diamond	Default	61.6	79.7	103.2	8996	501,452	4s	838MB
Diamond	Sensitive	57.4	72.0	95.9	11,580	762,236	25s	868MB
Diamond	Sensitive+	57.4	72.0	95.9	11,593	764,705	25s	870MB
Malt	Default	65	83	106	8305	433,846	57s	39,271MB
Blast	Default	56.2	70.1	93.2	11,277	760,515	1889s	414MB
Blast	Fast	57.4	71.6	95.1	10,832	703,998	1361s	699MB

Table A.3 Local aligner benchmarks; 100MB of Query-set I

Application		Bit-scores			Hits		Performance	
Name	Profile	Q_{25}	Q_{50}	Q_{75}	# query	Total	Runtime	Memory
lambda0	Default	65.5	80.1	84.7	193,070	4,469,891	126s	1135MB
lambda1	Default	65.9	80.1	84.7	192,983	3,824,099	147s	901MB
lambda2	Default	66.2	82.0	86.7	194,552	3,815,294	49s	2587MB
lambda3	Default	66.2	82.0	86.7	194,566	3,815,873	72s	2303MB
lambda3	Fast	68.6	82.4	87.0	187,546	3,182,364	25s	1583MB
lambda3	Sensitive	65.9	81.6	86.7	196,583	4,061,000	116s	3064MB
Diamond	Default	65.1	79.7	86.3	177,441	3,558,974	11s	1397MB
Diamond	Sensitive	63.9	79.3	86.3	180,964	3,960,232	52s	1437MB
Diamond	Sensitive+	62.8	79.0	86.3	184,810	3,994,875	64s	1785MB
Malt	Default	69	82	87	185,381	3,243,177	71s	31,367MB
Blast	Default	61.6	80.9	86.7	179,594	5,777,757	17760s	439MB
Blast	Fast	61.6	80.9	86.7	179,515	5,763,446	18169s	712MB

References

Afgan, E., Baker, D., Batut, B., van den Beek, M., Bouvier, D., Cech, M., Chilton, J., Clements, D., Coraor, N., Grüning, B. A., Guerler, A., Hillman-Jackson, J., Hiltemann, S. D., Jalili, V., Rasche, H., Soranzo, N., Goecks, J., Taylor, J., Nekrutenko, A., & Blankenberg, D. J. (2018). The Galaxy platform for accessible, reproducible and collaborative biomedical analyses: 2018 update. *Nucleic Acids Research, 46*(Webserver-Issue), W537–W544.

Altschul, S. F., Boguski, M. S., Gish, W., & Wootton, J. C. (1994). Issues in searching molecular sequence databases. *Nature Genetics, 6*(2), 119.

Altschul, S. F., Gish, W., Miller, W., Myers, E. W., & Lipmanl, D. J. (1990). Basic local alignment search tool. *Journal of Molecular Biology, 215*(2), 403–410.

Altschul, S. F., Madden, T. L., Schäffer, A. A., Zhang, J., Zhang, Z., Miller, W., & Lipman, D. J. (1997). Gapped BLAST and PSI-BLAST: A new generation of protein database search programs. *Nucleic Acids Research, 25*(17), 3389–3402.

Amstutz, P., Crusoe, M. R., Tijanić, N., Chapman, B., Chilton, J., Heuer, M., Kartashov, A., Leehr, D., Ménager, H., Nedeljkovich, M., et al. (2016). *Common Workflow Language, v1.0.* Technical report, Software Freedom Conservancy.

Aruoba, S. B., & Fernández-Villaverde, J. (2014). *A Comparison of Programming Languages in Economics*. Technical report, National Bureau of Economic Research.

Austin Common Standards Revision Group. (2014). POSIX.

Bahram, M., Hildebrand, F., Forslund, S. K., Anderson, J. L., Soudzilovskaia, N. A., Bodegom, P. M., Bengtsson-Palme, J., Anslan, S., Coelho, L. P., Harend, H., et al. (2018). Structure and function of the global topsoil microbiome. *Nature, 560*(7717), 233–237.

Baker, M. (2016). 1,500 scientists lift the lid on reproducibility. *Nature News, 533*(7604), 452.

Banerji, S., Cibulskis, K., Rangel-Escareno, C., Brown, K. K., Carter, S. L., Frederick, A. M., Lawrence, M. S., Sivachenko, A. Y., Sougnez, C., Zou, L., et al. (2012). Sequence analysis of mutations and translocations across breast cancer subtypes. *Nature, 486*(7403), 405–409.

Bannon, J. (2014). Heading for $100: The Declining Costs of Genome Sequencing & The Consequences.

Bazinet, A. L., & Cummings, M. P. (2012). A comparative evaluation of sequence classification programs. *BMC Bioinformatics, 13*, 92.

Berthold, M. R., Cebron, N., Dill, F., Gabriel, T. R., Kötter, T., Meinl, T., Ohl, P., Sieb, C., Thiel, K., & Wiswedel, B. (2007). KNIME: The Konstanz Information Miner. In *Studies in classification, data analysis, and knowledge organization (GfKL 2007)*. Springer.

Blischak, J. D., Davenport, E. R., & Wilson, G. (2016). A quick introduction to version control with Git and GitHub. *PLoS Computational Biology, 12*(1), e1004668.

Boccara, J. (2016). Strong types for strong interfaces.

© The Author(s), under exclusive license to Springer Nature Switzerland AG 2022   339
H. Hauswedell, *Sequence Analysis and Modern C++*, Computational Biology 33,
https://doi.org/10.1007/978-3-030-90990-1

Brooks, F. P. (1995). *The mythical man-month*. Boston, MA: Addison-Wesley, anniversary edition.

Brown, W. E., & Sunderland, D. (2019). *P1601: Recommendations for Specifying "Hidden Friends"*. Technical report, International Organization for Standardization.

Buchfink, B., Xie, C., & Huson, D. H. (2015). Fast and sensitive protein alignment using DIAMOND. *Nature Methods, 12*(1), 59.

Buffalo, V. (2015). *Bioinformatics data skills: Reproducible and robust research with open source tools* (1st ed.). O'Reilly Media.

Burkhardt, S., Crauser, A., Ferragina, P., Lenhof, H.-P., Rivals, E., & Vingron, M. (1999). q-Gram based database searching using a suffix array (QUASAR). In S. Istrail, P. A. Pevzner, & M. S. Waterman (Eds.), *RECOMB* (pp. 77–83). ACM.

Burkhardt, S., & Kárkkäinen, J. (2001). Better filtering with gapped q-grams. In A. Amir & G. M. Landau (Eds.), *CPM. Lecture notes in computer science* (Vol. 2089, pp. 73–85). Springer.

Burrows, M., & Wheeler, D. J. (1994). *A Block-Sorting Lossless Data Compression Algorithm*. Technical report, Systems Research Center.

Calabrese, M. (2018). *P1292R0: Customization Point Functions*. Technical report, International Organization for Standardization.

Camacho, C., Coulouris, G., Avagyan, V., Ma, N., Papadopoulos, J., Bealer, K., & Madden, T. L. (2009). BLAST+: Architecture and applications. *BMC Bioinformatics, 10*(1), 421+.

Chao, K.-M., Pearson, W. R., & Miller, W. (1992). Aligning two sequences within a specified diagonal band. *Computer Applications in the Biosciences, 8*(5), 481–487.

Cock, P. J., Fields, C. J., Goto, N., Heuer, M. L., & Rice, P. M. (2010). The Sanger FASTQ file format for sequences with quality scores, and the Solexa/Illumina FASTQ variants. *Nucleic Acids Research, 38*(6), 1767–1771.

Cock, P. J. A., Antao, T., Chang, J. T., Chapman, B. A., Cox, C. J., Dalke, A., Friedberg, I., Hamelryck, T., Kauff, F., Wilczynski, B., & de Hoon, M. J. L. (2009). Biopython: Freely available Python tools for computational molecular biology and bioinformatics. *Bioinformatics, 25*(11), 1422–1423.

Cogswell, J. (2015). *Adding an Easy File Save and File Load Mechanism to Your C++ Program*.

Collet, Y., & Kucherawy, M. S. (2018). Zstandard Compression and the application/zstd Media Type. *RFC, 8478*, 1–54.

Coplien, J. O. (1995). Curiously recurring template patterns. *C++ Report, 7*(2), 24–27.

Costanza, P., Herzeel, C., & Verachtert, W. (2019). A comparison of three programming languages for a full-fledged next-generation sequencing tool. *BMC Bioinformatics, 20*(1), 301:1–301:10.

Crockford, D. (2002). Introducing JSON.

Crosswell, L. C., & Thornton, J. M. (2012). ELIXIR: A distributed infrastructure for European biological data. *Trends in Biotechnology, 30*(5), 241–242.

Curcin, V., & Ghanem, M. (2008). Scientific workflow systems - Can one size fit all? In *2008 Cairo International Biomedical Engineering Conference*: IEEE.

Dadi, T. H., Renard, B. Y., Wieler, L. H., Semmler, T., & Reinert, K. (2017). SLIMM: Species level identification of microorganisms from metagenomes. *PeerJ, 5*, e3138.

Dadi, T. H., Siragusa, E., Piro, V. C., Andrusch, A., Seiler, E., Renard, B. Y., & Reinert, K. (2018). DREAM-Yara: An exact read mapper for very large databases with short update time. *Bioinformatics, 34*(17), i766–i772.

Dagum, L., & Menon, R. (1998). OpenMP: An industry-standard API for shared-memory programming. *Computing in Science & Engineering, 5*(1), 46–55.

Dehnert, J. C., & Stepanov, A. (2000). Fundamentals of generic programming. In *Generic programming* (pp. 1–11). Springer.

Deutsch, L. P. (1996). GZIP file format specification version 4.3. Internet RFC 1952.

Dezso, B., Jättner, A., & Kovács, P. (2011). LEMON - An open source C++ graph template library. *Electronic Notes in Theoretical Computer Science, 264*(5), 23–45.

Di Tommaso, P., Chatzou, M., Floden, E. W., Barja, P. P., Palumbo, E., & Notredame, C. (2017). Nextflow enables reproducible computational workflows. *Nature Biotechnology, 35*(4), 316–319.

Dodt, M., Roehr, J., Ahmed, R., & Dieterich, C. (2012). FLEXBAR—Flexible barcode and adapter processing for next-generation sequencing platforms. *Biology, 1*(3), 895–905.

Döring, A., Weese, D., Rausch, T., & Reinert, K. (2008). SeqAn An efficient, generic C++ library for sequence analysis. *BMC Bioinformatics, 9*, 1–9.

Driesen, K., & Hölzle, U. (1996). The direct cost of virtual function calls in C++. In *ACM Sigplan Notices* (Vol. 31, pp. 306–323). ACM.

Dröge, J., Gregor, I., & McHardy, A. (2014). Taxator-tk: Fast and precise taxonomic assignment of metagenomes by approximating evolutionary neighborhoods. Preprint, arXiv:1404.1029.

Duran, J. W., & Ntafos, S. C. (1981). A report on random testing. In S. Jeffrey & L. G. Stucki (Eds.), *ICSE* (pp. 179–183). IEEE Computer Society.

Duret-Lutz, A., Géraud, T., & Demaille, A. (2001). Design patterns for generic programming in C++. In *COOTS* (Vol. 1, pp. 14–14).

Dusíková, H. (2019). *P1433R0: Compile Time Regular Expressions.* Technical report, International Organization for Standardization.

Edgar, R. C. (2010). Search and clustering orders of magnitude faster than BLAST. *BioInformatics, 26*(19), 2460–2461.

Ewing, B., & Green, P. (1998). Base-calling of automated sequencer traces using Phred. II. Error probabilities. *Genome Research, 8*(3), 186–194.

Ewing, B., Hillier, L., Wendl, M. C., & Green, P. (1998). Base-calling of automated sequencer traces using Phred. I. Accuracy assessment. *Genome Research, 8*(3), 175–185.

Ferragina, P., & Manzini, G. (2000). Opportunistic data structures with applications. In *Proceedings 41st Annual Symposium on Foundations of Computer Science* (pp. 390–398). IEEE.

Fourment, M., & Gillings, M. R. (2008). A comparison of common programming languages used in bioinformatics. *BMC Bioinformatics, 9*(1), 82.

Free Software Foundation. (2002). GNU Lesser General Public License (LGPL).

Fritz, M. H., Leinonen, R., Cochrane, G., & Birney, E. (2011). Efficient storage of high throughput DNA sequencing data using reference-based compression. *Genome Research, 21*(5), 734–740.

Garousi, V., & Zhi, J. (2013). A survey of software testing practices in Canada. *Journal of Systems and Software, 86*(5), 1354–1376.

Geiger, R. S., Varoquaux, N., Mazel-Cabasse, C., & Holdgraf, C. (2018). The types, roles, and practices of documentation in data analytics open source software libraries - A collaborative ethnography of documentation work. *Computer Supported Cooperative Work, 27*(3–6), 767–802.

Gerlach, W., & Stoye, J. (2011). Taxonomic classification of metagenomic shotgun sequences with CARMA3. *Nucleic Acids Research, 39*(14), e91–e91.

Gesellschaft für Forschungssoftware. (2018).

Giancarlo, R., Siragusa, A., Siragusa, E., & Utro, F. (2007). A basic analysis toolkit for biological sequences. *Algorithms for Molecular Biology, 2*, 1–16.

Goble, C. A. (2014). Better software, better research. *IEEE Internet Computing, 18*(5), 4–8.

Gog, S., Beller, T., Moffat, A., & Petri, M. (2014). From theory to practice: Plug and play with succinct data structures. In J. Gudmundsson & J. Katajainen (Eds.), *SEA. Lecture Notes in Computer Science* (Vol. 8504, pp. 326–337). Springer.

Gogol-Döring, A. (2009). *SeqAn - A Generic Software Library for Sequence.* PhD thesis, Free University of Berlin.

Google. (2017). Abseil Compatibility Guidelines.

Gotoh, O. (1981). An improved algorithm for matching biological sequences. *Journal of Molecular Biology, 162*, 705–708.

Gremme, G., Steinbiss, S., & Kurtz, S. (2013). GenomeTools: A comprehensive software library for efficient processing of structured genome annotations. *IEEE/ACM Transactions on Computational Biology and Bioinformatics, 10*(3), 645–656.

Grüning, B., Dale, R., Sjödin, A., Chapman, B. A., Rowe, J., Tomkins-Tinch, C. H., Valieris, R., & Köster, J. (2018). Bioconda: Sustainable and comprehensive software distribution for the life sciences. *Nature Methods, 15*(7), 475.

Gschwind, M. (2014). OpenPOWER: Reengineering a server ecosystem for large-scale data centers. In *2014 IEEE Hot Chips 26 Symposium (HCS)* (pp. 1–28). IEEE.

Guéguen, L., Gaillard, S., Boussau, B., Gouy, M., Groussin, M., Rochette, N. C., Bigot, T., Fournier, D., Pouyet, F., Cahais, V., Bernard, A., Scornavacca, C., Nabholz, B., Haudry, A., Dachary, L., Galtier, N., Belkhir, K., & Dutheil, J. Y. (2013). Bio++: Efficient extensible libraries and tools for computational molecular evolution. *Molecular Biology and Evolution, 30*(8), 1745–1750.

Hauswedell, H. (2009). *BLAST-like Local Alignments with RazerS*. Bachelor's thesis, Freie Universität Berlin.

Hauswedell, H. (2013). *Local Aligner for Massive Biological Data*. Master's thesis, Freie Universität Berlin.

Hauswedell, H., Singer, J., & Reinert, K. (2014). Lambda: The local aligner for massive biological data. *Bioinformatics, 30*(17), 349–355.

Hedin, G. (1996). Enforcing programming conventions by attribute extension in an open compiler. In *Proceedings of the Nordic Workshop on Programming Environment Research (NWPER'96)*.

Henderson, P., & Morris, J. H. (1976). A lazy evaluator. In S. L. Graham, R. M. Graham, M. A. Harrison, W. I. Grosky, & J. D. Ullman (Eds.), *POPL* (pp. 95–103). ACM Press.

Henikoff, S., & Henikoff, J. (1992). Amino acid substitution matrices from protein blocks. *Proceedings of the National Academy of Sciences of the United States of America (PNAS), 89*, 10915–10919.

Herbig, A., Maixner, F., Bos, K. I., Zink, A., Krause, J., & Huson, D. H. (2016). MALT: Fast alignment and analysis of metagenomic DNA sequence data applied to the Tyrolean Iceman. In *BioRxiv* (pp. 050559).

Herstein, I. (1964). *Topics in algebra*. Xerox College Publishing.

Ho, T., & Tzanetakis, I. E. (2014). Development of a virus detection and discovery pipeline using next generation sequencing. *Virology, 471*, 54–60.

Hoberock, J., Garland, M., Kohlhoff, C., Mysen, C., Edwards, C., Brown, G., Hollman, D., Howes, L., Shoop, K., Baker, L., & Niebler, E. (2020). *P0443: A Unified Executors Proposal for C++*. Technical report, International Organization for Standardization.

Holtgrewe, M. (2010). *Mason: A Read Simulator for Second Generation Sequencing Data*. Technical report, Freie Universität Berlin.

Hoste, K., Timmerman, J., Georges, A., & Weirdt, S. D. (2012). EasyBuild: Building software with ease. In *SC companion* (pp. 572–582). IEEE Computer Society.

Hunt, A., & Thomas, D. (1999). *The pragmatic programmer: From journeyman to master*. Harlow, England: Addison-Wesley.

Huson, D. H., Auch, A. F., Qi, J., & Schuster, S. C. (2007). MEGAN analysis of metagenomic data. *Genome Research, 17*(3), 377–386.

Huson, D. H., & Xie, C. (2014). A poor man's BLASTX–high-throughput metagenomic protein database search using PAUDA. *Bioinformatics, 30*(1), 38–39.

IHS Markit. (2010). Intel and AMD Retain Market Share amid Fast Growth.

ISO. (1998). *ISO/IEC 14882:1998: Programming Languages — C++*. Technical report, International Organization for Standardization, Geneva, Switzerland.

ISO. (2003). *ISO/IEC 14882:2003: Programming Languages — C++*. Technical report, International Organization for Standardization, Geneva, Switzerland.

ISO. (2011). *ISO/IEC 14882:2011: Programming Languages — C++*. Technical report, International Organization for Standardization, Geneva, Switzerland.

ISO. (2014). *ISO/IEC 14882:2014: Programming Languages — C++*. Technical report, International Organization for Standardization, Geneva, Switzerland.

ISO. (2015). *ISO/IEC 19217:2015: Programming Languages — C++ Extensions for concepts*. Geneva, Switzerland: International Organization for Standardization.

ISO. (2017a). *ISO/IEC 14882:2017: Programming Languages — C++*. Technical report, International Organization for Standardization, Geneva, Switzerland.

ISO. (2017b). *ISO/IEC 21425:2017: Programming Languages — C++ Extensions for Ranges*. Geneva, Switzerland: International Organization for Standardization.

ISO. (2019). *ISO/IEC 14882:draft: Programming Languages — C++*. Technical report, International Organization for Standardization, Geneva, Switzerland.

Jackman, S., Birol, I., Jackman, S., & Birol, I. (2016). Linuxbrew and Homebrew for cross-platform package management. *F1000Res, 5*, 1795.

Järvi, J., Willcock, J., & Lumsdaine, A. (2003). Concept-controlled polymorphism. In *International Conference on Generative Programming and Component Engineering* (pp. 228–244). Springer.

Kahlert, B. (2015). *API-Usability der auf Templatemetaprogrammierung basierenden Software-bibliothek "SeqAn"*. PhD thesis, Free University of Berlin.

Kehr, B., Weese, D., & Reinert, K. (2011). STELLAR: Fast and exact local alignments. *BMC Bioinformatics, 12*, S15. BioMed Central.

Kent, W. J. (2002). BLAT–the BLAST-like alignment tool. *Genome Research, 12*(4), 656–664.

Kianfar, K., Pockrandt, C., Torkamandi, B., Luo, H., & Reinert, K. (2017). Optimum search schemes for approximate string matching using bidirectional FM-index. Preprint, arXiv:1711.02035.

Kiełbasa, S. M., Wan, R., Sato, K., Horton, P., & Frith, M. C. (2011). Adaptive seeds tame genomic sequence comparison. *Genome Research, 21*(3), 487–493.

Knorr, K. (2017). *Vergleich aktueller Aminosäure-Alphabet-Reduzierungen und ihr Nutzen für die Homologiesuche*. Bachelor's thesis, Freie Universität Berlin.

Koenig, A. (1988). *C traps and pitfalls*. Addison-Wesley.

Kosar, T. (2012). *Data intensive distributed computing: Challenges and solutions for large-scale information management*. IGI Global.

Kramer, D. (1999). API documentation from source code comments: A case study of Javadoc. In J. Johnson-Eilola & S. A. Selber (Eds.), *SIGDOC* (pp. 147–153). ACM.

Krause, L., Diaz, N. N., Goesmann, A., Kelley, S., Nattkemper, T. W., Rohwer, F., Edwards, R. A., & Stoye, J. (2008). Phylogenetic classification of short environmental DNA fragments. *Nucleic Acids Research, 36*(7), 2230–2239.

Kucherov, G., Salikhov, K., & Tsur, D. (2016). Approximate string matching using a bidirectional index. *Theoretical Computer Science, 638*, 145–158.

Langmead, B., Trapnell, C., Pop, M., & Salzberg, S. (2009). Ultrafast and memory-efficient alignment of short DNA sequences to the human genome. *Genome Biology, 10*(3), R25.

Lehman, M. M. (1980). On understanding laws, evolution, and conservation in the large-program life cycle. *Journal of Systems and Software, 1*, 213–221.

Leipzig, J. (2017). A review of bioinformatic pipeline frameworks. *Briefings in Bioinformatics, 18*(3), 530–536.

Lemire, D. (2012). On the quality of academic software.

Li, H., Handsaker, B., Wysoker, A., Fennell, T., Ruan, J., Homer, N., Marth, G. T., Abecasis, G. R., & Durbin, R. (2009). The sequence alignment/map format and SAMtools. *Bioinformatics, 25*(16), 2078–2079.

Li, T., Fan, K., Wang, J., & Wang, W. (2003). Reduction of protein sequence complexity by residue grouping. *Protein Engineering, 16*(5), 323–330.

Lipman, D. J., & Pearson, W. R. (1985). Rapid and sensitive protein similarity searches. *Science, 227*(4693), 1435–1441.

Liu, H.-Y., Zhou, L., Zheng, M.-Y., Huang, J., Wan, S., Zhu, A., Zhang, M., Dong, A., Hou, L., Li, J., et al. (2019). Diagnostic and clinical utility of whole genome sequencing in a cohort of undiagnosed Chinese families with rare diseases. *Scientific Reports, 9*(1), 1–11.

Mackelprang, R., Waldrop, M. P., DeAngelis, K. M., David, M. M., Chavarria, K. L., Blazewicz, S. J., Rubin, E. M., & Jansson, J. K. (2011). Metagenomic analysis of a permafrost microbial community reveals a rapid response to thaw. *Nature, 480*(7377), 368–371.

Maiden, M. C. J. (2019). The impact of nucleotide sequence analysis on meningococcal vaccine development and assessment. *Frontiers in Immunology, 9*, 3151.

Malloy, B. A., & Power, J. F. (2017). Quantifying the transition from Python 2 to 3: An empirical study of python applications. In *2017 ACM/IEEE International Symposium on Empirical Software Engineering and Measurement (ESEM)* (pp. 314–323). IEEE.

Mansfield, J. (2017). Copy-and-swap.

McConnell, S. (2004). *Code complete* (2nd ed.). Redmond, WA: Microsoft Press.

Meuer, H. W. (2000). The TOP500 Project of the Universities Mannheim and Tennessee. In A. Bode, T. Ludwig, W. Karl, & R. Wismüller (Eds.), *Euro-Par. Lecture Notes in Computer Science* (Vol. 1900, pp.43). Springer.

Meuer, H. W. (2008). The TOP500 project: Looking back over 15 years of supercomputing experience. *Informatik Spektrum, 31*(3), 203–222.

Mittal, S. (2019). A survey of techniques for dynamic branch prediction. *Concurrency and Computation: Practice and Experience, 31*(1), e4666.

Moore, G. E. (1965). Cramming more components onto integrated circuits. *Electronics, 38*(8), 114–117.

Murphy, L. R., Wallqvist, A., & Levy, R. M. (2000). Simplified amino acid alphabets for protein fold recognition and implications for folding. *Protein Engineering, 13*(3), 149–152.

Myers, E. W., & Miller, W. (1988). Optimal alignments in linear space. *Computer Applications in the Biosciences, 4*(1), 11–17.

Myers, G. (1999). A fast bit-vector algorithm for approximate string matching based on dynamic programming. *Journal of the ACM, 46*(3), 395–415.

Nattestad, M. (2017). For bioinformatics, which language should I learn first?

Needleman, S. B., & Wunsch, C. D. (1970). A general method applicable to the search for similarities in the amino acid sequence of two proteins. *Journal of Molecular Biology, 48*(3), 443–453.

Nickolls, J., Buck, I., Garland, M., & Skadron, K. (2008). Scalable parallel programming with CUDA. *Queue, 6*(2), 40–53.

Niebler, E. (2014). Customization point design in C++11 and beyond.

Niebler, E. (2019). *Range-v3 Quick Start Guide.*

Niebler, E., Carter, C., & Di Bella, C. (2018). *P0896: The One Ranges Proposal.* Technical report, International Organization for Standardization.

O'Dwyer, A. (2018). Customization point design for library functions.

Oehlert, P. (2005). Violating assumptions with fuzzing. *IEEE Security & Privacy, 3*(2), 58–62.

Okonechnikov, K., Golosova, O., & Fursov, M. (2012). Unipro UGENE: A unified bioinformatics toolkit. *Bioinformatics, 28*(8), 1166–1167.

Pheatt, C. (2008). Intel® threading building blocks. *Journal of Computing Sciences in Colleges, 23*(4), 298–298.

Philippe, O., Hong, N. C., & Hettrick, S. (2016). Preliminary analysis of a survey of UK Research Software Engineers. In *4th Workshop on Sustainable Software for Science: Practice and Experience.*

Piezunka, H., & Dahlander, L. (2015). Distant search, narrow attention: How crowding alters organizations' filtering of suggestions in crowdsourcing. *Academy of Management Journal, 58*(3), 856–880.

Pitt, W. R., Williams, M. A., Steven, M., Sweeney, B., Bleasby, A. J., & Moss, D. S. (2001). The Bioinformatics Template Library-generic components for biocomputing. *Bioinformatics, 17*(8), 729–737.

Pockrandt, C., Ehrhardt, M., & Reinert, K. (2017). EPR-dictionaries: A practical and fast data structure for constant time searches in unidirectional and bidirectional FM indices. In S. C. Sahinalp (Ed.), *RECOMB. Lecture Notes in Computer Science* (Vol. 10229, pp. 190–206).

Pockrandt, C. M. (2019). *Approximate String Matching: Improving Data Structures and Algorithms.* PhD thesis, Freie Universität Berlin.

Pollard, M. O., Gurdasani, D., Mentzer, A. J., Porter, T., & Sandhu, M. S. (2018). Long reads: Their purpose and place. *Human Molecular Genetics, 27*(R2), R234–R241.

Prause, C. R., & Jarke, M. (2015). Gamification for enforcing coding conventions. In *Proceedings of the 2015 10th Joint Meeting on Foundations of Software Engineering* (pp. 649–660). ACM.

Prechelt, L. (2000). An empirical comparison of seven programming languages. *IEEE Computer, 33*(10), 23–29.

Preston-Werner, T. (2013). Semantic Versioning.

Raemaekers, S., van Deursen, A., & Visser, J. (2014). Semantic versioning versus breaking changes: A study of the maven repository. In *2014 IEEE 14th International Working Conference on Source Code Analysis and Manipulation* (pp. 215–224). IEEE.

Rahn, R., Budach, S., Costanza, P., Ehrhardt, M., Hancox, J., & Reinert, K. (2018). Generic accelerated sequence alignment in SeqAn using vectorization and multi-threading. *Bioinformatics, 34*(20), 3437–3445.

Rahn, R., Weese, D., & Reinert, K. (2014). Journaled string tree - A scalable data structure for analyzing thousands of similar genomes on your laptop. *Bioinformatics, 30*(24), 3499–3505.

Rausch, T., Zichner, T., Schlattl, A., Stütz, A. M., Benes, V., & Korbel, J. O. (2012). DELLY: Structural variant discovery by integrated paired-end and split-read analysis. *Bioinformatics, 28*(18), 333–339.

Reinert, K., Dadi, T. H., Ehrhardt, M., Hauswedell, H., Mehringer, S., Rahn, R., Kim, J., Pockrandt, C., Winkler, J., Siragusa, E., et al. (2017). The SeqAn C++ template library for efficient sequence analysis: A resource for programmers. *Journal of Biotechnology, 261*, 157–168.

Reinert, K., Langmead, B., Weese, D., & Evers, D. J. (2015). Alignment of next-generation sequencing reads. *Annual Review of Genomics and Human Genetics, 16*, 133–151.

Röst, H. L., Sachsenberg, T., Aiche, S., Bielow, C., Weisser, H., Aicheler, F., Andreotti, S., Ehrlich, H.-C., Gutenbrunner, P., Kenar, E., et al. (2016). OpenMS: A flexible open-source software platform for mass spectrometry data analysis. *Nature Methods, 13*(9), 741.

Runeson, P. (2006). A survey of unit testing practices. *IEEE Software, 23*(4), 22–29.

Schäling, B. (2011). *The boost C++ libraries.* XML Press.

Schwabe, R. F., & Jobin, C. (2013). The microbiome and cancer. *Nature Reviews Cancer, 13*(11), 800–812.

Siragusa, E. (2015). *Approximate String Matching for High-Throughput Sequencing.* PhD thesis, Free University of Berlin.

Siragusa, E., Weese, D., & Reinert, K. (2013). Fast and accurate read mapping with approximate seeds and multiple backtracking. *Nucleic Acids Research, 41*(7), e78–e78.

Smith, R. (2019). *P1103: Merging Modules.* Technical report, International Organization for Standardization.

Smith, T. F., & Waterman, M. S. (1981). Identification of common molecular subsequences. *Journal of Molecular Biology, 147*(1), 195–197.

Soito, L., & Hwang, L. J. (2016). Citations for software: Providing identification, access and recognition for research software. *IJDC, 11*(2), 48–63.

Spinellis, D. (2012). Package management systems. *IEEE Software, 29*(2), 84–86.

Standard C++ Foundation. (2019). Serialization and Unserialization.

Stroustrup, B. (1993). A history of C++: 1979–1991. In *The Second ACM SIGPLAN Conference on History of Programming Languages April 20–23, 1993, Cambridge, United States*, Digital Library (pp. 271–297). New York: ACM Association for Computing Machinery.

Stroustrup, B. (2012). Foundations of C++. In H. Seidl (Ed.), *ESOP. Lecture Notes in Computer Science* (Vol. 7211, pp. 1–25). Springer.

Stroustrup, B. (2017). Concepts: The Future of Generic Programming.

Sullivan, J. M. (2005). Impediments to and incentives for automation in the air force. In *Proceedings. 2005 International Symposium on Technology and Society, 2005. Weapons and Wires: Prevention and Safety in a Time of Fear. ISTAS 2005* (pp. 102–110). IEEE.

Sutter, H. (2005). The free lunch is over: A fundamental turn toward concurrency in software. *Dr. Dobb's Journal, 30*(3), 202–210.

Sutter, H. (2019). *P0707: Metaclasses.* Technical report, International Organization for Standardization.

Takanen, A., Demott, J. D., Miller, C., & Kettunen, A. (2018). *Fuzzing for software security testing and quality assurance.* Artech House.

Tatusova, T., DiCuccio, M., Badretdin, A., Chetvernin, V., Ciufo, S., & Li, W. (2013). *The NCBI handbook.* Bethesda, US: National Center for Biotechnology Information.

Tauch, A., & Al-Dilaimi, A. (2017). Bioinformatics in Germany: Toward a national-level infrastructure. *Briefings in Bioinformatics, 20*(2), 370–374.

The UniProt Consortium. (2019). UniProt: A worldwide hub of protein knowledge. *Nucleic Acids Research, 47*(Database-Issue), D506–D515.

Thompson, S. (1991). *Type theory and functional programming*. Addison Wesley.

Thornton, K. (2003). libsequence: A C++ class library for evolutionary genetic analysis. *Bioinformatics, 19*(17), 2325–2327.

Trapnell, C., Pachter, L., & Salzberg, S. L. (2009). TopHat: Discovering splice junctions with RNA-Seq. *Bioinformatics, 25*(9), 1105–1111.

Turnbaugh, P. J., Ley, R. E., Hamady, M., Fraser-Liggett, C. M., Knight, R., & Gordon, J. I. (2007). The human microbiome project. *Nature, 449*(7164), 804–810.

Ukkonen, E. (1985). Finding approximate patterns in strings. *Journal of Algorithms, 6*(1), 132–137.

Ukkonen, E. (1993). Approximate string-matching over suffix trees. In *Annual Symposium on Combinatorial Pattern Matching* (pp. 228–242). Springer.

Urgese, G., Paciello, G., Acquaviva, A., Ficarra, E., Graziano, M., & Zamboni, M. (2014). Dynamic gap selector: A smith waterman sequence alignment algorithm with affine gap model optimization. In I. Rojas & F. M. O. Guzman (Eds.), *IWBBIO* (pp. 1347–1358). Copicentro Editorial.

Vahrson, W., Hermann, K., Kleffe, J., & Wittig, B. (1996). Object-oriented sequence analysis: SCL - a C++ class library. *Computer Applications in the Biosciences, 12*(2), 119–127.

Vakatov, D., Siyan, K., & Ostell, J. (2003). *The NCBI c++ toolkit*.

van Heesch, D. (2008). Doxygen: Source code documentation generator tool.

Veldhuizen, T. L. (2003). *C++ Templates Are Turing Complete*. Technical report, Indiana University Computer Science.

Visnovska, T., Biggs, P. J., Schmeier, S., Frizelle, F. A., & Purcell, R. V. (2019). Metagenomics and transcriptomics data from human colorectal cancer. *Scientific Data, 6*(1), 1–7.

Vroland, C., Salson, M., Bini, S., & Touzet, H. (2016). Approximate search of short patterns with high error rates using the 01*0 lossless seeds. *Journal of Discrete Algorithms, 37*, 3–16.

Wala, J., & Beroukhim, R. (2017). SeqLib: A C++ API for rapid BAM manipulation, sequence alignment and sequence assembly. *Bioinformatics, 33*(5), 751–753.

Warr, W. A. (2012). Scientific workflow systems: Pipeline Pilot and KNIME. *Journal of Computer-Aided Molecular Design, 26*(7), 801–804.

Weese, D. (2013). *Indices and Applications in High-Throughput Sequencing*. PhD thesis, Free University of Berlin.

Weese, D., Holtgrewe, M., & Reinert, K. (2012). RazerS 3: Faster, fully sensitive read mapping. *Bioinformatics, 28*(20), 2592–2599.

Westbrook, A., Ramsdell, J., Schuelke, T., Normington, L., Bergeron, R. D., Thomas, W. K., & MacManes, M. D. (2017). PALADIN: Protein alignment for functional profiling whole metagenome shotgun data. *Bioinformatics, 33*(10), 1473–1478.

Wojtczyk, M., & Knoll, A. (2008). A cross platform development workflow for C/C++ applications. In *2008 The Third International Conference on Software Engineering Advances* (pp. 224–229). IEEE.

Wootton, J. C., & Federhen, S. (1996). Analysis of compositionally biased regions in sequence databases. In *Methods in enzymology* (Vol. 266, pp. 554–571). Elsevier.

Ye, Y., Choi, J.-H., & Tang, H. (2011). RAPSearch: A fast protein similarity search tool for short reads. *BMC Bioinformatics, 12*, 159.

Yoon, Y., Ban, K.-D., Yoon, H., & Kim, J. (2016). Automatic container code recognition from multiple views. *ETRI Journal, 38*(4), 767–775.

Zhang, H. (2009). An investigation of the relationships between lines of code and defects. In *ICSM* (pp. 274–283). IEEE Computer Society.

Zhao, Y., Tang, H., & Ye, Y. (2012). RAPSearch2: A fast and memory-efficient protein similarity search tool for next-generation sequencing data. *Bioinformatics, 28*(1), 125–126.

Printed in the United States
by Baker & Taylor Publisher Services